Modernity and Destining of Technological Being

Temple Davis Okoro

Modernity and Destining of Technological Being

Beyond Heidegger's Critique of Technology to Responsible and Reflexive Technology

PETER LANG
EDITION

Bibliographic Information published by the Deutsche Nationalbibliothek
The Deutsche Nationalbibliothek lists this publication in the Deutsche Nationalbibliografie; detailed bibliographic data is available in the internet at http://dnb.d-nb.de.

Library of Congress Cataloging-in-Publication Data
Names: Okoro, Temple Davis, 1971- author.
Title: Modernity and destining of technological being : beyond Heidegger's critique of technology to responsible and reflexive technology / Temple Davis Okoro.
Description: 1 [edition]. | New York : Peter Lang, 2016. | Includes bibliographical references.
Identifiers: LCCN 2016025315 | ISBN 9783631672884
Subjects: LCSH: Heidegger, Martin, 1889-1976. | Technology.
Classification: LCC B3279.H49 O356 2016 | DDC 601—dc23 LC record available at https://lccn.loc.gov/2016025315

ISBN 978-3-631-67288-4 (Print)
E-ISBN 978-3-653-06903-7 (E-Book)
DOI 10.3726/978-3-653-06903-7

Peter Lang Edition is an Imprint of Peter Lang GmbH.

Peter Lang – Frankfurt am Main · Bern · Bruxelles · New York · Oxford · Warszawa · Wien

This publication has been peer reviewed.

www.peterlang.com

In Memoriam
Engr. John. C. OKORO
&
Mrs. Mathilda E. OKORO

Preface

The early twentieth century was an era of a sensational and incredible cultural revolution in art, literature, science and technology, as well as in philosophy. The twentieth century is rightly called the century of technology. But this century, however rich in innovation and technological progress, is also that of technical modernity's backlash, deluded as it is by its own power and limitless needs to exploit its environment.

Modern technology is altering the way we live our lives. Ingenious devices make daily undertakings stress-free and allow societies and individuals to use their time efficiently, thereby increasing human capabilities. Technological developments mean that more and more things are possible with the touch of a button. Through easy access to information and improved communication, through efficiency in productivity and conveniences in travelling and improved lifestyle, but equally through the availability of destructive weapons and increased loneliness, as well as a fuelling of hyper-consumerism and accelerated environmental destructions, modern science and new technologies have transformed the way we see ourselves and our society, as well as leaving us in an objectively dangerous situation. They have therefore presented us with challenging questions on how to employ science and technology most proficiently to meet human desires.

The problem with technology is that most inventions have inadvertent consequences, and those unintended outcomes are mounting daily, causing devastations and generating risks of existential enormity. Therefore, a questioning of technology became crucial and imperative for Heidegger, among others, because modern technology carried with it a novel way of ordering the world, which he saw as tainting man's authentic and essential relation to being, thereby signifying a certain catastrophe imminent in European industrial modernity. What in essence propelled Heidegger to write on technology lies in his concern that all over the place in Europe and the West, man remains enmeshed in and bound to, lost and forgotten in technological modernity, a condition wherein the more technology advances itself the more it "threatens to slip from human control."

Nevertheless, while Heidegger's input to the philosophy of technology is broadly recognized, his work is sometimes contentious, perplexing and problematic to understand. Bringing together Heidegger's early works and later works, Temple Davis has carefully examined Heidegger's thoughts in this work, in which he highlights how science and technology conceal the enigmatic and distinctive presencing of beings; how our way of being-in-the-world is fraught by tension, and how destiny always carries a danger, a conclusion which led Heidegger to end up in a set of Poetic ideals.

In response to such profound ambiguities in Heidegger's extra-mental formulations and ever-growing apprehension with technology in the contemporary social sphere; and in an attempt to open a space closed by Heidegger for scientific and technological innovation as well as scientific respect for the question of being, Temple Davis addresses such complex and fiercely contested issues as the place of science and technical knowledge in the political sphere, the question of responsibility and reflexivity and the place of individual choice and the ordinary citizen's opinion in political decision making in a technological society.

In this perspective, the weakness of modern anthropology is what immediately appears at the centre of the debate and which underscores the need for a new ontology of care for the other, one that is consistent with the common sense of the everyday world in which it is inscribed. By taking a step away from a mindset of instrumentalization and towards a perspective on how things are truly used, especially with reference to their social value, Temple Davis gives new meaning to the interpretation of the *Kehre*. It's about reclaiming the meaning of technique, which has been hijacked by the need-driven process of technological framing. The idea is to tie this march of need back up with the destiny that it keeps obscuring, all while imposing itself as the essence of destiny. Temple Davis thus sheds light on a double erasure: the erasure of the essence of destiny at play in every attempt at gasping being, and the erasure of this very question of destiny by the logic of need that overdetermines the technical appropriation of the world. It is due to this double erasure that nature's sole *raison d'être* has become the fulfilment of needs and the satisfaction of what's lacking.

We're left with the question of whether Heidegger can help us solidify a posture capable of going beyond reservation with respect to this double erasure in favour of a more responsible attitude within a technical relationship with the world of another nature. Temple Davis doesn't think so, since Heidegger's ethics of responsibility remain solely subordinate to a poetics of dwelling. This entails no limitation applicable to a form of dwelling within a universe that has become intrinsically socio-technical. This is why Temple Davis turns to Jonas and Levinas to determine the conditions of a responsible and reflexive approach to technique as praxis limited by its participation in humanity's shared destiny.

While Levinas' appeal to the beyond and Jonas's demonstration of responsibility appeals to moral consciousness and mysticism, which demand collective authority and individual responsibility, merit accolades of wisdom, neither is adequate to address the risk and danger faced by modernity and its capitalist and market-driven technological advancements. Therefore, Temple Okoro has sought to highlight the underlying insights to be derived from the social sciences, especially the works of Ulrich Beck on reflexivity as crucial for opening up political arenas and the participatory procedures and systems associated with technological decision-making, to enable the world to deal better with risky and uncertain technology.

A key interest of the present approach lies within the project of going beyond ethical solutions that seek to limit technological praxis with recourse to shared values. With Beck and Giddens, Temple Davis underlines the prevalent need for a politics of technological interest that defines norms for the prudential use of technological power by limiting it from within in two ways: first, through the participation of those communities affected by its concrete uses, in a way that reaches the operational level of sub-politics; second, through the discursive structures of evaluation wherein an enlarged form of political deliberation between experts, spokespeople and decision-makers must take place, in other words, an extension of democratic responsibility upheld by three constantly correlated pillars.

It is thus first and foremost a new form of public commitment that techno-democracy calls for today. On one hand, this includes the anticipation of needs and lacks with respect to local uses; on the other hand, it concerns the adaptation of deliberative structures in line with a form of renewed and triangular social dialogue in which public authorities can participate in evaluative interactions that also include stakeholders and shareholders.

I will not hesitate to recommend this present work by Temple Okoro which, against Heidegger's ontologico-poetic resignation in the face of technology, suggests through the insight from contemporary social sciences an alternative assessment of technology that might better account for the absorption of modern technology and political life. His call for interactive technology assessment perceives modern technology assessment hypothetically to be an essentially democratic, reflexively reflective and discursively participatory process. Such reflexivity of actors participating in technology assessment undertakings aims to advance improved technology in a healthier society, and accentuates the timely participation of a comprehensive array of actors, formal and non-formal, to expedite social learning about modern technology and probable effects. It generates a completely novel temporal framework for ethics, and one that conceivably embraces future environmental snags. Rather than giving assertions of the present on the future, it tries to inaugurate the desires of the future within the present. Essentially, Temple Okoro has equipped us with an extremely substantial work by bringing together a great number of renowned and pertinent texts, authors, arguments and ideas, all of which have combined together to underpin and open an avenue between Heidegger's destining and romanticism in the face of modern technology and our responsibility for technology which enables us to reject technological imperatives through a politics of collective identity. Hence, he deserves our thanks and appreciation. And whilst the teething troubles brought about by modern technology are becoming universal in nature, we are being asked specifically to take care of our attitude towards the environment, and therefore to one another. This call for a constructive, democratic, participatory and reflexive assessment of technology therefore merits our attention in no better moment than now.

Prof. Marc Maesschalck
Institut Supérieur de Philosophie
Université catholique de Louvain-la-Neuve (UCL)
Belgium

Table of Contents

Section Two Towards Unmaking of Metaphysics and Questioning Technology

Some Abbreviations Used

BDT:	Building Dwelling Thinking
BH:	Brief über den Humanismus
BT:	Being and Time
BW:	Basic Writing
CF:	Confer
ED:	Aus der Erfahrung des Denkens
EDS:	Editors
EGT:	Early Greek Thinking
EI:	Ethics and Infinity
EM:	Einführung in die Metaphysik
GL:	Gelassenheit
HD:	Erläuterungen zu Hölderlins Dichtung
HW:	Holzweg
IBID:	Ibidem
ID:	Identity and Difference (Identitat und Differenz)
IM:	Introduction to Metaphysics
IPCC:	Intergovernmental Panel on Climatic Change
NGOs:	Non-Governmental Organizations
OB:	Otherwise than Being
P.; PP:	Page; Pages
PLT:	Poetry, Language, Thought
QCT:	The Questions Concerning Technology
SD:	Zur Sache des Denkens
SG:	Der Satz vom Grund
SZ:	Sein und Zeit
TRANS:	Translation or Translated by…
TI:	Totality and Infinity
TO:	Time and the Other
UNEP:	United Nations Environmental Program
US:	Unterwegs zur Sprache
VA:	Vorträge und Aufsätze
WD:	Was heisst Denken?
WCT:	What is Called Thinking
WM:	Was ist Metaphysik

General Introduction

Heidegger saw his career as a philosopher to re-awaken the forgotten question of being. He tried to free being from the oblivion into which it has been downgraded by traditional metaphysical perception of being. His approach was two-dimensional: negatively, when he called for destruction of; and positively when he opted for a reformation of the Being-question.

In his book, *What is Called Thinking*, Heidegger repeatedly declared: "The most thought-provoking is that we are still not thinking."[1] To put it in another way, we are still involved in a strictly logical thinking that is in keeping with the metaphysical and technological culture of modern societies. The supremacy of human reason over Being reaches its zenith, Heidegger notes, in the science of logic, which as the 'science of thinking' has been the tribunal, before which Being must stand.[2]

The problematic decline of our understanding of Being, according to Heidegger, began in ancient Greek thought. The pre-Socratics had a deeper notion of Being than the Socratic and Platonic philosophers. Despite the fact that Plato was to some extent still conscious of the 'presencing' attributes of Being, it was he, however, who first interpreted Being as constant presence or *eidos*, the eternally unchanging form, and in doing so, inaugurated Western metaphysics. Heidegger interprets metaphysics as a science of being as Being. Its prime concern, in essence, is not being (*Sein*), but the beingness of beings. The beingness of being was assimilated in the highest entity, which Aristotle called the first cause and the unmoved mover, and subsequently, Medieval Christianity and later thinkers would further reinforce this perception of being as permanent presence, this time grounded in a creator identified as God. With the coming of Descartes and modernity, that all changed. Homo sapiens, as a final point, arrogate this creation and grounding to itself, as the emergence of consciousness and subjectivity, thus, taking the center stage of human existence. In this critical period, entities, or 'things

---

1 Heidegger, M. *What is Called Thinking*? Trans. J. Glenn Gray, New York: Harper and Row, Pub. 1968. Perennial Library edition, 2004, pp. 4, 6, 28 etc. Originally published as *Was Heisst Denken*?, Tübingen: Max Niemeyer Verlag, 1954. (Hereafter, WD)

2 Heidegger, M. *Introduction to Metaphysics*, trans. Gregory Fried and Richard Polt, New Haven & London: Yale University Press, 2000, p. 25. (Hereafter IM). Originally published as *Einführung in die Metaphysik*, Tübingen: Max Niemeyer Verlag, 6. Aufl. 1998. (Hereafter EM)

in themselves' were irreversibly consumed by individual subject who was now the foundation of all knowledge and value, thereby setting Being question aside.

Here, we would therefore be asking the question: what is Being? This is necessary because Heidegger was not able to answer this question. He wrote about Being, but he did not explicitly express what Being was since it was something that always withdraws its full connotation. But for us to search for Being, Heidegger tells us that there is one being (a small b) that is the best place to commence our exploration. We should begin with the being of human beings. This is because, although everything, both animate and inanimate has being, according to Heidegger, only human beings are 'concerned' about their Being. The human being is the only being for whom Being is an issue. So, to certain extent, Heidegger already knows the meaning of Being but was unable to get the right word for it. For him, Being is not an entity, it is not this object, nor is it a concept or a specific event. It is rather the "unfolding" of things around us, and most significantly, it is also the "unfolding" that happens in our very lives.

One of the mistakes of modernity which Heidegger was very concerned about, was the tendency to turn the self into a thing. It was Descartes, who created a 'subjective' philosophy, based on the individual, with the "*cogito ergo sum*" slogan. This, according to Heidegger, is a thoughtless, wrongheaded illusion, since we are not really self-determining and independent from society and from Being, as well as from others and the way in which we are "thrown" into this world. As an alternative to the self, Heidegger chose the word *Dasein*. Etymologically, the German *Da*, means 'there' and *Sein* means 'Being'. So, Dasein means being-there; explicitly, "being-there-in-the-world." Thus, it is the nature of Dasein, in its "average-everydayness" to be in the world. But Dasein is not in the world like water in a glass or like our body is in a chair. When Heidegger says Being-in-the-world, he mean to suggest that we are engaged, absorbed, connected, entangled, caught up in a concernful way.

But how does this Dasein address the question of Being and the truth of Being? According to Heidegger, as this work explains, Dasein handles the question of Being, not by asking theoretical questions, not by addressing being scientifically, not through concepts, impressions and hypotheses, but through existence. Consequently, he has to do an existential analysis, that is, an analysis of Dasein's existence, in order to answer the question, "What is Being?" In doing so, Heidegger examined Dasein as a "being-with" an inauthentic collectivity, a sign of his early critique of modernity. Being-in-the-world shows itself in average everydayness in the form of what Heidegger calls the "ready-to-hand" in opposition to the present-at-hand. The ready-to-hand mode of being-in-the-world is well understood through the use of 'equipment' in which Being presents itself in terms of "in-order-to" in connection with the function or use. This is where Heidegger first developed an understanding of (technological) experience of human existence as a "being-in-the-world" in which he discovered the everyday character of engagement with equipment as prior to any theoretical presence of objects.

In his quest for the attainment of the truth of Being, Heidegger realized that there is need for "the abandonment of the transcendental-horizonal approach to the question of being in the attempt to speak more inceptually, more originally, *from within* an authentic experience of being."[3] He therefore calls on the need for Dasein to dwell poetically as he develops a comprehensive interpretation of Holderlin's poem fragment: "*Poetically man dwells on the earth.*" The poem fragment names the earth as the place of man's dwelling. It is the place of certain fourfold, namely: Earth, Sky, Mortals and Gods. The fragment thus suggests that we, as Dasein in our thrownness, live, build, and shelter with and from earth. Most significantly, we dwell poetically. Heidegger also suggests that thinking is itself a kind of poetic dwelling because it emerges from and brings into its own the sheltering power of earth. Only a new thinking of Being outside metaphysics can lead us to a proper way of dwelling on the earth, a thinking in which Dasein expresses openness to Being, in which thinking find itself "appropriated" by Being and the truth of Being experienced in an event which Heidegger called *Ereignis*. As a result, Being must be understood in and through the realm of *Ereignis*; "that realm, vibrating within itself, through which man and Being reach each other in their nature, achieve their active nature..."[4] In his later thought (Heidegger II), he understood this *Ereignis* "to be the event of appropriation out of which epochs of being occur. These epochs of being are fundamental ways in which being occurs and humans relate to this occurrence (for instance, by being challenged to calculate and plan everything as in our current epoch of technology).[5]

The deteriorating historical and metaphysical understanding of Being in such epoch is what eventually paved the way for the modern, technological, nihilistic era. Meanwhile, Heidegger's ambition was to explain the modern world philosophically, to revamp the power of thinking for our time. But he has to undertake this in the midst of the enormous technological revolution that altered the old

---

3 Vallega-Neu, D. Ereignis: The Event of Appropriation, in Davis, W. Bret, (ed.), *Martin Heidegger: Key Concepts*, Durham, UK: Acumen Publishing Limited, 2010, p. 142.

4 Heidegger, M. *Identity and Difference*, trans. J. Stambaugh, Chicago: The University of Chicago Press bi-lingual edition, 2002, p. 37. Originally published by Verlag Günther Neska in Pfullingen under the title, *Identitat und Differenz*, 1957, p. 26, (Hereafter, ID).

5 Vallega-Neu, D., *Op. Cit*, pp. 151–152.

European civilization, with its rural and religious pedigree, into a mass metropolitan industrial order well anchored in science and technology. As a result, shifting from the attention on the meaning of Being that prevails in his earlier work, Heidegger's later reflection after the renowned "*Kehre*" (turn), advances, among other things, a more unmistakable and unique philosophy of technology. In *The Question Concerning Technology* ("*Die Frage nach der Technik*" [1954]),[6] he maintains that technology is not merely a pragmatic application of science to the world but a "*Bestand*" a revealing, a revelation or reality about the world. Heidegger questions how this "revealing" in modern technology transpires. This is the focus of his philosophy of technology. He asks in what way modern technology is a form of revelation. Paraphrasing it, in what way does modern technology participate in the unfolding process of reality, namely, *Being*? He sought to understand and explain the essence of technology.

To get the understanding of the essence of technology, Heidegger looks back to the ancient Greeks to locate *technē* as a form of *poiēsis*, that is, a form of bringing-forth. It is a way of bringing something out of concealment to unconcealment. Simply put, technology is a mode of revealing which brings something into presence. As a form of revealing, technology demonstrates itself fundamentally as a happening of truth—an occurrence referred to by the Greeks as αλεθεια.[7] Truth, in Heideggerian understanding, is not a matter of correct or incorrect replicating of the world; it is a question of experiencing the being of things as they show themselves to us through our relatedness to them. Truth as *aletheia* is a process from being concealed to being revealed. It is not simply truth as the logical correspondence of statements to reality but truth as a revealing process. Everything, including our lives undergoes the process of truth revelation as long as it opens itself to the world. He attributes the concept to the pre-Socratic philosophers, mainly Heraclitus, Parmenides and Anaximander, who, he maintains, considered the essence of truth to lie in the disclosure of entities.

Technology, when understood from its Greek understanding as *technē* presents to us the real essence of technology, contrary to the present technological predicament we encounter in modernity. The problem with modern technology, according to Heidegger, is that "the revealing that rules in modern technology is a challenging, which puts to nature the unreasonable demand that it supply

---

6 Heidegger, M., *The Question Concerning Technology and Other Essays*, translated and with an Introduction by William Lovitt, New York: Harper & Row, Publishers, Inc, 1977, (Hereafter QCT).

7 *Ibid.*, p. 13.

energy which can be extracted and stored as such."[8] So, while the essence of technology in its Greek undertone is a *bringing-forth*, that of modern technology is a *challenging-forth* which transforms nature into a standing-reserve. Modern technology basically changes nature into a source of energy which it controls and uses as it wills. Nature, at the sway of modern technology, is ordered to stand by, to be immediately at hand, what Heidegger calls *Bestand* (standing-reserve). Further than this challenging-forth, man is himself nevertheless ordered to challenge nature, a situation that Heidegger calls *Ge-stell* (enframing) whereby the essence of modern technology is placed beyond man's control. "Man, precisely as the one so threatened, exalts himself to the posture of lord of the earth. In this way the impression comes to prevail that everything man encounters exists only insofar as it is his construct."[9] Meanwhile, although human beings take themselves for the masters of being, being "challenges" them to challenge beings. Modernity is therefore the total mobilization of the world by humans who are themselves mobilized in the process.[10] The world of modern science does not allow being to hold the highest rank. "During the epochs of science and technology, the being of entities takes place as objectivity and standing reserve respectively."[11] All traces of the ontological difference are wiped out.

There is, therefore certain inevitability with regard to technological being. According to Heidegger's analysis, modern technology is not only a challenging-forth, it is a *destining* as well, which is something outside our control. Like any destiny, however, technology as Gestell carries with it opportunity as well as danger. This danger mainly comes from seeing technology as the primary and exclusive way to reveal in such a way that it threatens the possibility of all other revealing. Once modern technology becomes the exclusive mode of interpreting the world, which Heidegger describes as "calculative thinking," the world becomes the instrumental resources for technological uses. The technological mind-set is a more tangible and aggressive expression of scientific thinking. Thus for a scientist, nature is a 'object' to be studied and investigated with mathematical accuracy. For a technologist, on the other hand, nature is nothing but a huge source of energy, to be unlocked, transformed, stored up and to be distributed and switched about anew.[12] Such a

---

8 *QCT.*, p. 14.

9 *QCT.*, p. 27.

10 Feenberg, A. *Heidegger and Marcuse: The Catastrophe and Redemption of History*, N.Y.: Routledge, 2005, p. 21.

11 Bruce V. Foltz, *Inhabiting the Earth: Heidegger, Environmental Ethics and the Metaphysics of Nature*, Atlantic Highlands, NJ: Humanities Press International, Inc., 1995, p. 60.

12 *OCT*, p. 16.

destining was originally set into motion by the exalted founders of modernity. According to Zimmerman, "Industrialization (i.e., technological culture) arises only in a society which understands Nature solely as an object which can be known by, and manipulated for, man."[13] Calculative thinking is at its best in such a technological culture that persists in the illusion of leading man to happier life.

Heidegger admits that technology has its value. "We cannot, of course, reject today's technological world as devil's work, nor may we destroy it—assuming it does not destroy itself."[14] His appeal is not a return to the archaic, pre-technological era, but to the very essence of technology. Heidegger asserts equally that "the essence of technology is by no means anything technological."[15] In making this rather conflicting assertion, he rejects what he calls the instrumental or anthropological definition of technology, which condenses technology all together to a particular instant of it. In other words, to define technology instrumentally is to identify it as *a means to an end*.[16] Although individual technological devices are correctly defined as means, Heidegger warns that we should in no way confuse the essence of technology with the essence of any single technological thing.

Heidegger does not leave us simply with a pessimistic depiction of the Western metaphysical thinking; he leads us rather to take a 'leap' which he described as a 'step back' (*Schritt zurück*) from metaphysical and technological thinking to their very essence or ground. "We may venture the step back out of philosophy into the thinking of Being as soon as we have grown familiar with the provenance of thinking."[17]

While this extremely calculating and rationalistic epoch may not mark the end of history, Heidegger appeared to believe that a post-metaphysical era of being was imminent. This would not be a straightforward retreat to an antiquated way of life, but rather a new approachability to being as is epitomized in the work of art. He reminds us that the *technē* of art has been the sole revealing in Greece in ancient time and people should not neglect the approach to knowledge that it presents. By considering the same ancient root, Heidegger emphasizes the primal thinking of the poet, seeing art as the other mode of revealing which can open new possibilities of other knowledge or life style.

---

13 ZIMMERMAN, M. *Eclipse of the Self: The Development of Heidegger's Concept of Authenticity*, 2nd edn. Anthens, Ohio: Ohio University Press, 1986, p. 219.

14 *ID*, p. 40. *Identity and Difference*, p. 38.

15 *QCT*, p. 4.

16 *Ibid.*

17 HEIDEGGER, M. "The Thinker As Poet" in *Poetry, Language, Thought*, trans. Albert Hofstadter, New York: Perennial Classics edition, 2001, p. 10.

Although Heidegger's eloquence and boldness obviously raises many of the worries aroused by the contemporary technological and environmental crisis, his trepidation with the essence rather than with the concrete facts of technology might appear to result in a situation comparable to his position on religion, namely, a strategy of deliberate non-interventionist attitude in the 'exclusively' ontic, the level on which the daily decisions of society over and above individuals operate. Heidegger's argument, according to Feenberg, is developed at such a high level of abstraction that he literally cannot discriminate between electricity and atom bombs, agricultural techniques and the Holocaust.[18] Our exposition therefore leads us to point out major ambiguities in Heidegger's approach.

Beyond Heidegger

In so far as we do not reject Heidegger's invitation for us to reflect critically on the limits of technology by considering its essence, we do not subscribe to any strategy that does not have some relation to a concrete, pragmatic significance or application.

There is no suspicion about the creativity, ingenuity and inspiration of Heidegger in the arena of twentieth century philosophy. All the same, as that century was evaporating into history books, Emmanuel Levinas, a student of Heidegger, called for "a profound need to leave the *climate*" of Heideggerian philosophy with its obsession with the question of being, and he is convinced that "we cannot leave it for a philosophy that would be be pre-Heideggerian."[19] We feel, in our opinion, that Levinas did much to move us beyond Heidegger into a post-Heideggerian climate. While confronting Heidegger's question of being, Levinas identifies lack of any ethical philosophy in Heidegger, and also identifies a new space for such an ethics with his own idea of alterity or exteriority of the other. In a treatise entitled *The Search for a Heideggerian Ethics*, Michael Zimmerman reasons that "the burden of Levinas' critique of Heidegger" resides in observing how "Heidegger's fascination with the Greeks led him to discount Jewish and Christian insistence on the importance of personal responsibility for the concrete other"[20] Along these

---

18 Feenberg, A. *Questioning Technology*, p. 187.

19 Levinas, E. *Existence and Existents*, trans. Lingis Alphonso, Pittsburgh: Duquesne University Press, 2001. p. 4.

20 Zimmerman, M. *Ethics and Responsibility in the Phenomenological Tradition*. Ninth Annual Symposium of the Simon Silverman Phenomenology Center, Duquesne, 1992, (pp. 57–90), p. 81.

lines, Our thesis, through Levinas, challenges Heidegger from above, with an entreaty to the good or moral beyond being.

Also while Heidegger's critique of technology underscores the negative effects of technology which impinges on human essence, Hans Jonas, another student of Heidegger, equally, doubts technology's purportedly positive progressive effects. The consequences have evolved beyond our compass, and as a result, we cannot predict the problems coming from the invention and utilization of a new device and which may well surface in the remote future. What this means is that *"the natural is swallowed up in the sphere of the artificial."*[21] But instead of describing man's condition as a time of want "when the gods have fled" as Heidegger would suggest, Jonas called this situation "ethical vacuum"[22] wherein he stressed that changes in technology are to be followed by changes in ethics which calls for ethical responsibility. As he states: "The first and most general conditions of responsibility is causal power, that is, that acting makes an impact on the world; the second, that such acting is under the agent's control; and the third, that he can foresee its consequences to some extent."[23] Jonas was interested, among other things, with the responsibility of the politician, because in the public domain, the modern statesmen play a significant role in determining the existence and welfare of the coming generations. What Jonas takes for granted here is that man, of his own accord, chooses the burden of responsibility.

We would therefore argue in this thesis that the nature of the crisis confronting us today is so comprehensive, infiltrating through every rank of the society and culture that its solution cannot be left to politicians, the academics, the scientists and technologists alone. Since we find Heidegger's ontological account of the problem of technology acceptable, but with lack of concreteness, we will show how the integration of reflexivity into the theory of technology can help to open up the future of modernity to multiple prospects previously shut out by Heidegger's rigidity.

There is no doubt that the consequences of new technologies have been increasingly reviewed, analyzed, and regulated. Also, these consequences (e.g. dangers, risk, impacts etc.) are today dominating public and political debates. For this reason, Ulrich Beck, in his book *Risikogesellschaft* (1996), has argued that we have entered a new phase in the modernization process, a phase referred to as reflexive modernization. Reflexive modernization as we shall see "entails the 'self-confrontation' of modern society with negative consequences of modernization,

---

21 HANS, J. *The Imperative of Responsibility: In Search of an Ethics for the Technological Age*, Chicago: The University of Chicago Press, 1984, p. 32.

22 *Ibid.*, p. 22.

23 *Ibid.*, p. 90.

among which is the environmental crisis."[24] We are moving into a third stage of social development within modernity. Traditional society was supplanted by the industrial society (simple modernity). This epoch saw the advent of rationalization and 'differentiation' of society, utilitarian mentality and wealth accumulation, positivistic assumptions and a contingent scientific techno-economic development leading to the inauguration of industrial and capitalist society. This new society has to unravel the human-induced teething troubles which stem from the expansion and progress of industrial civilization; to examine, as Beck said, how the risk generated due to the quest to modernize can be "prevented, minimized, dramatized, or channeled".[25] In the last part of *Risikogesellschaft,* Beck talks of an "unbinding of politics,"[26] a trend towards a flexible, decentralized, and consensus approach to governance, a bottom-up method sometimes referred to as "political modernization" to the detriment of top-down institutionalized command-and-control regulation of the conventional politics. It is in this bottom-up approach that greater involvement of nonstate actors in politics and policy-making swings to new settings, namely, the so-called Sub-politics.

On a related development, Anthony Giddens, in *The Consequences of Modernity*, unambiguously strikes a similar note with Beck's proposal.[27] He matches Beck to a substantial degree in highlighting the shifting "risk profile" of modern society, wherein, although scientific and technological improvements have curtailed several premodern risks like drought and natural disasters, it has, at the same time step up new types of disaster and ecological risks beyond calculability. Nevertheless, Giddens balances Beck's gloomy risk depiction with emphasis on how present modernity, through social institutions, may transform itself through 'disembedding' and 'reembedding' so as to cope with these new risks. This calls for reflexivity in decision making about technology and its applications.

---

24 Thomas, J. Misa, Brey, P. & Feenberg, A. editors, *Modernity and Technology*, Cambridge, Massachusetts: The IMT Press, 2003, p. 308.

25 Beck, U. *Risk Society: Towards a New Modernity*, London: Sage, 1992, p. 19.

26 *Ibid.*, p. 185.

27 Giddens, A., *The Consequences of Modernity*, Stanford Calif: Stanford University Press, 1990.

Our three-part approach in this work will examine the following questions:

1. What, according to Heidegger, is the real meaning of Being in its ontological profundity?
2. What is Heidegger's interpretation of modernity and technology and how is his question concerning technology derived from his fundamental ontology, namely, the question of Being?
3. How can responsible and reflexive approach towards technology open a space for greater engagement with technology to counter its weaknesses?

In this presentation, we shall find that technology, as Heidegger sees it, has become the crucial way of our dealing with the world. That said, we will take seriously Heidegger's assertion that a certain destining is at the center of modern technology. This way of posing the question of technology is today criticized by authors influenced by the new sociology of sciences and the so called science studies. Their critique try to show how the substantial perspective of Heidegger is closing the door to a more pragmatic approach to technology where the central problem would be that of learning how to act within the frame of a technological society. This is the question we would like to cope with this presentation.

Heidegger's analysis shows that amid the advancement of modern technology, human beings have lost some mode of being that made them human or authentic. They have lost various critical component of humankind like poetry or a personified sense of self. But then, how can one anticipate to comprehend the complexities of modernity without a sufficient description of the technological advancement that make it possible; and how can one study particular technologies without a theory of the bigger wider society, described today by some as a risk society, in which they develop? These questions have not even been asked, much less answered convincingly, by most principal actors and contributors to the fields.

Our hypotheses charts a three dimensional paths: Firstly, that earlier Heidegger (Heidegger I), exemplified by *Being and Time*, presents a form of unsubstantiated subjectivism, in which an individual is "authentic" when he, she or it resolutely decides a course of action in the absence of any ethical norms that would guide and limit his decision. His examination of modernity derives from the supposition that what we think and what we are can be reclaimed simply by stepping back to the earlier times and precise opinion of things.

Secondly, that Heidegger's critique of technology in *The Question Concerning Technology,* (Heidegger II) and related texts was profoundly shaped by his question of being and his critique of modernity. There is disclosure of human being in the complexity of new technologies, which, opposed to modern technology, brings "being" forward and demonstrate the forgotten aspect of "Being" instituted by the enframing of modern technology. Therefore, it is almost not unexpected that Heidegger cannot understand the manner wherein technology soared and functions inside modernity since modernity itself, or at any rate what we understand by this word, functions for him as a plunge further than the earlier times and righter opinion of things.

Thirdly, that Heidegger's discourse leaves ominously and worryingly tiny accessible opening for pragmatic action. How can humanity influence his relationship with technology if it is a dilemma of collective perceptions and something already outside his control? Heidegger appears to portray modern technology as a distraction or abnormality on the route of human development, and as such, call for a rather particular re-thinking of man's relationship with technology. At best, this way of looking at the question about technology appears as if to discourage practical steps by describing technology as a substantial problem beyond the range of human intervention, consequently creating no space to change this relationship. Any action to correct this cultural aberration seems insignificant and any pragmatic approach meaningless. As Rose puts it, "the critics would serve society better by acknowledging that people are agents, not victims, of this cultural transformation."[28] We would therefore add that an emphasis on our conscious, reflexive and democratic activity, rather than unconscious participation, would empower those within technological society to examine their relationships to their technologies. This thesis takes the view that a philosophical approach to investigating technology is necessary but its sociological interpretation is more important as we take a number of things for granted in our interactions with technological objects.

Our methodology is primarily exegetical, critical and more or less tentative, and is therefore not an attempt to posit a comprehensive understanding or conclusive reading of the complexities embedded in Heidegger's philosophy of being and technology.

Structural Analysis

This work is divided into three parts of two chapters each:

Part I: In this part, we will examine the fundamental existential constitution of Dasein according to Heidegger with particular interest on his treatment of Dasein as being-in-the-world. This will lead us to highlight his early instrumentalist ontology as regards Dasein's relationship with the world as well as Dasein's environmental

28 Ellen, R., The errors of Thamus: An Analysis of Technology Critique. *Bulletin of Science, Technology, & Society*, *23*(3), 2003, (147–156), p. 150.

and communal undertones; In understanding why being is important to Heidegger, we will keep in mind the difference between being and entities. We will conclude with Heidegger's early critique of modernity and mass culture in *Being and Time* with the discussion of equipment as reading-to-hand and present-at-hand.

In this part also, chapter two looks at Dasein's attainment of the experience of Being with explication of being and thinking, namely; the essential thinking of Being, dwelling in the neighbourhood of being, poetic existence, etc. These will bring out Heidegger's change of term from traditional anthropological point of view to phenomenological perspective of human being giving meaning to real ontology. At the end of this part, we will use Emmanuel Levinas to criticize Heidegger's ontology. Here, we will highlight what Levinas sees specious with Heidegger, namely, that Heideggerian ontology, which surrogates the relationship with the Other to the relation with Being in general leads inevitably to imperialist supremacy, to oppression and subjugation. Levinas's insight here is that Dasein whose Being-in-the-world, substantiates the meaning of being, is itself substantiated by ethics, by the relation to the other. That is why ethics is prior to ontology. Therefore, we will conclude this part in terms of this Levinas's fundamental claim, namely, ethics (which has to do with our relation to the other and lacking in Heidegger) is prior to ontology. This is because, outside the other, we would not have a self.

In **Part II,** we shall examine how Heidegger became a normative thinker of technology. We will see that in modernity, according to Heidegger, the withdrawal of being has been intensified and made stronger by a widespread neglect or indifference to the question of being in the wake of modern technology. Modernity is an era characterized by the mentality and philosophy that nothing is again impracticable or unapproachable.

Chapter three x-rays the beginning of modernity as an age of technological being with the emergence of subjectivity and human consciousness, carried out through Descartes' project. Tracing modernity to its medieval background to locate where the dislocation occured, we will examine the question of subjectivity as enunciated by Descartes and then show Heidegger's critique of Descartes in which he upholds that the preeminence of science and technology in modernity is founded on the condition that human being is understood as a subject. This led the way for his critique of modernity and metaphysics, the precise idea of reductionist mentality we have of things in a technological society and thus, his call for destruction, overcoming and reformulation of metaphysical question.

Chapter four bears the nucleus of Heidegger's question concerning technology. After an initial exposition of the meaning of technology, we will examine, first of all, some trajectories in the philosophy of technology, some anti-modernist views before Heidegger. Here, by way of literature review, we will look at different schools of thought on the question of technology, ranging from the existentialist perspective, with particular interest in Romano Guardini, to the Critical Theory perspective, highlighting the "Frankfurt School," and the later thinkers of the theory.

Heidegger and the question of technology will take the central part of this chapter. After examining Heidegger's teaching on science which leads to calculative thinking, we are able then to see his question concerning the essence of technology. Here, we shall see a certain destining at work in modern technology, an ordering that Heidegger calls "enframing" which reduces everything, humans included, into standing reserve (Gestell), as resources to be ordered, challenged and put into use at will. The challenge next, according to Heidegger, is for us to find ancillary practices and the works of art, specifically, the art of poetry which will lead to the practice releasement toward modern technology.

In **Part III,** we shall move toward a technological rationality and more pragmatic approach to finding a way of dealing with collective social risks posed by modern technology.

Chapter five raises an ethical dimension lacking in Heidegger's interpretation. Here, we will use the work of Hans Jonas, *The Imperative of Responsibility*, to determine the possibility of ethical input with regard to technology as it affects modern society. According to Jonas as we shall see, modern technology is a threat to humanity, because technology puts us in a position where ethics do not seem to apply. In other words, ethics are no longer in compliance with its object, namely, human actions. Modern technology has made nature an object of human responsibility, and accordingly, we must approach her not only with more and more sophisticated ingenuity but also with ethics. Since modern technology has brought about unintended consequences of such enormity that old policies and ethics can no longer restrain them, we shall see that, to meet the new underlying scale of individual and collective actions, according to Jonas, a new long-range ethics of responsibility has to be built, one that is capable of coping with the up-to-the-minute global condition of human life; using new political ventures to reinstate the importance of the personal and the individual in preference to the materialistic and impersonal in society.

So this chapter shows that since the conditions for ethical theory has changed drastically, and actions that are collective and covers longer time spans must be included in the ethical thinking, then ethics in the age of modern technology seems to be subjugated to at least three new conditions, which according to Jonas requires the unfolding of three "dimensions of responsibility." This principle served as inspiration for the preventive and precaution principle.

In 1986, shortly after Chernobyl, the German sociologist Ulrich Beck explained how risk has become completely omnipresent in society, which he now calls "risk society." We will explain in this chapter that, according to him, prevention can take place if technological risks are known and proven whilst precaution aims to prevent risks that have not yet been determined. Thus, faced with the unknown, all technological innovation should prove its harmlessness before being put into service. Since these principles are not always totally palpable, we will suggest in Chapter six that, although technological risk assessment is attributable to experts, the decision on whether the risks should be taken or not is a political one. In democratic social orders, such decision is possible on condition that the general public and its legislative body or representatives are well-informed. This leads us to the question of reflexivity in technological undertakings. We will show that reflexivity will draw attention not only to the intended, but also the unintended consequences of the actions we take. Here the attention is moved from modernity to reflexive modernity which entails the capacity to question the very suppositions taken for granted by the excitement and eagerness toward technological development. We will see that the principal meaning of reflexivity has to do with modernity with its own implications and side-effects, the apparatus by which modern societies develop in sequences of producing problems and solutions to these problems which in itself, produces new problems *ad infinitum*.

The chapter observes that one fundamental component of reflexive modernization is the role of citizens relating to scientists and public administrators over technological risks. In the risk society, claims of facts and knowledge about risks ought to be deliberated among scientific experts, political stakeholders, professionals, interested parties and laypersons, basically, a deliberation amongst diverse epistemologies. This sets the stage for a new form of political action, an entirely individualized, spontaneous and amorphous politics from below. Here we will emphasize that this sub-politics and epistemic community, a deliberative approach opens the way for a new reflexive democratization of technology, in which different actors should be involved in decision making about technology as it affects the society, the environment and individuals. This perhaps can lead us to alternative model of development which we see as a subject of further research.

Finally, we will make a critical evaluation of Heidegger's critique of technology and his solution to the problem and a normative evaluation of modernity with a vision for an alternative society to ascertain how today's technological society might transform itself into a different kind of social structure without being less technological.

Section One
The Reality of Being: Heidegger's Fundamental Ontology

Chapter I: Experience of Being: Dasein's Being-in-the-World

Introduction

Ontology has as its primary task the introduction of certain principles that will offer a systematic and safe foundation for the analysis of the question of truth. But due to certain historical exigencies like the emergence of skepticism which challenged the religious dogmatic views of human life as a result of scientific and positivistic mentality of the modern era, Heidegger stresses that we must acknowledge the apparent idea that philosophy has no candid foundation upon which to progress apart from the finite presence of life on earth, namely, our Being-in-the-world. Thus, ontology has for its fundamental discipline, according to Heidegger, the analytic of the Dasein. "As ways in which man behaves, sciences have the manner of Being which this man himself possesses. This entity we denote by the term Dasein."[1]

This means that the manner of being which man has is quite different from the manner of being of other beings because "Dasein itself has a special distinctiveness compared with other entities."[2] Although, by his very nature, Dasein is a being-in-the-world, he is not just one of other entities. "In the Heideggerian perspective, the ultimate goal of Dasein's existence is the experiencing of Being. He is transcendence by his very nature and has a destiny that goes beyond his everyday concerns."[3] He is therefore not simply called upon to be a being-in-the-world, but is predestined to understand Being and to be a being that is open to Being.

> [...] Dasein is special because he understands Being. He understands Being in a way other beings do not. This is an ontological way. Ontological means the Being of beings.

---

1 Heidegger, M. *Sein und Zeit,* 17. Aufl. (Tübingen: Max Niemeyeer, 1993). p. 11. Although this *magnum opus* of Heidegger was first published in 1927, it was only in 1962 that the first translation in English appeared. All our citations are from the 2007 edition, *Being and Time*, trans. John Macquarrie & Edward Robinson, 29th edition, Oxford: Blackwell Publishing Ltd. 2007, p. 32. (Hereafter, SZ, BT).

2 "Dasein has a special distinctiveness as compared with other entities.... Dasein is a being which does not just occur among other entities.... Dasein is ontically distinctive in that it is ontological" (*BT*. P. 32).

3 Vensus, A. G. *The Experience of Being As Goal of Human Existence: The Heideggerian Approach*, Washington D.C: The Council for Research in Values and Philosophy, Series IIIB, South Asia, Vol. 2, 2000, p. 17.

> Thus Dasein surpasses beings in its capability to investigate and thereby understand Being. [...] Understanding of being is itself a definite characteristic of Dasein's Being.[4]

The epistemological concerns for Dasein "stem from the fact that he finds himself in the world," otherwise called Facticity; "understands the world and expresses his understanding in discourse."[5] Constitutively and constitutionally, Dasein is involved with others (*Dasein* and *Seinde*) in the world. He is an *être-avec*, a being-with, others. This phrase consists of two concepts, namely: "being-in" (*In-Sein*) and "in-the-world" (*in-der-welt*).[6] Here are involved the contrasting elements: entanglement and inventiveness, appropriation and assimilation. Consequently human existence is fundamentally a 'being-in'.

In this first part of our study, we will examine Heidegger's primordial thought, by tracing its development from his fundamental ontology through the analysis of the nature of Dasein, with particular reference to the way in which Dasein is in his 'there', i.e., his 'being-in. In elaborating this point, our case study will also focus on Dasein's relational concerns especially his environmental and communal relationship with the world. This part of the study will introduce us to the fundamental question that, *ab initio*, occupied Heidegger's philosophical career, and in which his question concerning technology will revolve.

1. The Question of Being

The task of philosophy is, among others, to reveal the last foundations of being and of our knowledge of being. Ever since its origin, Greco-European philosophy has in general taken the question, 'What is being'? as its central question. However, despite the fact that each person thinks that he knows what being is, and although for many centuries traditional philosophy has dealt with this fundamental question, it has to be asked again and again. This is due to the fact, according to Heidegger, that the preliminary insights gained by the Greeks were accepted as "self-evident" and taken for granted by traditional philosophy. The problem was not restated and rethought over and again *as a theme for actual investigation (als thematische Frage wirklicher Untersuchung)*.[7] Consequently the question of Heidegger's fundamental

4 Iroegbu, O.P. *Kpim of Time – Eternity*, Ibadan: Hope Publications Ltd., 2004, p. 70.

5 Vensus, G. *The Experience of Being As Goal of Human Existence: The Heideggerian Approach*, p. 55.

6 *SZ*, p. 53.*BT*, p. 79.

7 *SZ*, p. 2. *BT*, p. 21. According to Heidegger, subsequent philosophers, 'down to the logic of Hegel' did only some 'retouchings' (*Übermalungen*) to what Plato and Aristotle had achieved. (SZ., p. 2. BT., p. 21. Hodges sarcastically remarked: For centuries, "there was

ontology becomes the question of the *meaning of Being* and this formed the nucleus of *Being and Time* as an attentive reading of the book indicates. But as we shall see, Heidegger did not provide a clear answer to the question of Being, for in his view, the book can only bring us merely within the arena of logically and realistically asking the question in a productive way. As he states:

> The question of Being aims therefore at ascertaining the *a priori* conditiond not only for the possibility of the sciences which examine entities as entities of such and such a type, and, in so doing, already operate with an understanding of Being, but also for the possibility of those ontologies themselves which are prior to the ontical sciences and which provide their foundation.[8]

1.1. Aristotle: The Metaphysical Question

To be unacquainted with Aristotle of Stagira (384–322 BC) would constitute a serious defect in a person's philosophic culture. St. Thomas invariably refered to Aristotle as the philosopher; St. Bonaventure regards him as the greatest of the pre-Christian thinkers. Dante considered him to be the master of those who know. To be put into contact with the thoughts of Aristotle is to be in the presence of one of the greates thinkers of all time.

The case in point is the metaphysics of Aristotle. Aristotle did not use the word metaphysics at all. Indeed, the origin of the word metaphysics is rather casual. It was in the words of Andronicus of Rhodes in the 1st C of the Christian era that the word metaphysics first emerged. In editing the works of Aristotle, he inserted 14 books of Aristotle's writing immediately after the works on physics. And as these books came immediately after the works on physics, they were spoken of in general terms ad meta-ta-physika. By a stroke of good fortune, these metaphysical treatises occupy in this arrangement, that position which they ought to occupy in the hierarchy of knowledge.

---

little else in philosophy, but the attempt to dot the "I"s and to cross the "t"s of Plato and Aristotle. Cf. Hordern, W. *Speaking of God: The Nature and Purpose of Theological Language*, New York: The Macmillian and co., 1964, p. 19. In a related note, Whitehead unequivocally remarked that the whole Western philosophical tradition consists of a series of footnotes to Plato. Cf. Whitehead, A. N. *Process and Reality*, New York: Free Press, 1979, p. 39.

8 *SZ*, p. 11. *BT*. p. 31.

In the cause of his long reflection upon metaphysics, Aristotle stated the basic theme of metaphysics: "And indeed the question which was raised of old and is raised now and always and is always the subject of doubt, 'what being is'".[9]

The question within all questions is the question of being. Even those who do not consider themselves philosophers constantly, in the everyday speech, refer to being. Things are and they are affirmed to be or things are not and existence is denied of them. In that sense, human language is the first form of being, the sign in which being first presents itself. When the young child finally speaks of itself in the first person singular, a profound experience of being as an existing thing has already taken place. But the subject matter of metaphysics as the ultimate wisdom is not the first understanding of being attained by a child. Metaphysics disengages being in its proper intelligibility to make it an object of thought. Such a metaphysical abstraction wherein being as being is disengaged by the intellect for its own sake, is not a return to the Platonic *Ideas*. Aristotle explicitly and consistently rejected Plato's theories of Ideas: "He whose subject is existing things qua existing ...this is the philosopher."[10] "With Aristotle," Heidegger remarks, "the greatest philosophical knowledge of antiquity is expressed, a knowledge which even today remains unappreciated and misunderstood in philosophy."[11]

The philosopher is preoccupied not with the Platonic *Forms* and *Ideas* but with the existing thing precisely as the existing thing. For Aristotle, philosophy is concerned with reality and reality is always the particular and actually existing thing. The question for philosophy is to determine that which makes the existing thing to be actual and real.[12] The metaphysical question for Aristotle is to determine the being of the concretely existing thing. In opening up this question of reflection, Aristotle mapped out metaphysics from every special science:

---

9 Aristotle, *Metaphysics*, BK. 7, C. 1, 1028-b 1, in *The Complete Works of Aristotle: The Revised Oxford Translation,* ed. Jonathan Barnes, New Jersey: Princeton University Press, 1984.

10 Aristotle, *Metaphysics*, BK. 4, C. 3, 1005b, 10.

11 Heidegger, M. *Aristotle's Metaphysics θ 1–3: On the Essence and Actuality of Force*, trans. W. Brogan and P. Warnek, Bloomington and Indianapolis: Indiana University Press, 1995, p. 188.

12 In the ancient world, only the Hebrews and the Greeks had any sense of the real. The Hebrews derived it from Divine revelation. For a moment of great intellectual significance, Aristotle saw the real as the existing thing precisely as the existing. He saw it and attempted to reflect upon it in a metaphysics which is of enduring significance. For Aristotle "actuality... is the existing of a thing" (*Metaphysics*, BK. 9, C. 6, 1048 a, 30). In a decisive moment of intellectual history Aristotle beheld existence and asked the question about the being of this or that entity and attempted to answer the question.

There is a science which investigates being as being and the attributes which belongs to this in virtue of its own nature. Now this is not the same as any of the so called special science; for none of these others treats universally of being as being, they cut-off a part of being and investigates the attributes of this part…[13]

The mathematical sciences of arithmetic or geometry, for instance, isolate quantity, discrete and continium, and investigate that quantity in detail. But then, quantity is only one aspect of being and mathematical sciences based on quantity do not extend either to the existential order of being or to the universal order of being as being. The science of biology was limited to the sphere of living bodily being. Each of the Spatial Sciences has every right to be but none of them has being precisely as being as the object of their study. Methodologically, the Spatial Sciences isolate certain spheres of being for investigation and examined those spheres systematically and in great details. But metaphysics has for its concern that which is not the concern of any partial knowledge of Spatial Science—Being precisely as Being.

Metaphysics, therefore, is the science of being precisely as being. But the first difficulty is that 'there are many senses in which a thing may be said to be'. Aristotle listed many of the ways in which things are said to be:

Some things are said to be because they are substance, others because they are affections of substance, others because they are a process towards a substance, or destruction or privations or quantities of substance or productive or generative of substance or things which are relative to substance or negations of one or these or of substance itself.[14]

In the midst of these pluralities of ways in which things are said to be, traditional philosophy is right in stating that this concept of being is the most general one of all. "From the fact that "being" expresses the most universal concept it does not follow that the notion of being is clear. It is rather the darkest notion of all, and this is why the question of being has to be raised once again."[15] According to Heidegger, this concept is used in the most diverse meanings:

Within the range of basic philosophical concepts—especially when we come to the concept of 'Being'- it is a dubious procedure to invoke self-evidence, even if the 'self-evident' is to become the sole explicit and abiding theme for one's analytic—'the business of philosophy.[16]

13 Aristotle, *Metaphysics*, BK. 4, C. 1, 1003b 20.

14 *Ibid.*, BK. 4, C. 2, 1003b, 8.

15 Herman P. *Heidegger's Philosophy of Being: A Critical Interpretation*, Princeton, New Jersey: Princeton University Press, 1998, p. 34.

16 *SZ*. p. 4; *BT*., pp. 23–24.

Thus the concept of "Being" is far from being evident, and therefore it should be reexamined in a systematic manner.

1.2. Heidegger and the Question of Being

Heidegger is in accord with Aristotle that question of Being is the fundamental question of philosophy. However, he is of the opinion that Aristotle failed to deal with the question satisfactorily. It is therefore not surprising that he accused traditional metaphysics of the forgetfulness of Being. In the modern world, the question of Being has been forgotten. This forgetfulness of Being, Heidegger holds, is responsible for the decline of man's history on this planet. Western tradition has completely derailed from a contemplation of Being as such, which is the '*raison d'être*' of any philosophical system. Heidegger concedes that at the very dawn of philosophy, Greek thinkers like Anaximander, Heraclitus and Parmenides posed the question of Being correctly. In their wonder, which is the beginning of all philosophy, they sought to explain the 'Being' of beings. But with Plato, subsequent philosophy diverted its attention from the Being of beings and concentrated its effort on this being or that being. But Being, the ground that nourishes all beings is left forgotten. John Caputo observes that it is in Plato and Aristotle that we see where the move to think Being as presence was first made, to which everyone else from then on just subscribed without question.

> They raised the question of Being for the first time. They let Being break out in all its wonder, and they did so in terms of time. But they themselves missed the temporal clues that were functioning behind their backs. Hence they got as far as question of Being, but not as far as the temporal meaning of Being—which can only be flushed out in a hermeneutical reflection that focuses on how ontological theories come to be.[17]

For Heidegger, the fact that the question of Being has not been answered is a dilemma of the maximum magnitude. As he writes; "*Basically, all ontology, no matter how rich and firmly compacted a system of categories it has at its disposal, remains blind and perverted from its ownmost aim, if it has not first adequately clarified the meaning of Being, and conceived this clarification as its fundamental task.*[18]

Heidegger's philosophical inquiry then was guided by a singular purpose; all of his efforts were working towards the end of discovering what it meant "to be." In a series of seminars with French contemporaries during the years 1966–1973, he made the question of the meaning of Being the central topic of the discussion.

---

17 Caputo, D. J. *Demythologizing Heidegger*, Bloomington, Indianapolis: Indiana University Press, 1993, p. 13.

18 *SZ.*, p. 11; *BT.*, p. 31. (Italics in the text).

Here we see him returning time after time to the task of expounding the meaning of Being, and the blueprint text for one of the last sessions in *Zähringen* in 1973 sums up the matter plainly and categorically: "We must constantly emphasize that the only question that has even moved Heidegger is the question of Being: What does 'Being' mean?"[19] Also, in an open letter written shortly before he died in May 1976, he stated in unequivocal terms: "The question, with which I greet you, is the only question that even up to the present hour I seek to inquire into even more inquiringly. One knows this question under the title 'the question of Being."[20] Likewise Jeff Malpas made a far-reaching argument that all of Heidegger's thought can be construed as an attempt to articulate the "place of Being."[21]

The term "beings" translates *das Seiende,* more literally "that which is". Thus, "beings", and its synonym "entities", refer to anything at all that has existence of some sort. Undoubtedly atoms and molecules are beings. Humans and animals are beings, so also their properties and activities.[22] The problem arising from this universality is that when we attempt to study ontology, we find that Being, which initially seems the simplest of notions, is actually the most mysterious. It is therefore not surprising when, in an untitled page prior to Sec. 1, Heidegger started his *magnum opus, Sein und Zeit* with a quotation from Plato's *Sophist* 244a that presented the matter in a more complex and subtle way—"For manifestly you have been aware of what you mean when you use the expression "being". We, however, who use to think we understood it, have now become perplexed."[23] Heidegger

---

19 Heidegger, M. *Seminare, Gesamtausgabe*, 15, Frankfurt am Main: Vittorio Klostermann, 1986, pp. 267–421. The reference is the seminar in Zähringen, September 6, 1973, p. 37. Cf. Heidegger, M., *Four Seminars*, trans. Andrew Mitchell & François Raffoul, Bloomington: Indiana University Press, 2003, p. 67.

20 Heidegger, M., *Reden und andere Zeugnisse eines Lebensweges, Gesamtausgabe,* 16, Frankfurt am Main: Vittorio Klostermann, 2000, p. 747.

21 Malpas, J., *Heidegger's Topology: Being, Place, World,* Cambridge, MA: MIT Press, 2006, p. 306.

22 Pilot, R., *Heidegger An Introduction*, Ithaca, New York: Cornell University Press, 1999, p. 2.

23 In the *Sophist* 244a, the Eleatic Visitor tells the early Greek physicists that he does not know what they appear to know: what they mean when they use the term "being" *(on; Seiend).* They proceed to answer their own question: namely through "storytelling". They speak of warfare and love, and say that in such happenings beings come to be; that they arise from other beings and pass away into other beings. In other words, people grasp beings through their generation and corruption. Heidegger's intention is to pose the same question to his contemporaries. With this citation from Plato's Sophist, it has been remarked that Being and Time begins 'in the middle of a Platonic dialogue.'

situates *Being and Time* in continuity with Greek Antiquity rather than with modernity and contemporary philosophy. *Being and Time* is thus altogether retrieval, albeit in a complex way, of the question of the meaning of Being.[24] As Dreyfus puts it; "What Martin Heidegger is after in *Being and Time* is nothing less than deepening our understanding of what it means for something (things, people, abstractions, language, etc.) to be."[25] He states further that "Heidegger's primary concern is to raise the question of being—to make sense of our ability to make sense of things—and to reawaken in people a feeling for the importance of this very obscure question."[26]

Even though several philosophers earlier has tackled very similar question, the approach Heidegger took generated an exceptional and intricate dimension toward the question. Palmer Richard explains that "from the beginning, Heidegger sought a method of going behind and to the root of Western conceptions of Being 'hermeneutics' that would enable him to render visible the presuppositions on which they have been based."[27] Hermeneutics means interpretation. "Hermeneutics concerns theories for correctly interpreting texts."[28] As Heidegger explains:

> The expression "hermeneutics derives from the Greek verb *hermeneuein*. That verb is related to the noun *hermeneus*, which is related to the name of the god Hermes. Hermes is the divine messenger. He brings the message of destiny. *Hermeneuein* is that exposition which brings tiding because it can listen to a message.[29]

Heidegger perceived hermeneutics as no less than a required tool to accomplish his general ambition of interpreting the meaning of Being. Hans-Georg Gadamer, the first philosopher who in *Truth and Method* developed Heidegger's account of

---

Cf. SALLIS, J., *Delimitations: Phenomenology and the End of Metaphysics*, Bloomington: Indiana University Press, 1986, p. 99. See also, MULHALL, S., *Inheritance and Originality*, Oxford: Oxford University Press, 2001, pp. 185–96.

24 Critchley, S. & Schürmann, R., *On Heidegger's Being and Time*, ed. Steven Levine, London: Routledge, 2008, p. 58.

25 DRYFUS, L. H., *Being-in-the-world: A Commentary on Heidegger's Being and Time, Division I*, Massachusetts: MIT Press, 1991, p. 1.

26 *Ibid*, p. 10.

27 PALMER, E. R., *Hermeneutics*, Evanston, IL: Northwestern University Press, 1969, p. 124.

28 SCHMIDT, L. K., *Understanding Hermeneutics*, Stocksfield, UK: Acumen Publishing Ltd., 2006, p. 1.

29 HEIDEGGER, M., *Unterwegs zur Spache*, 11., durchgesehene Aufl. (Gesamtausgabe 12) Stuttgart: Günther Neske, 1997, p. 122 (Hereafter US). Originally published by Verlag Günther Neske, Pfullingen, 1959. A translation is provided by PETER, D. HERTZ in *On the Way to Language*, San Francisco: Harper and Row, 1982.

interpretation into general hermeneutics defines it as the philosophical enterprise for which the central question is, How is understanding possible?[30] As he states later in the book: "Heidegger entered into the problems of historical hermeneutics and critique...in order to explicate the fore-structure of understanding for the purposes of ontology."[31]

With hermeneutics as a tool, maybe the primary question that ought to be answered then is: How did Heidegger conceive Being? It is scarcely significant to endeavour to determine the meaning of Being if someone does not, from the onset, have at least a trivial understanding of what Being *is*. The essential intricacy of this task is conspicuously revealed in Heidegger's conception of Being. He asked:

> But is Being a thing? Is Being like an actual being in time? *Is* Being at all? If it were, then we would incontestably have to recognize it as something which is and consequently discover it as such among other beings. This lecture hall *is*. The lecture hall *is* illuminated. We recognize the illuminated lecture hall at once and with no reservations as something that is. But where in the whole lecture hall do we find the "is"? Nowhere among things we find Being. Everything has its time. But Being is not a thing, is not a time. Yet Being as presencing remains determined as presence by time, by what is temporal.[32]

The implausible complexity Heidegger had due to his assumption becomes apparent at this juncture. His initial reception of radical nihilist philosophy of Nietzsche restricted his outlook to merely space and time. Consequently, he was compelled to locate his respond to the question of being within the limitations of a space-time gamut. Therefore, Heidegger must virtually contradict or disagree with himself in his understanding of Being. Since he already concluded that a clear-cut or definite understanding of Being was unattainable, Heidegger wanted an alternative track. He projected that *Being* may possibly be recognized in the

---

30 Gadamer, H-G., *Truth and Method*, 2nd ed., trans. J. Weinsheimer and D. G. Marshall, New York: Continuum, (1989) 2002, p. xxx. This book was originally to be titled *Fundametals of Philosophical Hermeneutics*. The publisher thought that "hermeneutics" was not known well enough for that word to be in the title, so the other word was chosen. For Gadamer hermeneutics is the philosophical theory of knowledge that claims all cases of understanding necessarily involve both interpretation and application. Cf. Schmdit, L. K., *Understanding Hermeneutics*, p. 2.

31 *Ibid.* p. 265 For more on Heidegger and Hermeneutics see for example Hoy Couzens David, "Heidegger and the Hermeneutic Turn," in Guignon B. C., ed., *The Cambridge Companion to Heidegger*, Cambridge: Cambridge University Press, 1993, pp. 170–194.

32 Heidegger, M., *On Time and Being*, translated and with Introduction by J. Stambaugh, Chicago: University of Chicago Press edition, 2002, p. 3.

course of the reflection on *beings* which on certain stage should possess or contain Being. Richard Pilot explains:

> Being is not a being at all; it is what marks being out as being rather than nonbeings—what makes the difference, so to speak, between something and nothing...*Being is the difference it makes that there is something rather than nothing.* Even if we cannot find a cause for the totality of beings, we can investigate the meaning of Being, for it does make a difference that there are beings rather than nothing."[33]

The variation from this distinction makes the disparity apparent and gives Heidegger the escape route from the disagreement or contradiction about the character and nature of Being. Thus, he wanted to discern a means wherein he could arrive at the meaning of Being from an analysis of beings and 'how' of their beingness. This leads to the question of phenomenology. Since as a practical concept, phenomenology deals with the 'how' of what is to be analyzed, Heidegger's debt to Husserl cannot be overstressed. It is obvious that *Being and Time* was possible only on the basis of Husserl's work.[34] In a section on the preliminary conception of phenomenology, Heidegger remarked: "The following investigation would not have been possible if the ground had not been prepared by Edmund Husserl with whose *Logische Untersuchungen* phenomenology first emerged.[35] In fact, Palmer observed that:

> In the phenomenology of Edmund Husserl, Heidegger found conceptual tools unavailable to Dilthey or Nietzsche, and a method which might lay open the processes of being in human existence in such a way that being, and not simply one's own ideology, might come into view. For phenomenology had opened up the realm of the preconceptual apprehending of phenomena.[36]

---

33 Pilot, R., *Heidegger An Introduction*, p. 3. Erickson commented that on Heidegger's analysis, even though Being is not itself an entity, it is always the Being of an entity. See, Erickson, S. A., *Language and Being: An Analytic Phenomenology*, New Haven: Yale University Press, 1970, p. 18.

34 Pöggeler, O., *Der Denkweg Martin Heidegger's*, 2nd Auflage, Pfullingen: Günther Neske, 1983, p. 67.

35 *SZ.*, p. 38; *Being and Time*, p. 62. However, on this same page, Heidegger registered his opposition by stating that phenomenology is not to be practiced "in its *actuality* as a philosophical 'movement' (*Richtung*'). Higher than actuality stands *possibility*". Ibid.

36 Palmer, R. E. *Hermeneutics*, p. 124. One good introduction to Husserl is Joseph J. Kockelmans, *Edmund Husserl's Phenomenology*, West Lafayette, Indiana: Purdue University Press, 1994. On phenomenology in general, cf. H. Spiegelberg, *The Phenomenological Movement: A Historical Introduction*, 3rd ed., The Hague: Martinus Nijhoff, 1982. For Heidegger's personal account of the beginnings of phenomenology and the phenomenological problems that he viewed as decisive, see the "Preliminary Part" of

But then, phenomenology can only provide a correct 'way of access' if that to which such access is sought exists. This means that "Being (the existing) can only be seen in so far as it shows itself. But the self-manifestation of Being would be meaningless were it not for the fact that that very being for which Being is an issue is itself capable of a seeing which renders intelligible the self-manifestation of Being."[37] So Heidegger's concern was to overcome the individual beings in order to arrive at this self-manifestation of Being, which for him still remains vague.

> We do not know what 'Being' means. But even if we ask, 'What is "Being"?' we keep within an understanding of the 'is' though we are unable to fix conceptually what that 'is' signifies. We do not even know the horizon in terms of which that meaning is to be grasped and fixed. *But this vague average understanding of Being* is still a fact.[38]

The collapse which established ontology has experienced in this respect, and the vague ideas which thrive in it, generally, do not encourage us to have much self-confidence in its technique. Nevertheless, given that it is also not worthwhile to assume any technique *a priori* since we ought to let ourselves be directed by the things themselves which appear to us instantly, the fundamental principles of phenomenology could possibly present an appropriate technique or method for ontology. According to Heidegger;

> We must keep in mind that the expression 'phenomenon' signifies that which shows itself in itself, the manifest. Accordingly the φαινομενα or 'phenomena' are the totality of what lies in the light of day or can be brought to the light—what the Greeks sometimes identified simply with τα οντα (entities)[39]

---

his 1925 lecture course *History of the Concept of Time*. Dilthey formulates an empirically based methodology for the human sciences. The human sciences require their own unique methodology, which Dilthy terms understanding (*Verstehen*) as opposed to explanation (*Erklären*). Dilthy rarely uses the word "hermeneutics" in his discussions of understanding. He defines hermeunetics as "the theory of the rules of interpreting written monuments." Wilhelm Dilthy, *Selected Works*, vol. 4, Rudolf Makkreel & Frithjof Rodi (eds.), Princeton, NJ: Princeton University Press, 1996, p. 238. This definition shows that Dilthy's hermeneutics is to be understood in the broad sense as a philosophical theory that justifies the universal validity of historical interpretations.

37 Macann, E. C., *Four Phenomenological Philosophers: Husserl, Heidegger, Sartre, Merleau-Ponty*, London: Routledge, 1993 (Taylor and Francis edition, 2005), p. 70.

38 *SZ.* p. 5. *BT.*, p. 25.

39 *SZ.* p. 28. *BT.*, p. 51. Phenomenology indicates primarily a principle of method, which can best be formulated in Husserl's phrase: "Back to the things themselves." This expression does not mean that one should return to naïve realism; but it indicates that in philosophy one should renounce all principles and ideas that are insufficiently explained or incorrectly founded, all arbitrary ways of thinking and all prejudices, and

In phenomenology, Heidegger had established the ideal technique for his venture. As soon as he had decided how he would execute his investigation, he had to apply this technique in such a way that would correspond with his understanding of Being in concrete terms. Moran puts it briefly:

> Heidegger wanted to employ phenomenology as the proper mode of access to the phenomena of concrete human life, factical life, as he had initially called it in his early lecture courses, a way of thinking about human nature that remained faithful to the *historical, lived*, practical nature of human experience.[40]

Here Heidegger's idea of the essential connection between Being and time becomes crucial. He supposed, that Being was discernible in beings which are enmeshed in time. He concluded as a result, that the phenomenon he wanted to comprehend was how Being manifests itself in the phenomenon of *Dasein* (*the being-there*) which exists no more than in time. This means that the comprehending of Being is automatically existential, it is comprehended in the appraisal of the existence or experience of Dasein. This also means that, the comprehending of Being have to be historical because the simple things obtainable to Dasein are the experiences of "being-there" which it has previously had. Heidegger describes this as an "authentic understanding," of Being, namely, knowledge of what Dasein is. As Grondin Jean states categorically: "The goal is to give Dasein knowledge of itself."[41] Heidegger's line of reasoning can be summarized as follows: Being is inextricably connected to time and thus to history; if that is the case, then the comprehending of being lies in the discernment of the procedures by which one's history is presented to one's Dasein; as a result, in examining the process of how

be guided only by the things themselves. Cf. Kockleman, J. J., *Martin Heidegger, A First Introduction to His Philosophy*, Pittsburg, Pa: Duquesne University Press, 1965, pp. 18–22, see also Richardson, W. J., *Heidegger, Through Phenomenology to Thought*, New York. Fordham University Press, fourth edition, 2003, pp. 46–47. Originally The Hague: Nijhoff, 1963.

40 Moran, D., *Introduction to Phenomenology*, London: Routledge, 2000, pp. 227–8.

41 Grondin, J., *Introduction to Philosophical Hermeneutics*, New Haven: Yale University Press, 1991, pp. 98–99. The "authentic" understanding of what Dasein is is arrived at, through Dasein's awareness of its own possibilities and potentialities of being. As Charles Guignon explains; "In becoming authentic, Dasein recognizes the most binding possibilities for its life and takes them over as its own. But it also comes to see its own life as dovetailing into the wider project of a 'destiny' it shares with its community". Cf. Guignon, C., *The Twofold Task: Heidegger's Foundational Historicism in "Being and Time", Tulane Studies in Philosophy: The Thought of Martin Heidegger*, vol. 32, ed. M. E., Zimmerman, New Orleans: Tulane University Press, 1984, pp. 55–56.

the phenomena of being is presented or comprehended by Dasein, we can possibly grasp what Being is. Heidegger's next step, having arrived at both the method and the target of his submission, was to identify precisely how being was manifest to Dasein. To do this he first needed to recognize the activity of Dasein.

1.2.1. Priority of Dasein over being and entities

There are many beings. Even though all of them are correctly called being, each entity is a being in a particular sense. The question for philosophy is to ascertain which of these beings that can show the way to the recognition of the accurate meaning of Being in its appropriate sense. A closer look indicates that one being could provide a firm starting point for the question about the meaning of Being. This being is the being that can question *itself* about its own being—man himself. Given that every question presupposes some pre-knowledge of the questioned and since only the entity that exists, questions about its own existence, only the questioning entity has some comprehension of Being and gets its vital character from it. As a result, only by a profound analysis of man's being or *Dasein*[42] can we arrive at an insight into Being of beings. Heidegger puts it succinctly: "This entity which each of us is himself and which includes inquiry as one of the possibilities of its Being, we shall denote by the term 'Dasein'."[43] For us to pose the question of

42 We follow the practice of not translating the German word Dasein. If this word were always used in its traditional meaning of "actual existence," as in the standard expression *der Kamft ums Dasein* (the struggle for survival), there would be no objection against rendering it such in English. But in addition to using it in this sense, Heidegger employs it first and foremost in the sense of "human *Dasein*" and as such he gives it a new and unique meaning. The various attempts at translating it cannot be considered successful. Expression as "human being," "human existence," or even "thing-in-being called man," fail to capture the specific Heideggerian meaning of Dasein (cf. Spiegelberg, H., *The Phenomenological Movement*, vol. 1, The Hague, 1960, p. 285, note 1). *Dasein,* "There-being," according to Richardson, "is the There of Being among beings—it lets beings be (manifest), thereby rendering all encounter with them possible" (cf. Richardson, W. J., *Heidegger, Through Phenomenology to Thought*, p. 40.

43 *SZ.* p. 7; *BT.*, p. 27. In his Marburg-lectures, Heidegger dwelt at length on the reasons why he prefers the neutral expression '*Dasein*' rather than '*Mensch*' (man), "...Das Dasein in seiner Neutralität ist nicht indifferent Niemand und Jeder, sondern die ursprüngliche Positivität und Mächtigkeit des Wesens... Die Neutralität ist nicht die Nichtigkeit einer Abstraktion, sondern gerade die Mächtigkeit des *Urspunges*, der in sich die innere Möglichkeit eines jeden konkreten faktischen Menschtums trät." Cf. Heidegger, M., *Metaphysiche Anfangsgründe der Logik im Ausgang von Leibniz*, Gesamtausgabe, Abt. II, Bd. 26, Frankfurt am Main: Vittorio Klostermann, 1978, p. 172.

being reasonably and realistically, "we must first clarify the Being of Dasein; it is from our everyday understanding of our own Being that we must attempt to unfold a more profound understanding of the question of the meaning of Being."[44]

But first we have to understand the distinction between the concepts *Being* (*Sein*) and *entity* (*Seinende*). What does 'entity' mean? We find the concept already in the Greek origin: it is the *οῶ* (*ov*). In the Medieval era, the concept *ενσ* (*ens*) corresponds to it. God is an entity; the house is an entity; the table is an entity. In order to express the distinction between God who created all entities and man who is one entity among others, man and everything created is called a *finite* entity, an *ens creatum*, while God is an *ens infinitum*—more accurately, *the* infinite entity, since there are no others.[45]

A preliminary understanding of Heidegger's use of the expression "Dasein" lies in a contemplation of the relationship between the question of Being and the human being itself. In this sense, an analysis of our human "existence"[46] constitutes an indispensable condition for an authentic ontology. Heidegger maintains that the attempt to realize and articulate a fundamental ontological understanding cannot correctly begin with the third-personal being of substantial entity and the clear-cut structure in which it is spelt out per se. To do so is already to have interpreted the fundamental phenomena of experience in a unique way, but all such interpretations presupposes the structures of a more essential way of

44 MULHALL, S., *The Routledge Guidebook to Heidegger's Being and Time*, New York: Routledge, 2013, p. 14.

45 BIEMEL, W., "Heidegger's Concept of Dasein" in *Heidegger's Existential Analytic*, F. ELLISTON, (ed.), The Hague: Mouton Publishers, 1978, p. 125.

46 According to Spiegelberg, "Heidegger introduces the term 'existence' in a sense which clearly differs from all previous usages, Scholastic as well as Kierkegaardian". (SPIEGELBERG, Op. Cit, vol.1, p. 327) The scholastic meaning of this term was and is that of actuality as distinguished from essence. The meaning given to it by Kierkegaard, Jasper and others is indicated by Heidegger with the usual adjective *existenziell*. To characterize the specific sense which this word has in his own philosophy, he coined the neologism existenzial (Spiegelberg, op. cit. p. 301). An adequate translation of both terms will render the English "existentiel" for *existenziell* and "existential" for *existenzial*. The fact that the English "existential" has been used to translate existentiell" undoubtedly leads to much confusion (Cf. BOCHENSKI, I. M., *Contemporary European Philosophy*, Berkeley, 1957, p. 163). "existential" refers to the ontic dimension of existence; "existentiell" is used in reference to the analysis of the structural features of *Dasein's* existence (Cf. *SZ*. p. 12, *BT*. p. 33). In *On theEssence of Truth*, Heidegger introduced a new spelling *Ek-sistenz*. This has led to the translation of Dasein by the English neologism "Ek-sistent" (Cf. LANGAN, T. *The Meaning of Heidegger*, New York, 1959, p. 11.

being—that of an interpreter.[47] The first indication that this is so, he opines, may be found in the central complexity in traditional metaphysics—the relation of what a thing is to the fact that it is. "The fundamental characters of man's being are not properties and qualities, but ways in which it is possible for him to be. *Da-sein* expresses being, and nothing else. It is a translation into German of the word *existentia*, and its usual meaning is simply *real existence*."[48]

The fundamental nature of Dasein lies therefore in its existence.[49] While Dasein exist, other entities are. The German word '*Existenz*' means 'to stand out from'. Dasein 'ex-sists'[50]; that is to say, Dasein stands out from all other entities in the world. In asserting this, Heidegger intends to draw a distinction between the way in which Dasein exists and the manner in which other "beings" exist. Dasein is not merely disinterestedly present in the world as an extended thing. Its potentialities are what distinctively define it as "who" it is. Dasein is outstanding amongst all other beings because, first of all, it is able to raise the question of its own being, and secondly, it can decide to relate itself authentically or inauthentically to its potentialities. This second characteristic is why Heidegger states that "the being which is an issue for this entity in its very Being, is in each case mine.[51] Trees, Rock, sand, even Monkeys and dogs are incapable of this kind of existence since their essence lies not exclusively in their potentialities, but in their actuality. Heidegger further underscores this position along these lines: "Man alone exists. The rock is, but it does not exist. The tree is, but it does not exist. The horse is, but it does not exist. The angel is, but it does not exist. God is, but he does not exist."[52]

---

47 *SZ*. pp. 312–316; *BT*., pp. 360–364.

48 Magda, K., *A Guide to Heidegger's Being and Time*, edited by John Llewelyn, New York: State University of New York, 2001, p. 47.

49 *SZ*., 42; *BT*., p. 67.

50 Heidegger very often uses hyphenated words for the sake of emphasis. Here, the word 'ex-sist' (*ek-sistert*) points to Dasein's peculiar character of standing out among other entities.

51 *SZ*, p. 24; *BT*., p. 67. We can compare this to the treatment given by Max Stirner in his book *The Ego and Its Own* (Cambridge University Press, 1995). There, Stirner rejects the left Hegelian approach to 'Man'. He considers this approach an abstraction which ignores the dynamic and unique qualities possessed by each individual man. Instead, he speaks of "the Unique One" or of "Ownness" or of "the Un-man". The upshot is that, like Heidegger's Dasein, Stirner's Un-man is more like an activity than a thing. In its most uncorrupted existence, it genuinely faces its own possibilities with honesty and gusto.

52 Heidegger, M., *Was ist Metaphysik*? Frankfurt am Main: Vittorio Klostermann, 1975, p. 16. In this analysis, Heidegger does not necessarily reject the reality of entities like rock or tree, but only tries to emphasize the distinctive category of being of Dasein as existence.

They all have distinct natures which can be emphatically summed up. Hence Heidegger used the word *"Categories"* to differentiate entities whose character is not that of Dasein.[53] Dasein, however, is purely potential, and its characteristics are distinguished from those of objectively present things by the term "existential."[54] It distinguishes itself in the course of existing by constantly running prior to any categorical outline of its characteristics. Whereas things that are neutrally present are referred to as "whats," Dasein is referred to as a "who." It is, generally speaking, what we think of when we think of "human being."

The above considerations are indicative that "Dasein" is first introduced as a terminological maneuver in order to refer to the being that we ourselves are. However, Friedrich-Wilhelm von Herrmann warns that there lies in the use of the term Dasein an ambiguity which, when not considered, is subject to inadvertent misinterpretation.[55] The danger consist in the fact that on the one hand, Heidegger introduces the word "Dasein" as a terminological formulation for the being that we are, and on the other hand, chooses the term "Dasein" as "purely an expression of Being to designate this being.[56] Herrmann elucidates as follows:

> "Dasein" is the term for the "human" being. But what is expressed in it is precisely not the "human" being *as* a being, but rather the *constitution of Being* of this being (Dasein as a pure expression of Being). If the human being is defined as Dasein in *Being and Time*, it is done so in view of its constitution of Being. But to take account of this is not yet enough to steer clear of the greatest danger of misunderstanding to which *Being and Time* is exposed. Thus, what is characteristic of Dasein as a constitution of Being is grasped from the outset only when one sees in the "Da," in the disclosedness, not *only*

Dasein as existence "is set apart in the being of Dasein as the only existing being which can undertake an inquiry into Being in terms of his peculiar existence". Cf. ALDERMAN, H., "Heidegger on Being Human", *Philosophy Today*, no. 15, 1971, p. 20; HEIDEGGER, M., "The Way Back into the Ground of Metaphysics," in *Existentialism from Dostoevsky to Sartre*, Edited by KAUFMANN, W., New York: New American Library, 1975, p. 272.

53 *SZ.*, p. 45; *BT.*, p. 70.Since the nature of Dasein is characterized by existence, Heidegger calls Dasein's existence-structure 'existential' and distinguishes it from categories which are characteristic of beings other than Dasein. 'Existential' refers to the 'who' (Dasein), while 'categories' refers to 'what' (the present-at-hand-entities in the broadest sense). Cf. WERNER, M., *Heidegger and Tradition*, trans. THEODORE KISIEL, T. & MURRAY G., Evanston: Northwestern University Press, 1971, pp. 87–88.

54 *SZ.*, p. 42; *BT.*, p. 67.

55 FRIEDRICH-WILHELM VON HERRMANN, *Subjekt und Dasein: Interpretationen zu "Sein und Zeit,"* 2. Stark erw. Aufl. Frankfurt a. M.: Vittorio Klostermann, 1985, p. 21.

56 *SZ.* p. 12; *BT.*, p. 33. Der Title Dasein (ist) als reiner Seinsausdruck zur Bezeichnung dieses Seienden gewählt.

the character of Being of the human being but the *essential connection* between the character of Being of the human being and the Being as such of beings as a whole.[57]

The magnitude of Herrmann's precaution cannot be overestimated. We have already seen that what Heidegger raises most is the question of Being, not this or that being, but Being itself. The distinction between Being and beings is what Heidegger refers to as the ontological difference. Ontological in the sense that Dasein "is ontically distinguished by the fact that, in its very Being, that Being is an *issue* for it."[58] Keeping this ontological difference in mind, which can be taken to be same as the *Seinsfrage* (Being question) itself, we can then see why Heidegger maintained the use of the word "Dasein" and does not use the traditional and more familiar anthropological designations such as "I," "ego," "Consciousness," Subject," and "animal rationale" to refer to beings which we ourselves are.[59]

Therefore, Heidegger declares that Dasein has a little comprehension of Being, prior to his asking the question of Being. Other beings are open to Dasein in his association with them and he is capable of knowing what they are as well as how they are. Specifically, Dasein is capable of comprehending what makes them what they are, namely, their being. If Dasein did not have this primary understanding of Being (*Seinsverständnis*), he would by no means be capable of raising the question of Being in any way.

This feature of Dasein's understanding of Being makes Dasein ontically different from every other entities, although like any other entity, Dasein is an entity within the world. Hence, Heidegger concludes; "Dasein is ontically distinctive in that it is ontological."[60] Both ontically and ontologically, then, the human being

57 Herrnmann, F. W., *Subjekt und Dasein*, pp. 21–22.

58 *SZ*. p. 12; *BT*., p. 32.

59 Remmon, E. B., *Heidegger and a New Possibility of Dwelling*, Frankfurt/M: Peter Lang, (European Universities Studies, Series xx), 2003, p. 10. All these designations fail to capture what is decisive about the beings that we are. It is extremely important to be able to understand the radical character of Heidegger's designation of the Being of the human being as Dasein, a designation that is made possible by the fundamental insight of the ontological difference and which breaks through the entire philosophical tradition since antiquity.

60 *SZ*. p. 12; *BT*., p. 32., Here it is necessary to make some clarifications on the conflicting but related terms in use. ***Ontology*** is a philosophical investigation of Being. ***Ontological*** means pertaining to Being. ***Ontical*** means pertaining to particular facts about entities, without regard to their Being. For instance, "How old is the sun?" is an ontical question, while "What is the way of Being of stars?" is an ontological question. Ontical questions can be answered by experimental science, but ontological questions call for philosophy. Cf. Pilot, R., Heidegger, *An Introduction*, p. 34.

has priority because it is the only being that maintains a relationship with its existence (Existenz). "Existence", according to Heidegger, "is only possible on the grounds of the understanding of Being."[61] Thus our prior description of Dasein as existence is anchored in his understanding of Being.

Another feature of Dasein which is to be seen in connection with its 'existence' is 'mineness' (*Jemeinigkeit*). The interpretation of Being begins with the interpretation of the Being of the individual, because, through its own self consciousness, individual Dasein can be aware of Being itself and be able to question itself in its own Being.[62] Human existence cannot be grasped as an instance or special case of some genus of the things present-at-hand.[63] In a more accurate reading of Heidegger, Mulhall writes, "as long as Dasein exists, it can never achieve wholeness; it will always be ahead of itself, essentially related to a possibility, to something that it is not yet."[64] Dasein is steadily already engrossed in the *possibilities* of its being; it is always already *being-ahead-of-itself* as possibilities which are still to be realized. To other present-at-hand entities, their own being is never an issue. But Dasein is "that entity which in its Being has this very Being as an issue, comports itself towards its Being as its ownmost possibility."[65] Since human existence by its very nature is one's own and cannot be treated as a sampling of a class, "one must always use a *personal* pronoun when one addresses it: 'I am', 'you'."[66]

Dasein, therefore, has a choice or possibility to understand itself in terms of its ownmost Self, i.e., authentically (*eigentlich*) or in terms of the world and entities therein i.e., inauthentically (*uneigentlich*), which is the way it does proximally and

61 HEIDEGGER, M., *Kant and the Problem of Metaphysics*, fifth Edition, Enlarged, trans. RICHARD TAFT, Indianapolis: Indiana University Press, 1997, p. 159. Here, Heidegger explains that if understanding of Being did not occur, man could never be as the being which he is, and this would be so regardless of the wonderful faculties with which human beings have been equipped. Moreover, man is a being in the midst of beings in such a way that for man the being which he is himself and the being which he is not are always manifest. We call this mode of the human being existence.

62 DEREK, R. M., *Heidegger's Philosophy and Theory of the Self*, Aldershot, England: Ashgate Publishing Limited, 2001, p. 120.

63 SZ. p. 42; *BT.*, p. 68.

64 MULHALL, S., "Human Mortality: Heidegger on How to Portray the Impossible Possibility of Dasein" in *A Companion to Heidegger*, p. 298. As Heidegger puts it, Dasein's mode of being is such that something is always left outstanding, or say incomplete; but if Dasein cannot bring its own existence into view as a whole, then how could I produce an existential analytic of its own kind of being that might bring it into view as a whole.

65 SZ. p. 42; *BT.*, p. 68.

66 SZ. p. 42; *BT.*, p. 68.

for the most part in everydayness.[67] "As modes of Being, authenticity and inauthenticity...are both grounded in the fact that any Dasein whatsoever is characterized by mineness."[68] Thus authenticity is the standing out of individual Dasein from the Dasein of others by the disclosure of the world through the application of the projects of the individual Dasein in a world it finds already in place.[69] It is not merely "a matter of finding a better way of life", but "a matter of finding out what it is to be really, fully human."[70]

In a nutshell, Heidegger speaks of a threefold priority of Dasein over the other entities. George Vensus summarized these under the following: (i), *Ontically*, Dasein has priority because only Dasein can be characterized in terms of existence. He exceeds other beings, above and beyond his openness to Being. (ii), *Ontologically*, Dasein has priority over entities because he is capable of understanding Being (*Sein-Verstand*). (iii), *Fundamentally*, Dasein has priority, which Heidegger describes as an *ontico-ontological* priority, for the simple reason that by his understanding of Being, Dasein equally comprehends his own being, the being of other Daseins and that of entities. In this way, Dasein supplies the ontico-ontological

---

67 Heidegger calls this factor that leads to inauthentic collectivity of everydayness '*das Man*', which is rendered in English as the 'they' or the 'they-self'. Cf. *S.Z.*, pp. 126–127; *BT.*, pp. 163–164. Also see VYCINAS, V., *Earth and Gods: An Introduction to the Philosophy of Martin Heidegger*, The Hague: Martinus Nijhoff, 1961, pp. 30–31.

68 *SZ.* p. 43; *BT.*, p. 68. Authentic existence is that mode of relation to the other that promotes existence in the full sense. It lets the human stand out as human in freedom and responsibility because according to Heidegger, "freedom is the realization of the uniqueness of man" (Heidegger, M., *The Question of Being*, trans. W. KLUBACK & J.T. WILD, London: Vision Press Ltd. 1956, p. 15). To realize this uniqueness is not something automatic. Rather it involves a gradual development whereby man gradually rebels against the inauthentic collectivism of the 'they'. He extricates himself from the crowd in order to be fully himself. Rebellion in man is his refusal to be treated as an object. According to Heidegger, man is able to rebel against his inauthentic state only through the phenomena of dread/anxiety, conscience and death. Through these native traits in man, he becomes authentically himself. "Authenticity is clearly not a matter of growing into an ideal identity. It means rather deciding to remain affirmatively alert to what no one can change." (Scott, E. Charles, Care and Authenticity, in Davis, W. Bret, (ed.) *Martin Heidegger: Key Concepts*, Durham, UK: Acumen Publishing Limited, 2010, p. 65.

69 DEREK, R. M., *Heidegger's Philosophy and Theories of the Self*, p. 132.

70 GUIGNON, C., Authenticity and the Question of Being, in MCMANUS, D., ed., *Heidegger, Authenticity and the Self: Themes From Division Two of Being and Time*, New York: Routledge, 2015, p. 16.

environments for the likelihood of all additional ontology.[71] Thus, "*fundamental ontology* from which alone all other ontologies can take their rise, must be sought in the *existential analytic* of Dasein."[72] And so, to understand well the *Dasein* is a good prerequisite to understanding the *Sein*. In avoiding the traditional distinctions of reality in the lines of immanent-transcendent, effect-causality, object-subject, and noumenal-phenomenal (or Ideal-Real in Plato); and in positively returning to the existential concreteness of the here and now human being, in his historicity and humanity, ontology according to Heidegger, would pave a better way towards and a more realistic perception of Being.[73] Therefore, in Dasein we have the right access to the question about the meaning of Being and in so doing, generates a cavity that transcendentally grounds every additional spheres of investigation. Graeme Nicholson puts it neatly: The architecture of our science and thought is threefold. There is:

> Ontic knowledge that has been prepared by,
> An antecedent ontological knowledge; But the latter for its part has been prepared by,
> A certain understanding of being that is central to our life.[74]

While ontic means particular, specific, individual beings, ontological means the Being of those particular Beings. Thus Dasein, while being ontic, relates to the ontological: the Being of beings. It interrogates. It is also interrogated.[75] That interrogation will bring Heidegger to unravel the various structures and nature of Dasein as a being-in-the-world. That is our next preoccupation.

---

71 Cf. VENSUS A. G., *Authentic Human Destiny: The Paths of Shankara and Heidegger*, Washington, D.C.: C.R.V.P. Publication, 1988, p. 113.

72 SZ., p. 13; *BT.*, p. 32. Heidegger's theory of fundamental ontology is stated vividly in his *Kant und das Problem der Metaphysik*, 2. Aufl Frankfurt am Main: Vittorio Klostermann, 1951, pp. 13 and 209. According to Heidegger, "the unveiling of the constitution of the Being of Dasein is Ontology. Insofar as the ground for the possibility of metaphysics is found therein—the finitude of Dasein as its fundament—it is called Fundamental Ontology." *Kant and the Problem of Metaphysics*, p. 163. (see also pp. 164–173).

73 IROEGBU, P. O., *Metaphysics the Kpim of Philosophy*, Owerri, Nig.: International Universities Press, Ltd. 1995, p. 227.

74 GRAEME, N., *Illustrations of Being: Drawing upon Heidegger and upon Metaphysics*, London: Humanities Press International, Inc., 1992, p. 90.

75 *SZ.* p. 14; *BT.*, p. 35.

1.3. The Nature of Dasein as Being-in-the-World

From the very beginning, Heidegger wanted to make it a maxim worthy of our retention that Dasein is not merely a conscious subject. He demonstrates his resentment with the thought of Descartes, who inaugurated the subjective turn in modern philosophy with his attempt to prove that the ideas in the human mind correspond to the reality of the external world. Although Heidegger dedicated *Being and Time* to Husserl as his mentor, the subject/object dichotomy is manifestly obvious in his thought. Husserl's treatment of Intentionality and Consciousness, Experience and Meaning bears the unmistakable stain of Cartesian tradition.[76] In conceptualizing the human relation to the world as one in which subject contemplates objects, the Cartesian-Husserlian submission reduces all human experiences to the distinction between the perceiver and the perceived. To distance himself from this Cartesian and Husserlian conception, Heidegger maintains that Dasein is not a subjective ego detached from it's worldly surrounding.

Consequently, it is not surprising that Heidegger entitled the second chapter of Division One of *Sein und Zeit* "Being-in-the-World in General as the basic state of Dasein."[77] This means that Dasein is primordially and essentially directed to the world and therefore each manifestation of his being-man is a way to relate himself actively to the world. In the words of Gilles Vannier, "being-in-the-world is the *a priori* human condition."[78] Dasein is primordially and above all an intentional and self-transcending being. Only through his familiarity with the world does Dasein become himself. His being is being-in-the-world. There are three major terms in the expression of Being-in-the-world. These are "in-the-world" (*in-der-Welt*); "being-in" as such (*In-Sein*) and "the being of the being that is in the world (*das Seiende*).[79]

---

76 Phenomenology in Husserl's sense was elaborated into a distinctive philosophical position according to the pattern set up by Descartes, Kant and Fichte. On this point see Brisart's excellent paper: 'L'intentionalité comm. "titre d'un probleme central" selon Heidegger', *Cahiers du centre d'études phénoménologiques* (CEP 2) Louvain-la-Neuve: Cabay, 1982, pp. 32–84.

77 *SZ*. p. 53; *BT*., p. 78. "Das In-der-Welt-Sein überhaupt als Grundverfassung des Daseins."

78 Vannier, G., *L'existentialisme: Littérature et philosophie*, Paris: L'Harmattan, 2001, p. 29. "Etre et Temps pose les jalon d'un "philosophie qui replace les essences dans l'existence et ne pense pas qu'on partir de leur facticité" (Merleau-Ponty). Autrement dit, l'être-au-monde est un *a priori* de la condition humaine".

79 *SZ*., p. 53; *BT*., p. 78–79.

Since this being-in-the-world is the foremost and fundamental trait which we encounter as we look at Dasein, the analysis of human existence has to commence with the explanation of being-in-the-world. Having already made a consideration of the being of Dasein, we shall in this section dwell on the nature of Dasein by analyzing the manner in which Dasein is in his 'there', i.e., his being-in and the 'world' as archetypal manner of Dasein's 'being-in'.

1.3.1. Meaning of Dasein's 'Being-in' (In-Sein)

This is the fundamental state or condition of the Dasein as always in-something. It is in reality both as a discovery and in its ontological definition. The proposition "in" generally indicates a relation of the contained to the container; for instance, the bed is in the room and the room is in the house or the vestment is in the cupboard. The word "in" here has a spatial connotation, that is, a spatial relation between two or more beings. Heidegger gave a lucid clarification of the term in the following words:

> This expression [...] Being-in [...] designates the kind of Being which an entity has when it is 'in' another one, as the water is 'in' the glass, or the garment is 'in' the cupboard. By this 'in' we mean the relationship of Being which two entities extended 'in' space have to each other with regard to their location in that place. Both water and glass, garment and cupboard, are 'in' space 'at' a location, and both in the same way [...]. All entities whose Being 'in' one another can thus be described have the same kind of Being—that of Being-present-at-hand—as things occurring 'within' the world [...] they are of such a sort as to belong to entities whose kind of Being is not of the character of Dasein [...]. Being-in, on the other hand, is a state of Dasein's Being; it is existential [...]. "Being-in" is thus the formal existential expression for the Being of Dasein, which has Being-in-the-world as its essential state.[80]

The "in" demonstrates the way in which we are related to the world. We are in the world such that our being is immersed in it, enclosed in it, dwelling in it.[81] Thus in the expression "being-in-the-world," the particle "in" signifies that man's being can be understood simply in the course of Dasein's fundamental relationship to the world. Rather than referring to a spatial relation, the proposition "in" points out an affiliation "with" and a being "with." "In" as well as "with" expresses that *Dasein* is acquainted with something, is accustomed to something, is conversant with something, and that it takes this "something" to heart or in the words of Richardson, "to be "at home" or "to sojourn" in, to be entrusted with a privileged 'familiarity

---

80 *SZ.*, p. 54; *BT.*, pp. 79–80.

81 Iroegbu, P. O. *Kpim of Time: Eternity*, p. 71.

with the world-about".[82] Thus to be or being in the world is to enter into reality, world-reality, and to take part in its offers: positive and negative. We manifestly share in its joys and sorrows. Devoid of the structured existential reality, which we tag "the world," *Dasein* can neither be nor be understood. The assertion being-in-the-world does not simply express an actual location or state of affairs or position, but as Heidegger puts it; "The expression '*bin*' is connected with '*bei*', and so '*ich bin*' (I am) means in its turn "I reside" or "dwell alongside" the world, as that which is familiar to me in such and such a way.[83] Thus, 'being-in' denotes a private and existential 'in-hood' which entails the association of dwelling and participation. Therefore, 'being-in' means Dasein being familiar with his environment and in association to the entities of his environs.

Dasein's "being-in" can take various and diverse forms, all of which could be examined phenomenologically. "It consists in having to do with something, producing something, consuming something, abandoning something, interrogating something, considering and determining."[84] All these modalities seem to be actual ways of a basic form of "being-in" which Heidegger calls "concern" (*Besorgen*). "Because Being-in-the-world belongs essentially to Dasein, its Being-towards-the-world (*Sein-zur-Welt*) is essentially concern."[85] The term 'concern' is used here by Heidegger as an ontological expression and according to Michael Inwood, it ordinarily has two senses: first, 'caring, worrying' about something, and, secondly, 'taking care' of things. 'Care' (*Sorge*), as Heidegger uses the word, involves both these senses, but its meaning is more fundamental than either. Even one, who is, in the ordinary senses of the words uncaring or carefree, or careless, is in Heidegger's sense, caring or careful. It is because Dasein's being-in-the-world is care that we speak of its concern (*Besorgen*) about the ready-to-hand, such as shoes and hammers, and its solicitude (*Fürsorge*) for other people. But again, Inwood explains, concern and solicitude are compatible with neglect, contempt, and hatred; the only entities that lack care, concern, and solicitude are those that are wholly incapable

---

82 Richardson, W. J., *Heidegger: Through Phenomenology to Thought*, p. 52.

83 *SZ.*, p. 54; *BT.*, p. 80. "…ich wohne, halte mich auf bei…der Welt, als dem so und so Vertrauten." The preposition '*bei*', like '*an*', does not have quite the semantical range of any English preposition. Our 'alongside', with which we shall translate it when other devices seem less satisfactory, especially in the phrase 'Being alongside' ('*Sein bei*'), is often quite misleading; the sense here is closer at that of 'at' in such expression as 'at home'or 'at my father's, or that of the French '*chez*'.

84 Vensus A. G., *Authentic Human Destiny: The Paths of Shankara and Heidegger*, p. 105.

85 *SZ.* p. 57; *BT.*, p. 84.

of them, such as stones, trees, and animals. Care is distinct from specific attitudes such as willing, wishing, striving, or knowing.[86]

The transcendental relationship of Dasein to the world thus assumes primarily and primordially the form of "being concerned." Dasein *Sorge* was Heidegger's phrase for concern and caring, concerning the self and its existence. When faced with the world and other beings, the human being feels anxiety and dread. The world appears multifarious and dangerous. Consequently, the human being, Dasein, must care for itself as no one else can or will. Taking care of the self is a sign that the individual recognizes dangers in the life. Recognizing threats demonstrates an appreciation of the physical self. It is worthwhile to conclude that concern with the physical self precedes the consciousness of concern for the emotional self. Whereas an infant might want human contact unconsciously and unknowingly, it only understands the need for food and other basic physical needs. Care is correlative to the significance of the world. Only if Dasein is care can it dwell in a significant world, and only if it dwells in a significant world can Dasein be care.[87] So care-taking or 'being-in' is the essence of Dasein's being.[88]

The most central feature of Dasein's 'being-in' is what Heidegger refers to as "*Erschlossenheit*", (disclosedness) of Dasein. "*Dasein is its disclosedness*."[89] The "There" (*Da*), which belongs to the expression "Dasein" (Being there) means disclosure, or more literally, disclosedness (*Erschlossenheit*). Concern is a form of disclosedness in the sense that in our engagement with things, things are disclosed. But it is also a form of disclosedness of Dasein in the sense that Dasein discloses itself to itself in its engagement with things.[90] Dasein is unavoidably connected to the world in which he lives. He is inescapably related to it in various aspects of his being-in-the-world. This makes him open up to the world. He is uncovered in the process of opening-up. Hence, Dasein's 'being-in' is the clearing (*Lichtung*) wherein the world is revealed or unveiled (Heidegger takes up this

86 Inwood, M., *Heidegger, A Very Short Introduction*, Oxford: Oxford University Press, 2000, p. 58.

87 *Ibid.*, p. 59.

88 *SZ.* p. 57; *BT.*, p. 84. For more on this cf. Also Gilbert Ryle, "Heidegger's Sein und Zeit", in *Heidegger and Modern Philosophy: Critical Essays*, ed. M. Murray, London: Yale University Press, 1978, pp. 58–59.

89 *SZ.* p. 133; *BT*, p. 171.

90 Gorner, P., *Heidegger's Being and Time An Introduction*, Cambridge: Cambridge University Press, 2007, pp. 70–71. For reasons of clarity, Heidegger uses the terms 'uncover' or 'dis-cover' (*Entdecken*) and uncoveredness or dis-coveredness *(Entdecktheit)* for the disclosing and disclosedness of entities other than Dasein.

question further in his explanation of the Truth of Being as we shall see in chapter two). He is disclosed to the various elements that exist: to nature, to fellow Daseins, to the society, and to the particular community in which he lives, and subsequently to the entire world. According to Heidegger, "to say that Dasein is 'illuminated' ("*erleuchtet*") means that as being-in-the-world it is cleared (*gelichtet*) in itself, not through any other entity, but in such a way that it *is* itself the clearing."[91] It is the clearing of Dasein towards all entities of the world and is the basis of his familiarity with the world; it enables Dasein to encounter entities and be involved with them.[92]

This open relationship to the world likewise opens the world for more discoveries. Dasein searches, researches, explore and in the process discover more about the world. There is mutual exchange, however. While Dasein discovers more about the world, he is also discovered. As Heidegger states: "To say that in existing, Dasein is its "there", is equivalent to saying that the world is 'there'; its *Being-there* is Being-in."[93]

This state of being can affect Dasein negatively. It imposes itself on Dasein. Unable to surpass his limited nature, the Dasein caves into the pressure of weakness, fear and oppression from the nature of the cosmos. This is what Heidegger calls *Verfallenheit* (Fallenness).[94] Dasein falls! He tends towards less being, with less dynamism and with more and more submission to the cosmic forces that are often oppressive. He struggles with his emotions and passions. This leads Heidegger to two fundamental modes of the disclosedness of Dasein's being. These are what he refers to as 'Being-in-a-mood' or 'State-of-mind' (*Befindlichkeit*) and Understanding (*Verstehen*). State-of-mind and understanding are characterized equiprimordially by *discourse*.[95]

1.3.2. "Moodness" (Befindlichkeit)

Befindlichkeit, as coined by Heidegger, is the first focal consciousness of oneself as 'being-in-the-world'. It is the ontological characterization of what we are ontically familiar with as 'moods' (*Stimmung*)[96]. Derived from the German phrase,

---

91 *Ibid.*

92 *Ibid.* See also Richardson, J., *Existential Epistemology: A Heideggerian Critique of Cartesian Project*, Oxford: Clarendon Press, 1986, p. 25.

93 SZ., p. 133; *BT.*, p. 171.

94 *SZ.*, p. 134; *BT.*, p. 172.

95 *SZ.*, p. 133; *BT.*, p. 172.

96 Richardson, W. J., *Heidegger, Through Phenomenology to Thought,* p. 65.

"*Wie befinden Sie sich?*," it refers to the mode or manner Dasein is 'placed' or 'to be found' (*Sich finden*) in life and in the world. Richardson describes it as "the already-having-found-itself-affectively-there-ness of There-being."[97] It is the ontological possibility of having particular mood. It is the existential structure of affectivity. It is a primary way of being human.[98] No matter what we do, we always find ourselves in a certain mood. In *The Basic Problems*, Heidegger defines *Befindlichkeit* thus: "To be affectively self-finding is the formal structure of what we call mood, passion, affect and the likes, which are constitutive for all comportment towards beings."[99]

Being in a mood is the state of mind in which Dasein is adjusted in his everyday dealing with the world. A particular mood reveals how one is and how one is faring in any given moment. So, what Befindlichkeit designates "is our always finding ourselves in a situation (in a world), with a particular orientation to that situation, constituted by such things as moods, disposition, inclinantion, beliefs, past experience and so on."[100] It tells the story by saying something about Dasein. As such, often Dasein is overshadowed by moods and he does not decide the particular situation in which he finds himself. The 'there' which Dasein always has is its 'situation.' The 'there' or human situation is constituted by *Situatedness* together with *Understanding* and *Discourse*. Because of this basic mood, man realizes his own situation in the world: "Every mode of historical man's comportment—whether accentuated or not, whether understood or not—is attuned, and by this attunement is drawn up into beings as a whole."[101] Attunement is one way Dasein discloses or uncovers itself to itself in its being-in. It is precognitive and prior to any psychological explication.[102] It cannot therefore be denied that mood is something primordial which is characteristic of our being man. Dasein is always in some mood or other, and shows itself; "even in the most indifferent and

---

97 *Ibid.*, p. 66.

98 Ballard, W. B., *The Role of Mood in Heidegger's Ontology*, Lanham, Maryland: University Press of America, 1991, p. 26.

99 Heidegger, M., *The Basic problems of Phenomenology*, trans. Hofstadter A., Bloomington: Indiana University Press, 1982, First Midland Book Edition, 1988, p. 281.

100 CERBONE, R. David, *Understanding Pheomenology*, Routledge, 2014, p. 54.

101 Heidegger, M., *Vom Wesen der Wahrheit*, First published in 1943, Frankfurt a. M, p. 9 in *Wegmarken (Weg)* 3, durchgesehene Aufl. Frankfurt a. M.: Vittorio Klostermann, 1996, 177–202. This translation is from *On the Essence of Truth* in *Basic Writings*, Revised and Expanded Edition, D. F. Krell, (ed.), New York: HarperCollins Publishers, 1993, p. 129.

102 Schmidt, L. K., *Understanding Hermeneutics*, p. 69.

inoffensive everydayness the Being of Dasein can burst forth as a naked 'that it is and has to be'."[103] This means that moods indicate not only that we are thrown into the world, but that we *are* and *have to be*.[104] As a result, Dasein does not initiate his existence; he rather finds himself delivered over already as existing, but of which he is responsible.

> The 'there' of Dasein is something to which Dasein is delivered over. I am responsible for what I make myself, how I exist, which possibilities of being I realize, but I am not responsible for having this responsibility. I find myself existing and with the responsibility of existing.[105]

Undoubtedly, the mood tells us something about our own being in its relation to other beings, but it is very difficult to specify why one is in a certain mood and what this mood tells us about our own being. These character of the unavoidable, the unknowability of his origin and destiny—the 'whence' (*woher*) and the 'whither' (*wohin*) of Dasein—his sheer thatness, which remain concealed is called "thrownness" (*Geworfenheit*).[106]

In *Befindlichkeit* therefore, Dasein comes to express its mode or passion arising from its being affected in his thrownness into the world. This thrownness, this situatedness of Dasein must be conceived as the "facticity (*Faktizität*) of its being delivered over."[107] According to Heidegger, "the mood brings Dasein before the "that-it-is" of its 'there', which, as such, stares it in the face with the inexorability of an enigma."[108] Without wanting it and without having freely chosen to be, man *is*. Therefore, moods reveal that Dasein's Being is something of which Dasein has been handed over and which, in its existential constitution, Dasein has to be. That Dasein is, and that it has to be in the mode of existence manifest itself, but the whence and wither of his Being remains shrouded in darkness. In this respect, Dasein's Being is a paradox. Dasein may come to feel assured that "it knows where it is going," and in scientific and technological enlightenment, it may achieve some understanding of its origins, but this sense of purpose and scientific knowledge "counts for nothing" as compared with the more fundamental enigmatic truth disclosed in moods. Thus, facticity refers to the inevitable and unalterable feature of Dasein's thrownness. Not even our genetic lineage and decent can respond to

103 *SZ.*, p. 134; *BT*, p. 173 (als nacktes "Daß es ist und zu sein hat).
104 Critchley, S. & Schürmann, R., *On Heidegger's Being and Time*, p. 87.
105 Ballard, W. B., *The Role of Mood in Heidegger's Ontology*, p. 73.
106 *SZ.*, p. 135; *BT.*, p. 174.
107 *Ibid.*
108 *SZ.* p. 136; *BT.* p. 175.

the question of whence and wither we came into Being. And most strikingly, we have no idea to what destination our existence is headed for, except for our position with regard to death. Nevertheless for Heidegger, it is this dual ambiguity that makes the 'thrown' condition of human life the more total and concrete.

On another note, Heidegger uses the term *factical* rather than *factual* when describing the characteristics of a person or group because *persons* do not have properties in the same way that things do. Bruce Ballard gave a practical example: the factical characteristic of *being a student* is a state that simply does not occur of itself, nor is it caused in a way a natural event is caused. Being a student is a role which is acted out. It depends on an agreement between pupil, institution and faculty. It is a state requiring continual decision and renewal. In these and numerous other ways, characteristics of persons (factical) should be distinguished from properties (factual) of inanimate things.[109]

However, if Dasein is essentially being-in-the-world, then mood has to disclose to man not only his "being thrown," but also the world, other Daseins and things.

> Having a mood is not related to the psychical in the first instance, and is not itself an inner condition which then reaches forth in an enigmatical way and puts its mark on things and persons. It is in this that the second essential characteristic of state-of-mind shows itself. We have seen that the world, Dasein-with, and existence are equiprimordially disclosed; and state-of-mind is a basic existential species of their disclosedness, because this disclosedness is essentially Being-in-the-world.[110]

Dasein is not alone in this world into which he is thrown and existence of others is absolutely essential to it's factically Being-there. Therefore, to understand others within the world and to understand the thereness of others is to exist.

1.3.3. Understanding, Interpretation and Assertion

'**Understanding**' (*Verstehen*), therefore, is the second indispensable and existential structural aspect of Dasein's disclosedness. Conventionally speaking, 'understanding' has an allusion to the rational, coherent or articulate grasp of things outside oneself. However in Heideggerian usage, it is the manner wherein Dasein can surpass the factical thrownness of his state-of-being there.

Dasein not only possesses an existential possibility to be always in a mood, but it is primordially understanding. "Dasein is the possibility of Being-free for its ownmost potentiality-for-Being....Understanding is the Being of such

---

109 BALLARD, W. B., *The Role of Mood in Heidegger's Ontology*, p. 11.

110 *SZ.*, p. 137; *BT.*, p. 176.

potentiality-for-Being."[111] This primordial understanding is all the time already present in mood and all understanding in its turn is connected with a particular mood. To Dasein, as being-in-the-world, its own being and the world are already disclosed to some degree, and this being-disclosed is essentially a sort of understanding. *"Understanding is the existential Being of Dasein's own potentiality-for-Being; and it is so in such a way that this Being discloses in itself what its Being is capable of."*[112] Dasein is what it can be, it is its possibilities. Understanding discloses Dasein to itself in the sense that it discloses its possibilities. Heidegger calls this potentiality and disclosing possibilities of being "projection" (*Entwurf*). Primordial understanding always moves in the range of possibilities. "With equal primordiality the understanding projects Dasein's Being both upon its 'for-the-sake-of-which' and upon significance, as the worldhood of its current world."[113] As Gorner explains, "What thrownness (*Geworfenheit*) and projection (*Entwurf*) have in common is throwning (*werfen*). In affectedness I am thrown, in understanding I am throwing. I am throwing myself ahead of myself into possibilities, projecting myself into them"[114] Dasein then is thrown into a mode of being which is eventually determined by a project. By projecting itself, Dasein is its possibilities and therefore is what it becomes.

Understanding discloses Dasein's possibilities of being, and since comportment to entities other than itself is an indispensable feature of Dasein's being, understanding also discloses the possibilities of entities other than Dasein. "Not only is the world, *qua* world, disclosed as possible significance, but when that which is within-the-world is itself freed, this entity is freed for *its own* possibilities. That which is ready-to-hand is discovered as such in its service*ability*, its us*ability*, and its detriment*ality*."[115] But Dasein constantly goes beyond 'what-he-is' towards 'what-he-is-not-yet'. "As projecting, understanding is the kind of Being of Dasein, in which it *is* its possibilities as possibilities."[116]

---

111 *SZ.*, p. 144; *BT.*, p. 183.

112 *SZ.*, p. 144; *BT.*, p. 184. (Italics in the original)

113 *SZ.*, p. 145; *BT.*, p. 185.

114 Gorner, P., *Heidegger's Being and Time: An Introduction*, p. 80.

115 *SZ.*, p. 144; *BT.*, p. 184. "...Dien*lichkeit*, Verwend*barkeit*, Abträg*lichkeit*..." Heidegger italicized the part of these words which constitutes them in their possibility to such and such. For instance, when the hammer is disclosed in its 'usability', it is disclosed in its possibility to be 'usable' as a hammer.

116 *SZ.* p. 145; *BT.*, p. 185. "Das Verstehen ist, Entwerfen, die Seinsart des Daseins, in de res seine Möglichkeiten als Möglichkeiten ist."

The projective nature of understanding involves Dasein as being-in-the-world in its entirety. For this reason, primordial understanding implies fundamentally a certain view, a "sight," of things, of fellow beings and of Dasein itself. "In its projective character, understanding goes to make up existentially what we call Dasein's "sight" (*Sicht*).[117] Heidegger wants to state here that being is disclosed through primordial understanding in what it is and as it is. This "sight" or view consequently has nothing to do with the operations of any of the senses as such. The precise view ("sight") of Dasein which constitutes Dasein as such is circumspection (*Umsicht*), attention, (*Rüchsicht*) or transparency (*Durchsichtigkeit*), according to Dasein's fundamental ways of being, that is, insofar as Dasein's view is concerned with equipment, fellow beings and Dasein itself respectively.

> Primordial understanding is always marked by a mood, and a mood is also always a bringing to light. In its primordial understanding marked by a mood, Dasein must always of necessity choose one of the three modalities of knowing namely, circumspection, attention or transparency.[118]

Interpretation (*Auslegung*): The projecting of understanding has the prospect of evolving itself or "making *explicit* of what is understood" which Heidegger refers to as interpretation (*Auslegung*).[119] Interpretation does not draw any new information about what is understood, but to a certain degree consists in working out the possibilities which are already projected in understanding. In the words of Heidegger, "in it (i.e., interpretation) the understanding appropriates understandingly that which is understood by it. In interpretation, understanding does not become something different. It becomes itself."[120]

Heidegger highlights two moments involved in the act of interpretation, namely, 'as-structure' (*Als-Struktur*) and the 'fore-structure' (*Vor-Struktur*).[121] In his interpretative circumspection, Dasein understands or sees entities as being for such

117 *SZ.* p. 146; *BT.*, p. 186.

118 Kockelmans, J. J., *Martin Heidegger, A First Introduction to his Philosophy*, Pittsburgh, Pa: Duquesne University Press, 1965, p. 71.

119 Gorner, P. *Heidegger's Being and Time: An Introduction*, p. 82. Heidegger made use of two German terms for the English term "Interpretation", viz., "*Interpretation*" and "*Auslegung*". While "*Auslegung* is used in the broader and primordial sense of any activity or better, Dasein's activity that 'lays-bare' or 'laying-out' (*aus + legen*) something as something, *Interpretation* is more technical and theoretical, as in the exegetical interpretation of a text.

120 *SZ.* p. 148; *BT.*, p. 188. "In ihr eignet sich das Verstehen sein Verstandenes verstehend zu."

121 *Ibid.* See also Vensus A. George, *Authentic Human Destiny*, p. 111.

and such a purpose, i.e., 'in-order-to'. Interpretation is this explicit understanding. "What is explicitly understood or interpreted has the structure of *something as something* (*Etwas als Etwas*)."[122] For instance, he sees a table "*as* a table, a door *as* a door, a chair as a chair, a bridge *as* a bridge and so on."[123] The 'as' that constitute the structure of the explicitness of what is understood, Heidegger calls 'as-structure.' In interpreting something, Dasein makes clear, what is *already there*. In the case of the entities of the world, what is *already there*, is the mode and purpose for which Dasein makes use of them as entities.[124] Hence, the 'as-structure' is made explicit in interpretation. It is the making explicit of what is already there in the entity as something within-the-world.

The explication or interpretation of the 'as-structure' is grounded in the 'fore-structure' (*Vor-Struktur*). Since interpretation does not necessarily involve *working out* the significance of something, but involves *presuppositions*, interpretation of something as something always essentially involves 'fore-structure' which according to Heidegger consists of 'fore-having' (*Vorhabe*), fore-sight (*Vorsicht*) and fore-conception (*Vorgriff*).

Fore-having consists of Dasein's understanding of his world, in its entirety, purpose and participation. It has to do with what Dasein has beforehand of any interpretation. As Philipse Herman explains; "The word *Vorhabe* is derived from the German verb *vorhaben*, which means to intend, to have planned. The term *Vorhabe* calls attention to the fact that in pro-jecting our Dasein, we 'have in advance' (*vor-haben*) a referential structure of instruments, institutions, and possibilities (*Bewandtnisganzheit*) that derives its point from Dasein as the ultimate 'for the sake of which', and that functions as a background for interpreting entities or texts."[125]

Fore-sight is the point of view (*Hinsicht*) which fixes that with regard to which what is understood is to be interpreted. It brings limits on fore-having by seeing something from a certain perspective. "Literally it means a previous looking towards…The movement of understanding from what is still unclear to expliciteness

---

122 GORNER, P., *Heidegger's Being and Time, An Introduction*, p. 82.

123 *Ibid.*

124 *SZ.*, pp. 148–150; *BT.*, pp. 188–190. Cf. also, PUTHENPURACKAL J. J., *Heidegger Through Authentic Totality to Total Authenticity*, Leuven: Leuven University Press, 1987, p. 30; VENSUS A. GEORGE, *Authentic Human Destiny*, p. 111.

125 HERMAN, P., *Heidegger's philosophy of Being: A Critical Interpretation*, New Jersey: Princeton University Press, 1998, p. 55.

'is always done under the guidance of a perspective which fixes that with regard to which what has been understood is to be interpreted'"[126]

Fore-conception (*Vorgriff*) is the conceptuality (*Begrifflichkeit*) in terms of which the interpretation is framed. "It denotes the conceptual structure that we beforehand have decided (*entschieden*) to use in order to understand something."[127]

Interpretation gives rise to "**assertions**" (*Aussage*). This means that the entity that is interpreted is expressed in an assertion. Heidegger considers assertion (judgement) as a derivative form of interpretation, which in the final analysis is grounded in understanding.[128] In expounding the full constitution of assertion, he structures three significations or characteristics of assertion.[129] The assertion does not necessarily, nor even primarily, has to be an expression in words; but it includes everything in which Dasein makes its primordial, previously interpreted understanding known, no matter in what way this is done. If this expression is made known through language, the reason is that "*discourse is existentially equiprimordial with state-of-mind and understanding*."[130] The understanding becoming conscious of the "as-structure"; namely, understanding a thing *qua* thing, Heidegger refers to as "meaning".

126 Schmidt, L. K., *Understanding Hermeneutics*, p. 72: *SZ*. 150; *BT*., p. 191.

127 Herman, P., *Heidegger's philosophy of Being: A Critical Interpretation*, p. 55.

128 *SZ*. p. 154; *BT*., p. 195.

129 *SZ*., p. 154–5; *BT*., pp. 196–97. Here Heidegger mentioned that the preliminary signification of assertion is "pointing out" (*Aufzeigen*) in the sense of 'exhibiting', or 'drawing attention to'. In this sense, Heidegger is referring to the original meaning of the Greek term *"logos"* ((λογος) as "*apophanasis*" (αποφανασισ). This means letting entity be seen in itself. For instance, the statement 'the hammer is too heavy' is not a representation of the hammer but the hammer itself. Secondly, assertions means no less than "*predication*" in which a subject is given a specific character by attributing to it a predicate, and thus determine the subject by the predicate. By virtue of predication it is not just the entity that is uncovered (the hammer) but the entity in its determinate character (the too heavy hammer). Lastly, assertion means "*communication*" (*Mitteilung*) or speaking forth (*Heraussage*). What is shared through communication is a common mode of concernful dealing with an entity. This sharing of being is not a case of somehow transferring a private mental content from one subject to another. As being-in-the-world Dasein is always already outside itself. In a nutshell the, "*assertion*" can be defined as "*a pointing-out which gives something a definite character and which communicates*" *Cf. SZ*., p. 156; *BT*., p. 199. Cf. Also, Vensus A. George, *Authentic Human Destiny*, p. 111ff; Gorner P., *Heidegger's Being and Time An Introduction*, p. 85.

130 *SZ*. pp. 161–62; *BT*., pp. 203–204.

Meaning (*Sinn*) is not a quality of a thing, but it is everything that a being is and means to Dasein.[131] It is that wherein the intelligibility (*Verständlichkeit*) of something maintains itself. When the entities within the world have come to be understood, we say that they have 'meaning. As Heidegger states it:

> When entities within-the-world are discovered along with the Being of Dasein—that is, when they have come to be understood—we say that they have *meaning* (*Sinn*). But that which is understood, taken strictly, is not the meaning but the entity, or alternatively, Being. Meaning is that wherein the intelligibility (*Verständlichkeit*) of something maintains itself. That which can be articulated in a disclosure by which we understand, we call "meaning". The *concept of meaning* embraces the formal existential framework of what necessarily belongs to that which an understanding interpretation Articulates. *Meaning is the "upon-which" of a projection in terms of which something becomes intelligible as something; it gets its structure from a fore-having, fore-sight, and a fore-conception.* In so far as understanding and interpretation make up the existential state of Being of the "there", "meaning" must be conceived as the formal-*existential* framework of the disclosedness which belongs to understanding. Meaning is an existentiale of Dasein, not a property attaching to entities, lying 'behind' them, or floating somewhere as an 'intermediate domain'.[132]

When a thing is accessible to us, the thing "has meaning". Meaning is the context that gives us access to the thing.[133] Given that this context comes from our own capacity to project possibilities, we can say that meaning "is an *existentiale* of Dasein."[134] It is suggestive of a mode of Being in the world. As Galven explains:

> I can at times focus upon the way I make use of the world. When such focusing occurs, the specific manner in which that part of the world becomes available to me is explicit. When such explication occurs, I say that the way in which I make use of the world is *meaningful*. Meaning, then, is a mode of my Being here in a world.[135]

An entity, therefore, has meaning only in relation to Dasein; outside this relation, it is not meaningless (*sinnlos*) but "unmeaningful" (*unsinniges*).[136]

Since the question of Dasein's 'being-in' has to do with his concernful dealing with the world and his being-affected (*Betroffenheit*) with his disclosedness

---

131 Kockelmann, J. J., Martin Heidegger, *A First Introduction to his Philosophy*, p. 72.

132 *SZ.* p. 151; *BT.*, p. 193. 'Sinn ist das durch Vorhabe, Vorsicht und Vorgriff strukturierte Woraufhin des Entwurfs, aus dem etwas als etwas verständlich wird'.

133 Polt, R., Heidegger An Introduction, p. 72.

134 *SZ.*, p. 151; *BT.*, p. 193.

135 GELVEN, M., *A Commentary on Heidegger's Being and Time*, Northern Illinois University Press, 1989, p. 98.

136 *SZ.*, p. 152; *BT.*, p. 193.

to the entities that exists in the world, we will now examine the last part of that unitary expression which is the 'world.'

1.4. The World

If "being-in" is the formal existential expression of the Being of Dasein in its dwelling, its being familiar with, its staying in close proximity with things and places, what then is the "world" "in" which Dasein always already finds itself? Heidegger used a range of variants to characterize the world. In *The Fundamental Concepts of Metaphysics*, he explains:

> The naïve concept of world is understood in such a way that world basically signifies beings, quite undifferentiated with respect to 'life' or 'existence,' but simply beings. In characterizing the way and manner in which the animal lives we then saw that if we can speak meaningfully of the world and world-formation of man, then world must signify something like the *accessibility of beings*. But we also saw in turn that with this characterization we get caught up in an essential difficulty and ambiguity. If we determine world in this way, then we can also say in a certain sense that the animal has a world, namely has access to something that we, *for our part*, experience as beings. But then we discovered that while the animal does have access to something, it does not have access to beings as such. From this it follows that world properly means *accessibility of beings as such*. Yet this accessibility is grounded upon a *manifestation of beings as such*. Finally, it was revealed that this is not a manifestation of just any kind whatsoever, but rather manifestness of beings as such as a whole.[137]

Since according to Heidegger, human beings found themselves always 'thrown' into the world from which they were inseparable, the world could not be bracketed from man. Existence is always a Being-in-the-world. The world is our context.[138] Here we shall first consider Heidegger's notion of the world in relation to Dasein and then, Dasein's relational concerns, i.e., the worldly character of the environment (*Umwelt*) and equipments, the wider world beyond the immediate 'world around us'; and finally the communal world (*Mitwelt*) or what Heidegger called 'Being-with'(*Mitsein*). All these will enable us to have a better understanding of Heidegger's interpretation of Being which is constitutionally in the world.

---

137 HEIDEGGER, M., *The Fundamental Concepts of Metaphysics: World, Finitude, Solitude*, trans. W. MCNEIL & N. WALKER, Bloomington and Indianapolis: Indiana University Press, 1995, p. 284. Published in German as *Die Grundbegriffe der Metaphysik. Welt—Endlichkeit—Einsamkeit*, Frankfurt am Main: Vittorio Klostermann, 1983, 1992.

138 HANS, W. COHN, *Heidegger and the Roots of Existential Therapy*, London: Continuum, 2002, p. 25.

1.4.1. *Dasein and World*

The 'world' which is the last part of the unitary word under consideration (i.e. Being-in-the-world) indicates the stark existential reality of our cosmos. It is this earth seen as our dwelling, our place of being, thinking and acting. Dasein, whether in its average everydayness or otherwise, is in the world. In what he titles 'the worldhood of the world', Heidegger explains the world as the phenomenon of the material contents of our universe. These are the entities that abide in the world. These include stones, trees, animals, mountains, hammers etc. These are the ontical (concrete beings) realities of the world. Dasein too is in the world the way they are. But Dasein is also in the world in another sense; he is aware of and familiar with the world, and aware of other entities in the world and of itself. So, at the level of ontology, we can pass over and discuss not these individual concrete entities, but their being *autant que tel* (as such).[139] This means that the "world" is not to be taken as a sum of things, but as another structural element of Dasein's definition which has an ontologico-existential undertone.

Furthermore, by saying that the concept "world" has an ontologico-existential meaning, Heidegger wants to point out that he does not intend to limit his search to a description of what man ordinarily calls "world," but wants to ascertain its essential structure, its being. He also wants to convey that this structure is founded on human existence: "Ontologically, 'world' is not a way of characterizing those entities which Dasein essentially is not; it is rather a characteristic of Dasein itself."[140] In non-Heideggerian terminology, a world is a system of purposes and meanings that organizes our activities and our identity, and within which entities can make sense to us.[141] The search for the worldhood of the world is an ontological preoccupation. In a précis form, Heidegger outlines different meanings which the term "world" can assume, namely:

---

139 Iroegbu, P. O., *Kpim of Time: Eternity*, p. 71. The distinction between ontological and ontic is derived from the distinction between Being and being. One can regard a being simply as it is. This is the ontic standpoint: it has to do with the Greek *on*, the *ens*, being. But one can only try to understand the Being of beings, that which makes this being be what it is, its fundamental and constituent structure. In this case, one does not stop with being as it is immediately given, but one tries to understand being *as* being, that is, the proper being of this being, in short, its Being. This is the ontological order. In his later works Heidegger also distinguishes between the being of beings and Being, and refers to the former as "beingness" (*Seiendheit*) of beings.

140 *SZ*. p. 64; *BT*., p. 92.

141 Pilot, R., *Heidegger An Introduction*, p. 54.

As an *ontical* concept, the totality of entities in the world. "World" signifies the totality of beings which can be present within the "universe."

As an *ontological* term, the Being of the beings of the world. In a derivative form "world" can also be used in this sense in reference to a particular realm encompassing a particular group of beings; for example, the "world" of the mathematician or the "world" of the scientist.

As the "wherein" the Dasein can live. It is the public, the open world; my personal world or our common world "'in which' a factical Dasein 'lives'". Accordingly, this sense of the world is *pre-ontological* and existentiell. Here, 'world' stands for the 'we-world' (*Wir-Welt*) with others and one's own closest world of the environment (*Umwelt*)[142]

As the *ontologico-existential* concept of the world-hood, that is, the being of the world of (*Weltlichkeit*) Dasein's 'wherein'; one that "designates the ontological and existential concept of *worldliness*." This meaning embraces in itself the *a priori* character of world-hold in general.[143]

Heidegger tells us that he would restrict himself specifically to the third sense of the world, that is, the 'world' taken in the pre-ontological and pre-thematic sense. World, "*Umwelt*", is that place, the wherein where Dasein dwells. There he is. There he inhabits. It is his world of abode. "*Umwelt* means a world that is round-about us, a world that is nearest, first at hand."[144] Understood in this sense, world has environmental and communal dimensions. Our surrounding world is the world that is nearest to us. "What first shows itself to us, as nearest and pre-ontological, is Dasein in everyday life and everydayness....What everydayness shows is how Dasein 'initially and for the most part' is in the world."[145] In everyday experience, we are involved with the world in a fundamentally practical way; we operate in the world using the entities we find there. In this unreflective way of being we encounter and use objects we find in the world according to the purpose we have. We behave as if we exist in a world of entities with which we can interact according to our choice....We know entities according to what we use them to do,

142 *SZ*. p. 65; *BT*., p. 93. What Heidegger means by 'we-world' is nothing but the 'communal world' (*Mitwelt*). '*Umwelt*' literally means 'the world around'. In his characteristic manner, Heidegger maintained the same prefix 'um' later when he speaks of the related notions, such as '*Umwelt*' (environmental world), '*Umgang*' (dealings or going about), '*Um-zu*' (in-order-to) etc.

143 *SZ*. p. 65; *BT*., p. 93.

144 Madga, K., A *Guide to Heidegger's Being and Time*, p. 68. Cf. SZ., p. 66; *BT*., p. 94. "The world of everyday Dasein which is closest to it is the *environment*."

145 Mugerauer, R., *Heidegger and Homecoming: The Leitmotif of the Later Writings*, Toronto: University of Toronto Press, 2008, pp. 28–29.

and the ways in which they impact on our existence.[146] Dasein is in its world, then, through engaged, concerned dwelling. Any human being is involved in a world, engaged in a sphere of concerns and issues.

Various human disciplines, namely, anthropology, psychology and biology have all botched to present an ontological sound treatment of human being because they erroneously treat humans as "whats" rather than as "whos". Dasein is a *subject* not an *object* and any treatment that neglects this existential characteristic fail to get to the primordial "truth" about human-being. Heidegger mentions Descartes by way of illustration in this respect.

In spite of the fact that Descartes effectively discovered "thinking being" (*rex cogitans*) as different from "extended being," (*rex extensa*) he spent most of his energies investigating "thinking" and abandoned the prior question of "being" itself. So, Heidegger concludes that Being remains unclarified in Descartes' thought.[147] He points out the rather contradictory nature of Descartes' modern ontology by posing the question: If the being of the thinking subject (*res cogitans*) is fundamentally different from the being of other things (*res extensa*), then is it possible to say that there is such a thing as being in general? His anxiety here is that, on the one hand, he is being critical of modern ontology insofar as it understands the being of both *res cogitans* and *res extensa* as present-at-hand, which means that both have essentially the same way of being; on the other hand, he asks how it is possible to draw such an inflexible distinction between the thinking subject and other beings and yet approve a unitary concept of being. If the subject is indeed so different from other entities, then it must have a different mode of being.

Heidegger, in fact suggests that Descartes passes right over the phenomenon of the world when he presumes that it can be known through knowing (*Erkennen*), i.e., "*intellectio* in the sense of the kind of knowledge (*Erkenntnis*) we get in mathematics and physics."[148] In reducing the world to "extended matter," Descartes radically split humans off from that world. "A world that one does not inhabit is a world in which one is essentially implicated and by which one is not essentially constrained."[149] Only by God's grace is a connection between the two permissible, and consequently Charles B. Guignon writes, "…the quest for certainty that

---

146 Derek, R. M., *Heidegger's Philosophy and Theories of the Self*, p. 133.

147 *SZ*. p. 95; *BT*., p. 128.

148 Ibid.

149 Mulhall, S., *Heidegger and Being and Time*, second Edition, London and New York: Routledge, 2005, p. 40.

motivates Cartesian foundationalism ends in frustration."[150] Descartes never asks about the being that is universal to both thinking and extended "matter" and so leaves humans estranged from their own environment. "The world in itself consists only of particles and energy; the world for us includes subjective values that we project on to things…The result of these assumptions is a technological approach to the world."[151] Furthermore, Descartes did not find the proper method of access to Being, because for him, Being itself remained determined by entities.[152] Heidegger is resolute that such a means of considering Being shuts off Dasein's Being as Being-in-the-world.[153]

A key element of Heidegger's mission will be to restore this fracture by indicating that the powerful legacy of Cartesian Dualism in anthropology, psychology and biology must be overcome. He does not so much want to demonstrate that this dualistic way of conceptualizing the world is false so much as he wants to show that it is inadequate and incomplete. He wants to "get behind" Cartesian thinking in order to reveal how this way of seeing the world is possible and grounded in a more primordial mode of existence. So, having challenged these modern assumptions, we can equally expect Heidegger to confront our technological mind-set to life as well as we shall see later in this work. At the moment let us look at the various aspects from which Heidegger viewed the world with particular reference to its environmental and communal undertones.

1.4.2. *The Worldhood of the World* (The Environmental World of Dasein)

In our everydayness in the surrounding world, we cannot live in isolation but are enmeshed in its cares. Our going about (*Umgang*) *in* the world does not take place without our having to do *with* things. It is a world, in which, Dasein is related to the entities and other Daseins, in circumspective concern and reverential solicitude respectively. "Dasein goes about in the world and goes about his business with the things he meets within it. *Umgang* basically means the practical, using and

---

150 Guignon, B. C., *Heidegger and the Problem of Knowledge*, Hackett Publishing Co., 1983, p. 29.

151 Pilot, R., *Heidegger, An Introduction*, p. 56. The subjective values in questions could be judgments about good and bad, beautiful and ugly which are "value judgments" that merely reflect our own desires, instead of saying something about the world. The right way to live, we presume, is to spell out our own desires or create new ones and then enforce our will on the external world.

152 *SZ.* p. 98; *BT.*, pp. 130–131.

153 *SZ.* p. 105; *BT.*, p. 140. "In Dasein there lies an essential tendency towards closeness."

handling way of taking care of things."[154] According to Heidegger, Greek thought was more primordial than ours because the Greeks were closer to that which primordially appeared to man. The old Greeks spoke about πραγματα *(pragmata)* in reference to that which one has to do with in one's πραξισ (praxis), one's concernful dealing with things.

Heidegger calls the beings that we go about with in everyday life "*equipment*"[155] or "useful things" (*das Zeug*).[156] In those dealings, Dasein is not primarily interested in mere knowledge but in action, in manipulating things and putting them in use. What could be more trifling than knowing that in our daily concern we deal with things? However, this objection takes for granted that we know what specifically makes a thing a thing. In what does the thingness of a thing actually reside? Some think that it lies in the reality or materiality of things, which could be further explained through extension; for others, things are "objects of value". Yet the things which we bump into in our everyday concernful dealing clearly do not appear to us as material objects or as objects of value.

Heidegger's own phenomenological concern with useful, everyday things (electric gadgets, tools and equipments, paraphernalia etc.) can be misleadingly simple, and we may fail to see its profound significance. "Dasein encounters are usable, employable in the pursuit of its purposes: in Heidegger's terms, they are not just present-at-hand, the object of theoretical contemplation, but handy or ready-to-hand."[157] Heidegger's key insight is that we never come across things initially as objectively present (*vorhanden*), but rather as something that is useful or handy, something that is ready to hand or as he puts it, "readiness-to-hand" (*zuhandenheit*).[158] Thus the distinction is made between entities present-at-hand and ready-to-hand:

---

154 Magda, K., *A Guide to Heidegger's Being and Time*, p. 69.

155 Heidegger, M., *The Basic problem of Phenomenology*, p. 163. What is given to us is never an object in isolation. Rather, what is primarily given is a contexture, a contextual whole of equipmentality, the nearest things that surround us "we call equipment."

156 "*das Zeug*": The word 'Zeug' has no precise English equivalent. While it may mean implement, instrument, or tool, Heidegger uses it for the most part as a collective noun which is analogous to our relatively specific 'gear' (as in 'gear for fishing') or the more elaborate 'paraphernalia', or still more general 'equipment'.

157 Mulhall, S., *Heidegger and Being and Time*, p. 41. The expressions '*present-at-hand*' and '*ready-to-hand*' corresponds to distinctions between 'theory' and 'practice' respectively. Heidegger does not reject the importance or role of theoretical knowledge but his submission is that 'theory' is a founded mode of 'practice'.

158 SZ., p. 69; *BT.*, p. 98.

> Presence-to-hand is neither a super-property nor a formal structure common to everything existent. Instead, it is one of the several ways in which we can *encounter* entities. It is to be contrasted, for example, with "readiness-at-hand" (*Zuhandenheit*), in which we encounter entities in terms of their usefulness (or uselessness) to our practical projects. Crucially, because presence-to-hand, readiness-to-hand are just different *ways of encountering* what Heidegger calls "intrawordly entities"—a term coextensive with "physical objects"—they are *not* different kinds of *entities*. For the same entity, a hammer, for example—could in principle be encountered in different ways of being: once as a present-to-hand object weighing two kilograms, and another time as a ready-to-hand item of equipment useful for hammering.[159]

However, Heidegger pointed out that there is no such thing as equipment:

> Taken strictly, there 'is' no such thing as an equipment. To the Being of any equipment there always belongs a totality of equipment, in which it can be this equipment that it is. Equipment is essentially 'something in-order-to...' *("etwas um-zu...")*. A totality of equipment is constituted by various ways of the 'in-order-to', such as serviceability, conduciveness, usability, manipulability.[160]

In this sense, we still speak of equipment for cutting and hammering in reference to the things used for cutting and driving nails. "We encounter our environment by 'looking around', and find equipment as 'in-order-to' do this or that."[161] A cup is there in order to drink, a knife in order to cut, a hammer in order to drive nails into the wood, the telephone, in order to talk etc. Equipment is by no means by itself, but is always and exclusively in allusion to other equipment. The question then is; what is distinctive, appropriate and characteristic of equipment, what is the "equipmentness" of equipment, that which makes equipment equipment? What constitutes this equipment manifold specifically as unity or its focal point?

The various pieces of equipment are connected by their "what...for" and this "what...for" gives the equipment manifold unity.[162] But this process does not go on indefinitely because all these in-order-to leads back to an ultimate

---

159 Boedeker, E., Phenomenology, in H. Dreyfus, & M. Wrathall, (eds.), *A Companion to Heidegger*, Malden, MA: Blackwell Publishing, 2005, p. 159. See also Puthenpurackal, J. J., *Heidegger Through Authentic Totality to Total Authenticity*, pp. 15, 19. For a detailed study and analysis of present-at-hand and ready-to-hand, see Gerold, Prauss, *Erkennen und Handle in Heidegger 'Sein und Zeit'* Freiburg: Karl Alber, 1977; Harman, G., *Tool-being: Heidegger and the Metaphysics of Objects*. Open Court Publishing, 2011.

160 *SZ.*, p. 68; *BT.*, p. 97.

161 Greaves, T., *Starting with Heidegger*, London: Continuum International Publishing Group, 2010. p. 39.

162 *SZ.*, p. 68; *BT.*, p. 97.

'for-the-sake-of-which' in Dasein where there is no further involvement. Thus, Dasein is the ultimate focal point to which the whole involvements are focused.

> But the totality of involvements itself goes back ultimately to a "towards-which" in which there is no further involvement: this "towards-which" is not an entity with the kind of Being that belongs to what is ready-to-hand within a world; it is rather an entity whose Being is defined as Being-in-the-world, and to whose state of Being, worldhood itself belongs.[163]

The ready-to-hand constantly manifests itself as an entity belonging to the world. That, which determines the ready-to-hand, has one way or another affiliation to the world and to its worldhood. "An entity is discovered when it has been assigned or referred to something, and referred as that entity which it is."[164] Within this world, Dasein goes about with equipment in its diverse cares and concerns (*Besorgen*). In such a world, Dasein always already understands, whether explicitly or not.[165] Concern is inseparable from 'world' in the sense of the structure of significance from out of which the ready-to-hand shows itself. A world-less subject would not be able to comport itself to *any* entities.[166] In this state of concern equipment and its being matter to Dasein. Equipment becomes equipment by mere things being given functional properties by human beings, in order words, its "serving to," its "being good for," or its usability or pragmatic value. This view which Dasein has regarding equipment, Heidegger calls circumspection (*Umsicht*),[167] the sight

---

163 *SZ.*, p. 84; *BT.*, p. 116.

164 *SZ.*, p. 84; *BT.*, p. 115.

165 Remmon, E. B., *Heidegger and New Possibility of Dwelling*, p. 20. The term "to understand" refers here to the knowledge which is still completely and immediately related to our being-concerned itself. For instance, one makes use of a hammer in the right way without explicitly understanding the proper mode of being of this piece of equipment or the 'hammerness' of the hammer. By using the hammer in the right way within a certain equipment manifold, Dasein has appropriated it in the most suitable way, for a hammer is not there to be looked at, but to hammer with. Thus, by using the hammer, Dasein, in its everyday concernful dealing with things, has to submit to the assignment that is constituent of this piece of equipment, namely, "what…for". Cf. Kockelmans, J. J., *Martin Heidegger, A First Introduction to His Philosophy*, p. 32.

166 Gorner, P., *Heidegger's Being and Time, An Introduction*, p. 46.

167 *SZ.*, p. 69; *BT.*, p. 98. The word 'Umsicht', which is translated by 'circumspection', is here presented as standing for a special kind of 'Sicht' ('sight'). Here, as elsewhere, Heidegger is taking advantage of the fact that the prefix 'um' may mean either 'around' or 'in order to'. 'Umsicht' may accordingly be thought of as meaning 'looking around' or 'looking around for something' or 'looking around for a way to get something done'. In ordinary German usage, 'Umsicht' seems to have much the

that occurs in the *Umwelt*, or environment. So our speaking of involvement of equipments towards Dasein amounts to Dasein letting them be involved with him. In order words, the letting-be-involved of entities, by Dasein, is the ontological condition for an entity being encountered in its readiness-to-hand. The 'letting-be-involved', of which we speak, is the *a priori* condition for any ontic involvement.[168] When this is understood ontologically, the 'letting-be-involved of entities by Dasein means disclosing the totality of involvements as the world within which entities can have their involvement.[169]

In this sense, any single ready-to-hand object, however isolated or self-contained it may seem, is encountered within a world of work.[170] It is the work to be produced that makes man go to the aforementioned "referential totality within which the equipment is encountered."[171] A bed is made of wood or iron and as such refers us to "nature." Finally the work produced has a reference to the "for-the-sake-of-whom", that is, the person who would use it.[172] This complex of references does not stand in isolation from each other; rather, they are closely interconnected which results in an equipmental-referential totality "for the world is defined, in effect, as the totality of all those roles in their essential interrelations."[173] Hence, the world of everyday Dasein "is an intersubjectively constituted referential totality, in which it becomes possible for Dasein to encounter other entities, both of its kind and of the ready-to-hand."[174] Sometimes however, the complexities of this equipmental interconnectedness remain unnoticed, overlooked or are

---

same connotation as our 'circumspection'—a kind of awareness in which one looks around before one decides just what one ought to do next. But Heidegger seems to be generalizing this notion as well as calling attention to the extent to which circumspection in the narrower sense occures in our ever-day living (cf. footnote).

168 Puthenpurackal, J. J. *Heidegger Through Authentic Totality to Total Authenticity*, p. 23.

169 Sallis, J., *The Concept of the World: A Study in the Phenomenological Ontology of Martin Heidegger*, Michigan: University Microfilms International, 1977, p. 90.

170 Mulhall, S., *Heidegger and Being and Time*, p. 49. Work '*Werk*' here does not refer to the act of working but instead to the 'end' of the act. It refers to the product achieved by working rather than to the process of working as such.

171 *SZ.*, p. 70; *BT.*, p. 99.

172 *SZ.*, pp. 70–71: *BT.*, pp. 99–100.

173 Crowell, S. & Malpas, J., (eds.), *Transcendental Heidegger*, Stanford, California: Stanford University Press, 2007, p. 95.

174 Puthenpurackal, J. J., *Heidegger Through Authentic Totality to Total Authenticity* p. 24.

taken for granted by Dasein in his everydayness until the smooth operation of the system is dislocated or disturbed.

> When equipment cannot be used, this implies that the constitutive assignment of the "in-order-to" to a "towards-this" has been disturbed. The assignments themselves are not observed; they are rather 'there' when we concernfully submit ourselves to them (*Sichstellen unter sie*). But *when an assignment has been disturbed*—when something is unusable for some purpose—then the assignment becomes explicit.[175]

With the dislocation in the complex system, the equipmental character of entities within the world is disclosed. According to Heidegger, this can happen in three ways: First, a piece of equipment can become unusable, damaged or un-readiness-to-hand so that it is no longer good or suitable for what it was originally meant to serve. Heidegger calls this inability of the equipment to be equipment "*Conspicuousness*" (*Auffälligkeit*).[176] Here, the equipment, having lost its equipmentality, lies before us as a present-at-hand (*Vorhandenheit*) entity and asks for repair or replacement so that it can be "ready-to-hand" again. Secondly, when particular equipment needed to perform certain task appears to be missing, the equipments that are there become '*obtrusive*'. "The more urgently (*Je dringlicher*) we need what is missing, and the more authentically it is encountered in its un-readiness-to-hand, all the more obtrusive (*um so aufdringlicher*) does that which is ready-to-hand become—so much so, indeed, that it seems to lose its character of readiness-to-hand.[177] This can happen, for instance, when the need to hammering nails arises, Dasein cannot find a hammer. So, the deficiency of the equipment to accomplish a task is called obstrusiveness (*Aufdringlichkeit*). "This causes all ready-to-hand objects within the work-world that are given to become present-to-hand, as Dasein cannot deal with anything until a hammer is also given."[178] Thirdly, an equipmental system is disclosed when a piece of equipment that was lost suddenly reappears and by its unruly presence, 'stands in the way' of our concernful dealing with an equipment and thus becomes disturbingly '*obstinate*' (*Aufsässigkeit*) where it "stands in the way of our concern".[179]

In all these cases of 'conspicuousness', 'obtrusiveness' and 'obstinacy', a being that strictly speaking, should be "ready-to-hand" appears to us as "merely given" as

---

175 *SZ.*, p. 74; *BT.*, p. 105.
176 *SZ.*, p. 73; *BT.*, p. 102.
177 *SZ.*, p. 73; *BT.*, p. 103.
178 Dotray, Z., "Heidegger's Unzuhandenheit as a Fourth Mode of Being," in *Macalester Journal of Philosophy*, Vol. 19, Iss. 1, Article 12, 2010, (205–208), p. 213.
179 *SZ*, p. 73: *BT*, p. 102.

"merely there."As such, Heidegger maintains that Dasein come across equipment present-at-hand alongside ready-to-hand.

> Un-readiness-to-hand [...] implies that what cannot be used just lies there; it shows itself as an equipmental Thing which looks so and so, and which, in its readiness-to-hand as looking that way, has constantly been present-at-hand too. Pure presence-at-hand announces itself in such equipment, but only to withdraw to the readiness-to-hand of something with which one concerns oneself [...]. This presence-at-hand of something that cannot be used is still not devoid of all readiness-to-hand whatsoever; equipment which is present-at-hand *in this way* is still not just a Thing which occurs somewhere.[180]

If the world can emerge in concern, then it must have been revealed formerly to Dasein in some way. For, if the reunion of Dasein with what is "ready-to-hand" is to be possible, then the world appears explicitly as that "wherein" Dasein already was and to which it later also can unequivocally return. Dasein cannot encounter equipment except insofar as the equipment belongs to Dasein's world, for the "ready-to-hand" would be worthless if it were not interwoven with this worldly constitution. In all this Dasein exercises a spatializing function. It is purely with respect to Dasein's spatiality, that that of the entities ready-to-hand could be understood. Hence, in Heidegger's examination of space, he glimpses into the spatiality of Dasein as well as entities.[181]

The spatiality of Dasein, which is essentially being-in-the-world, is not to be compared with scientific space; the distances of concernful living cannot be compared with objective measurements but are defined only by that activity of Dasein in which it concernfully deal with things and by the directing intention that created them.[182] The world which Dasein discovers in its concernful dealing with things is the real world.

In a nutshell, under the worldhood of the world, Heidegger scrutinizes the world of Dasein in its everydayness in contrast to the imitative world of science. We can speak of many worlds as there are different meaningful and referential totalities. We often talk of the world of science and technology, the academic world, the medical world etc. The world of Dasein has its hub in Dasein itself

---

180 *SZ.*, p. 73; *BT*, p. 103.

181 Heidegger follows a phenomenological, two-structural appraoch—the noetico-noematic structure—in his analysis of spatiality which is the same as the spatiality of the ready-to-hand entities within-the-world. Cf. Puthenpuackal, J., *Heidegger Through Authentic Totality to Total Authenticity*, p. 18.See also Didier, F., *Heidegger et le probleme de l'espace*, Paris: Editions de Minuit, 1986, p. 115; Ricour, P., *Temps et Récit III*, Paris: Seuil, 1985, chap. 3 "Temporalité, Historisité, intra-temporalité."

182 Kockelmans, J. J., *Martin Heidegger, A First Introduction to his Philosophy*, pp. 48–49.

and originally it coincides with our own environment (*Umwelt*) to the extent that this environment is experienced in our lived experiences. Heidegger show splendidly how the things within this world are given not as material objects which simply occur as obviously present-at-hand (*Vorhanden*), but as serviceable things or pieces of equipment (*Zeuge*) which refer to probable applications within our "practical" world and are thus ready-to-hand or handy ((*zuhanden*). "The true result and the decisive outcome of the analysis of the world is…the referential structure of the involvement and significance as a possibility, which Dasein ever projects."[183] Things of this kind allude to one another and create reference -systems and only within these systems does the significance of these things become manifest. Taken in this sense, the world is not a creation of Dasein, but Dasein's way of giving meaning to the existential interpersonal complex in which he finds himself. It is that through which beings are "determined" as beings (tools). In essence, "Heidegger wants to identify the world or its worldhood not with being, but rather with being as understood in a certain manner. This understanding of being, in contrast to understanding as presence-at-hand, is described as 'readiness-to-hand' (*Zuhandenheit*), an expression that captures the serviceability or usability connotations that belongs to the very being of implements."[184]

1.4.3. The Communal World of Dasein

In this section, we pay attention to the other constitutive structures of being-in-the-world. Two of these structures are indicated by "being-with" (*Mitsein*) and "Dasein with others" (*Mitdasein*) which are just as primordial as being-in-the-world. In Dasein's being-in-the-world, he is not only concerned with entities, but also connected to other Daseins. Heidegger comments: "These entities are neither present-at-hand nor ready-to-hand; on the contrary, they are *like* the very Dasein which frees them, in that *they are there too, and there with it*."[185] As Magda, observed, "Just as Dasein is never a worldless subject, but in advance refers himself to the possible presence of things within a world, so he is never isolated, otherless "I," but in advance understands himself as I-myself-with (possible other selves).

---

183 Wiplinger, F., *Wahrheit und Geschichtlichkeit: Eine Untersuchung über die Frage nach dem Wesen der Wahrheit im Denken Martin Heideggers*, Freiburg/München: Karl Alber, 1961, p. 190. "Daseigentliche Resultat und der entcheidende Ertrag der Weltanalyse ist…die Verweisungsstruktur der Bewandtnis und Bedeutsamkeit als einer möglichkeit, die Dasein je entwirft."

184 Stapleton, T., "Dasein as being -in-the-world" in Bret, W. D. (ed.), *Martin Heidegger: key Concepts*, Durham, UK: Acumen Publishing Limited, 2010, pp. 50–51.

185 *SZ.*, p. 118; *BT.*, p. 154. (Emphasis in the original)

The "with" already refers him to the other as a self; that is, as one who exists in the same way as he himself and yet is the "other" *with* whom he can be together in the same world."[186] Only by considering these structures will it be possible to understand being-in-the-world in its profundity.

1.4.3a. "Being-With"

Dasein is not in isolation in the world into which it was thrown. The existence of others is absolutely unavoidable to its facticity in the world. Others are as equiprimordially present to Dasein as equipment is: in our dealing with entities the presence of others is discovered at the same time because they also are involved in the pieces of equipment. A car belongs to someone; it is driven by someone and built somewhere by another. It is therefore bizarre to think that Dasein first encounters equipment and only afterwards, by abstraction and reflection, the others. "In clarifying Being-in-the-world we have shown that a bare subject without a world never 'is' proximally, nor is it ever given. And so in the end an isolated "I" without Others is just as far from being proximally given."[187]

Being-with is an aspect of Dasein whether ontically another person is there or not. We did not mention the others explicitly in the preceding analyses, but they were implicitly present; for the world cannot possibly be understood without any relation to them. The individual can never be understood in isolation. Dasein is not solitary self-contained being. As being-in-the-world, our existence is already a being together with others. Heidegger emphasizes this distinction between the ontological and the ontic even more clearly in a later passage. "Even if the particular factical Dasein does *not* turn to Others, and supposes it has no need of them or manages to get along without them, it *is* in the way of Being-with."[188] What makes "being-with" possible is not the spatial proximity of two beings but their mutual relation. Where no mutual relation is possible there can be no "being-with". In other words, "being-with" is proper to Dasein. "Being-with others is a basic structure of each Dasein's self, for the sake of which he exists: Dasein therefore exists essentially for the sake of others…His world is in advance a world he shares with others; his being-in-the-world is in itself a being-with-others-in-the-world."[189] The term "with" is reminiscent of a togetherness which is not an isolation. To be

186 Magda, K., A *Guide to Heidegger's Being and Time*, p. 78.
187 *SZ.*, p. 116; *BT.*, p. 152.
188 *SZ.*, p. 123; *BT.*, p. 160.
189 Magda, K., *A Guide to Heidegger's Being and Time*, p. 76.

with someone suggests that there is a communion between the one and the other. What they have in common bind them together. In the words of Heidegger;

> This 'with' is something of the character of Dasein; [...]. By reason of this with-like (*mithaften*) Being-in-the-world, the world is always the one that I share with Others. The world of Dasein is a with-world (*Mitwelt*). Being-in is Being-with Others. Their Being-in-themselves within-the-world is Dasein-with (*Mit-dasein*)[190]

Being-with, then, is 'something which belongs essentially to my mode of being. It is not a property which I possess by virtue of the fact that there happen to be other people.[191] Seen from my standpoint, the other's way of being is "*Dasein with*" me and others. I can discover the others as co-existent because I myself am "being with," i.e., I share with them my openness to things and the world. Even in being alone, Dasein is being-with in the world. "The Other can *be missing*, only *in* and *for* a Being-with."[192] Being alone is what Heidegger refers to as a 'deficient mode of being-with'. Being aggressive to others, being timid before them or ignoring them are just permutations of Being-with. The likelihood of being alone itself confirms that the being of Dasein is being-with. For Heidegger, these imperfect human relations are symptomatic of everyday life. He is not merely recounting the phenomenal character of our experience of others. He is, he believes, unfolding a structural characteristic of Dasein. Dasein alone is incomplete, it has no nature of its own in which to laze about, but has to decide how to be. "Virtually everything Dasein does or is cries out for others, as suppliers of its raw materials, as purchasers of its products, as hearers, or as readers. Dasein's world is essentially a public world, accessible to others as well as itself."[193] The principal understanding of the interconnectedness of Dasein with the Other is summed up by W.R. Schroeder as follows:

> The materials one work on are made by others; the tasks one performs are taught by others; the products one makes are destined for others; the functional use of things come to them collectively— everyone understands what things are for; one's own understanding is a specification of the common sense. One's existence is always articulated in a world
> [...]
> with others.[194]

190 *SZ.*, p. 118; *BT.*, pp. 154–155.
191 GORNER, P., *Heidegger's Being and Time, An Introduction*, p. 60.
192 *SZ.*, p. 120; *BT.*, p. 157.
193 INWOOD, M., *Heidegger*, p. 35.
194 SCHROEDER, W. R., *Sartre and his Predeccessors: The Self and the Other*, p. 133.

What this means is that in the everyday environment, I always experience things in relation to other people. So, even while no other people happen to be around, I admit their importance merely by using things. My ways of using the thing and the thing itself as a tool *refer* to my human community.[195] It is only because every Dasein is essentially being-with that each can experience the other Dasein in his own being-with. Heidegger summed it up this way: "Only so far as one's own Dasein has the essential structure of Being-with, is it Dasein-with as encounterable for Others."[196] Given that others are first found in the area of our practical everyday concerns, we are likely involved in shared projects and have parallel goals and values. As a result, rather than being considerably different from us, they are 'those from whom one mostly does not distinguish oneself, those among whom one is, too. Although "being-with", then, presupposes a parity of nature, co-existence is possible only because this equality of nature applies to beings which by their nature are open to anything that manifests itself to them and which subsequently can share the world that is universal to them.

1.4.3b. Modes of Dasein's Being-with (Early Critique of Modernity)

From the point of view of care (*Sorge*), whilst Dasein's 'being-in' with the ready-to-hand entities is described by 'concern' (*Besorgen*), Heidegger characterizes the primordial connection of Dasein and other Daseins or the way men act toward-one-another as "solicitude" (*Fürsorge*).[197] From Heideggerian perspective, solicitude may show itself in different modes: "Being for, against, or without one another, passing one another by, not "mattering" to one another—these are possible ways of solicitude."[198] Thus, to neglect someone, to be against someone, and to hate someone are the possible forms of solicitude. In other words, solicitude

---

195 Pilot, R., *Heidegger, an Introduction,* p. 60.

196 *SZ.*, p. 121; *BT.*, p. 157. '…Mitdasein characterisiert das Dasein anderer, sofern es für ein Mitsein durchdessen Welt freigegeben ist. Das eigene Dasein ist, sofern es die Wesensstruktur des Mitseins hat, als für Andere begegnend Mitdasein.'

197 Dieses Seiende wird nicht besorgt, sondern steht in der Fürsorge.' There is no good English equivalent for 'Fürsorge', which is here translated by 'solicitude'. The more literal 'caring-for' has the connotation of 'being fond of', which is not applicable here; 'personal care' suggests personal hygiene; 'personal concern' suggests one's personal business or affairs. 'Fürsorge' is regularly used in context where we would speak of 'welfare work' or 'social welfare; this is the usage which Heidegger has in mind in his discussion of 'Fürsorge' as 'a factical social arrangement'. Cf. *SZ.* p. 121*; BT.*, p. 157, footnote.

198 *SZ.* p. 121; *BT.*, p. 158.

could be either *negative* or *positive*. The negative or deficient mode of solicitude is referred to as "'indifference' modes that characterize everyday, average Being-with-one-another."[199] Existentially expressed, being-with-one-another has the character of *distantiality*. The more unassuming this kind of being is to everyday Dasein itself, all the more obstinately and primordially does it work itself out. As Heidegger states:

> These modes of being show again the characteristics of inconspicuousness and obviousness which belong just as much to the everyday Dasein-with of Others within-the-world as to the readiness-to-hand of the equipment with which one is daily concerned. These Indifferent modes of Being-with-one-another may easily mislead ontological Interpretation into interpreting this kind of Being, in the first instance, as the mere Being-present-at-hand of several subject.[200]

What is crucial here is simply the inconspicuous power and dominance by others which Dasein in his being-with has now been enmeshed unconsciously. The everyday constitutes the situation of Dasein, such that it is the there of its being-there, but Heidegger also largely sees escape from the confines of the everyday as the only means of 'authentic appropriation'. "In Heidegger, the self of the everyday is mechanically dispersed into the 'they,' the term Heidegger uses to denote the structures of modern public disclosure, interpretation and communication."[201]

As such, Dasein loses the 'practical' concern or 'solicitude' with all Other Dasein. He is enmeshed in the inauthentic collectivity of the anonymous '*They*' which Heidegger refers to as '*das Man*'.[202] This is indicative of Heidegger's early critique of modernity in *Being and Time*. Individuals are constituted by social and cultural relationships that can never, hypothetically, be completely comprehended. Charles Guignon observed that, "Far from being an autonomous and isolated subject, the self is pictured as the 'Anyone' (*das Man*), a 'crossing point' of cultural systems unfolding through history. To be human, in Heidegger's view, is to be a place-holder in a network of internal relations, constituted by public language, of the communal world into which Dasein is thrown."[203]

---

199 *SZ*. p. 121; *BT*., pp. 157–158.

200 *SZ*. p. 121; *BT*., p. 158.

201 Robert, J., *Philosophizing the Everyday: Revolutionary Praxis and the Fate of Cultural Theory*, Ann Arbor, MI: Pluto Press, 2006, p. 48.

202 SZ. p. 126; *BT*., p. 164.

203 Guignon, C., *Heidegger and the Problem of Knowledge*, Indianapolis: Hackett Publishing Co., 1983, p. 86.

The "das Man" has its own ways to be. These ways to be which Heidegger called "'distantiality' is grounded in *averageness*, which is an existential characteristic of the 'they.'"[204] As he explains:

> Thus the "they" maintains itself factically in the averageness of that which belongs to it, of that which it regards as valid and that which it does not, and of that to which it grants success and that which it denies it. In this averageness with which it prescribes what can and may be ventured, it keeps watch over everything exceptional that thrusts itself to the fore. Every kind of priority gets noiselessly suppressed. Overnight, everything that is primordial gets glossed over as something that has long been well known. Everything gained by a struggle become just something to be manipulated. Every secret loses its force. This care of averageness reveals in turn an essential tendency of Dasein which we call "leveling down" (*Einebnung*) of all possibilities of Being.[205]

Distantiality, averageness, and leveling down, as ways of being of the 'they', create what we know as "publicness" (*Öffentlichkeit*). Publicness obscures everything, and then claims that what has been so covered over is what is familiar and reachable to everybody. In it, all genuineness and specialty that essentially belongs to Dasein is annihilated and obscured. Only the superficial in things is touched upon. The 'they' controls the way, in which, the world is construed. It presents each judgment and decides upon it and takes away Dasein's liability. This is a critical evaluation of everyday life in industrial-urban society ushered in by modernity. As he states further:

> In utilizing public means of transport and in making use of information services such as the newspaper, every Other is like the next. This Being-with-one-another dissolves one's own Dasein completely into the kind of Being of 'the Others', in such a way, indeed, that the Others, as distinguishable and explicit, vanish more and more. In this inconspicuousness and unascertainability, the real dictatorship of the "they" is unfolded. We take pleasure and enjoy ourselves as *they* (*man*) take pleasure; we read, see, and judge about literature and art as *they* see and judge; likewise we shrink back from the 'great mass' as *they* shrink back; we find 'shocking' what *they* find shocking. The "they", which is nothing definite, and which all are, though not as the sum, prescribes the kind of Being of everydayness.[206]

What is negative about this condition is that indifference "alienates Dasein from its ownmost non-relational potentiality-for-Being; that is, it drifts along towards an alienation (*Entfremdung*) in which its ownmost potentiality-for-Being is hidden

---

204 *SZ*, p. 126; *BT.*, p. 164.
205 *SZ*, p. 127; *BT.*, p. 165.
206 *SZ*, pp. 126–127; *BT.*, p. 164.

from it."[207] This attitude is the explication of 'inauthenticity'. This is condition where Dasein has relinquished the potentialities to 'Others', or to the 'anonymous 'they'. Here Dasein becomes so estranged in itself in its oblivious immersion in its everydayness. In such a society, nothing actually original can be revealed because language has relapsed to the state of idle talk (*Gerede*), the passing around of ungrounded opinions. Heidegger says that "this 'movement' of Dasein in its own Being, we call its '*downward plunge*' (*Absturz*). Dasein plunges out of itself into itself, into the groundlessness and nullity of inauthentic everydayness. But this plunge remains hidden from Dasein by the way things have been publicly interpreted, so much so, indeed, that it gets interpreted as a way of 'ascending' and 'living concretely.'"[208]

Idle talk frees the 'they' from having to understand anything in a genuine, original way. In other words, one is not completely involved with the content of the talk, but fairly sketchily and ambiguously hears what is said. "Idle talk [...] is the kind of Being which belongs to Dasein's understanding when that understanding has been uprooted."[209] It is Heidegger's conviction that his countrymen had been uprooted and made homeless. This homelessness is manifested in the 'big-city' Germans who were distracted by curiosity about new places, faces and products. In curiosity, Dasein allows himself to be carried away by the looks of the world. The things looked for are merely for the sake of seeing and not in order to understand. In curiosity Dasein is agitated about innovations, relentlessly looking for excitement and varying encounters. This, in turn, leads to incessant interruption and dissipation, thereby losing touch with its highest possibilities. Thus the curiosity-stricken German Dasein is "everywhere and nowhere. This mode of Being-in-the-world reveals a new kind of Being of everyday Dasein—a kind in which Dasein is constantly uprooting itself."[210] This point will be highlighted more when we treat Heidegger's critique of technology in part II of our study.

Now let us return to the ***positive*** mode of solicitude which has two extreme possibilities. The first is that which can "*leap in* (*einspringen*) for him" and take away 'care' from the Other and takes over his worries, "that with which he is to concern

---

207 *SZ.*, pp. 178, 254; *BT.*, pp. 222, 298. It should be noted that one of the most difficult aspects of understanding Heidegger lies in his use of 'authenticity'. Just as 'Dasein' does not name our subjectivity, 'authenticity' does not mean the subjective, existential aspects of individual humans but indicates the ontological dimension. Cf. R. Mugerauer, *Heidegger and Homecoming*, p. 37.

208 *SZ.*, p. 178; *BT.*, p. 223.

209 *SZ.*, p. 170; *BT.*, p. 214.

210 *SZ.*, p. 173; *BT.*, p. 217.

himself."[211] The Other is subjugated and dependent, and thus the 'solicitude' here is condensed to the level of 'concern' for the ready-to-hand. The other is reduced to mere equipment for my use as if he were an entity.

In contrast to this, the second positive solicitude has also the possibility of *leaping ahead* (*vorausspringen*) of the Other. This type does not impede or take away the liberty of the other rather "this kind of solicitude pertains essentially to authentic care—that is, to the existence of the Other, not to a '*what*' with which he is concerned; it helps the Other to become transparent to himself *in* his care and to become *free for it*."[212]

Everyday being-with-one-another regularly sustain itself between the two extremes of solicitude, namely, the negative mode, or one of the two types of positive solicitude; Or between that which leaps in and dominates and that which leaps forth and liberates (vorspringend-befreienden).[213] As we have already seen, Dasein is related to what is "ready-to-hand" through its everyday "concern" (*Besorgen*) and this concern's own vision of things is called "circumspection" (*Umsicht*). With respect to other Daseins, Heidegger said that solicitude (*Fürsorge*) is guided by "considerateness" (*Rücksicht*) and "forbearance" (*Nachsicht*).[214] 'Considerateness' and 'forbearance', just as 'solicitude' can have different modes of expression and has to be taken in a very broad sense so that they can also indicate the more negative modes of "regard". Modes of Dasein such as 'being-for-oneself', 'being-oneself', 'being-alone' and 'being-away' are all expressions of Dasein's being-with in the negative sense. These expressions point to Dasein's lack of awareness of his primordial existential being-with.[215]

One may perhaps object that Heidegger's analysis of "being-with" leaves no more space for the phenomenon of "being-alone." His reply is that "being-alone" is fundamentally an extrication of oneself from the others in such a way that "Being-alone is a deficient mode of Being-with" and therefore is not possible without a certain understanding of the other. Segregation and isolation presuppose the

211 *SZ.*, p. 122; *BT.*, p. 158.

212 *SZ.*, p. 122; *BT.*, p. 159. See also the excellent work by Pierre Kaufmann, *L'Experience emotionnelle de l'espace,* Paris: Vrin, 1967.

213 Ibid.

214 *SZ.*, p. 123; *BT.*, p. 159 "Wei dem Bedorgen als Weise des Entdeckens des Zuhandenen die *Umsicht* zugehört, so ist die Fürsorge geleitet durch die *Rücksicht* und *Nachsicht*." Heidegger is here calling attention to the etymological kinship of the three words which he italicizes, each of which stands for a special kind of sight or seeing ('Sicht') (Cf. footnote).

215 Vensus, A. G., *Authentic Human Destiny*, p. 136.

existence of others so that "being-alone" is possible only on the basis of a previous "being-with."[216]

Concluding Remarks

What we have seen so far from *Sein und Zeit* is Heidegger's conception of the world from the point of view of Dasein. In inauthentic projection, the world is totalized by an unending accumulation of the same, in dispersion and distraction; and this happens in such a fashion that the distinction between equipment ready-to-hand, objects present-at-hand, and any other Dasein, gets blurred.[217] In the later phase of his writings, it is interpreted in terms of the 'truth of Being'. This project that unifies all beings according to utter manipulation and accumulation will be called "technology", whose essence is enframing (Ge-stell). In his essay, "*Das Ding*", Heidegger speaks of the 'worlding' of the world (*das Welten der Welt*), and explains it by means of the 'fourfold' (*das Geviert*). The examination of this truth of Being will be our concern in the next chapter.

216 *SZ.*, p. 121; *BT.*, p. 157.

217 Critchley, S. & Schürmann, R., *On Heidegger's Being and Time*, p. 119.

Chapter II: Heidegger's Search for the Truth of Being

2. Attaining the Experience of Being

Introduction

Heidegger's two main philosophical fascinations are the distinction between Being and being and the nature and idea of Truth and how it became hostage to metaphysics.

Since Heidegger's philosophical voyage may be regarded as an unmitigated meditation on the problem of the meaning of Being and its difference from beings, we started with the introductory analysis of the Being of Dasein as that being which understands Being. So, in the previous chapter, we have examined the nature of Being and the being of Dasein as Being-in-the-world. Dasein's being-in-the-world is characterized by care. At this level, he is caught up in a concernful involvement with entities and other Daseins which he encounters in his everydayness as being-in-the-world. In developing the term Dasein, Heidegger is seeking to avoid the subject-object replica which interprets human being in a way which dislocates the relational manner in which we find ourselves within the world: as part of the world of one another, not apart from it. Dasein accentuates the basic openness to world and others out of which we live and through which we become who we are. The world is not a thing at all, and is not outside Dasein in the way that we would say that I am looking at a tree outside my window or the street light is just opposite my house.

We have so far seen also how Heidegger tries to answer the question of the meaning of Being. In a sense, he is digging through the levels and blockades that have been set up by preceding thinkers and by the history of ontology since the beginning of philosophical reflection on the topic of what it means to be. Heidegger finds that the pre-Socratics like Parmenides and Heraclitus offer truer, more "essential" accounts of Being—but these have been lost in the course of the metaphysic of Plato and Aristotle. The Cartesian Dualism smartly set up the structure of Being as rather simple and atomistic, a mechanical reproduction of the scientific method. The dualism sets reality apart, and covers over the essential nature of being. This means that no one is really right and that the search for meaning must take the prime importance in one's life if one is to live authentically in the world. However, Heidegger presents a phenomenological analysis that is planned to undermine the assumptions that has given rise to the common problems of post-Cartesian philosophy (e.g., knowledge of the external world, the mind-body problem, etc.). To do this efficiently, "he takes as his starting point neither a self-contained subject, nor an objective account of the world derived from science, but rather what he calls being-in-the-world."[1] For Heidegger, this means getting out of the subject/object relationship that has contaminated and covered over the real meaning of what it means to be.

Heidegger explains in *Being and Time* that what he was doing as "fundamental ontology" which is the expression of the conditions for comprehending the meaning of Being and "an existential analytic of Dasein". He, nonetheless, saw himself as continuing and perfecting the "transcendental" projects of Kant and Husserl devoid of approving their transcendentalism.[2] However, while Heidegger acknowledges his previous thought, he can no longer use a transcendental-hermeneutic method, and has to adopt different strategy in his ensuing philosophy.

Therefore, in the quest for the truth of Being, Heidegger abandoned the ontological approach and turn to an "ontopoetical" approach. In his renowned *Kehre* ('turning'), he altered his erstwhile 'ontic-ontological' history of Being into a recollative thinking of Being.

In this chapter, we shall therefore see how Heidegger abandoned metaphysical project of attempting to ground beings as such, in favour of a poetic and meditative thinking which tries to retrieve a non-metaphysical, poetic idea of Being as evoked in the event of appropriation. This is Heidegger's transition from Being to *Ereignis* which takes place in two stages. The first stage of this transition is what we will consider in this chapter. The second stage is equivalent to the move from completing to overcoming metaphysics as we shall see in part two of our study.

In this first stage, we shall examine how Dasein can attain the experience Being by going back to the Greek origin to unearth the ontological difference between Being and what-is, which Heidegger opines was the missing point of modern thought. This Greek path will lead us to examine the essential thinking of Being and Dwelling in the nearness of Being which results from essential thinking. Poetic dwelling eventually leads us to the question of the truth of Being which Dasein can only arrive at by a leap to a level of *Ereignis* or the event of appropriation.

---

1 RUDD, A., *Expressing the World, Skepticism, Wittgenstein, and Heidegger*, Chicago: Open Court Publishing Company, 2003, p. 56.

2 Blattner has an excellent discussion of "Heidegger's Trancendental Idealism" and its close relation to Kant's, based on a reading of the passage from S.Z 255. See BLATTNER, W. *Heidegger's Temporal Idealism*, Cambridge: Cambridge University Press, 1999, pp. 233–254.

But going beyond this Heideggerian idea of ontological and poetical representation, we shall see how Levinas, a man with much affinity with Heidegger, prevailed the supposed apathy and ethical indifference and coldness of Heidegger's ontology in favour of ethics, and eventually split with Heidegger's whole assumption of social being with the question of responsibility.

2.1. The Way: Greek Origin

The image of paths or way is frequently encountered in the writings of Heidegger. One of his books is titled "Forest Paths." At the end of *Being and Time*, the image of the way emerges;

> One must seek a *way* of casting light on the fundamental question of ontology, and this is the way one must *go*. Whether this is the *only* way or even the right one at all, can be decided only after one has gone along it.[3]

In a reflection on the meaning of philosophy, Heidegger reiterated the theme of the way:

> ...Nor the word philosophy is speak in Greek. The word as a Greek word is a path. This path [...] lies before us for the word has long since been spoken, i.e., set for [...] it lies behind us for we have always heard and spoken these words accordingly. The Greek word philosophy is a path along which we are travelling.[4]

The reality of philosophy can be understood and attained only by entering upon these paths which leads back to Greek philosophy. Heidegger located his Golden Age in Pre-Socratics, where poets and thinkers lived in openness to Being that was covered up when thinking started to take a wrong turn with Plato.[5] This return along this philosophic way is not undertaken with any expectation or generating a renaissance of pre-Socratic philosophy for its own academic sake.

> If we so stubbornly insist on thinking Greek thought in Greek fashion, it is by no means because we intend to sketch a historical portrait of Greek antiquity, as one of the great past ages of man [...]. We search for what is Greek neither for the sake of the Greeks themselves nor for advancement of scholarship.[6]

---

3 *SZ.*, p. 437; *BT.*, p. 487. (Emphasis in the text).

4 Heidegger, M., *What is Philosophy*? Trans., William Kluback and Jean T. Wilde, New York: Twayne, 1958, p. 29.

5 Rudd, A., *Expressing the World*, p. 205. This is very similar to the picture that Nietzsche presents in the book *The Birth of Tragedy*, although for Heidegger, Nietzsche's "inverted Platonism" represents not an escape from the tradition, but its nihilistic culmination.

6 Heidegger, M., *"Anaximanders Fragment" in Early Greek Thinking*, Trans. David F. Krell and Frank Capuzzi, New York: Harper and Row, 1975, p. 50. (Hereafter, EGT)

Greek antiquity was understood by Heidegger philosophically and not culturally nor politically. Greece was the location were being has apprehension and apprehension "is the happening that has the human being."[7] The retrieval of the Greek origin therefore is the recovery of that which lies at the source of man's spiritual being and that which lies at that source is being and the call of being. It is the event in which being calls man to thinking. "Philosophy does not spring from myth. It arises solely from thinking and in thinking. But thinking is the thinking of being. Thinking does not originate: it is when being presences."[8] When being calls man to thinking, being comes to language. Within the Greek language, a fragment from Anaximander constitutes the first written philosophical text. It is the first extant written response to being's call of man to thinking as distinct from myth making poetry or even praying. Predictably, there arc many rendering of this fragment—Diels, Jaeger, Nietzsche, Burnet, have each offered different translations and reconstitution of the text. Heidegger accepted the construction and translation of Burnet. He accepted "these as the immediate genuine works of Anaximander."[9] While he accepted these as the authentic works of Anaximander, Heidegger did not lose other renderings of the text. He accepted Burnet rendering "with the proviso that the preceding parts of the text…are positively retained."[10] These preceding paths possess strength and eloquence and they constitute 'a secondary testimony concerning Anaximander's thinking.'

The paths which Heidegger regarded as inauthentic but still wished to retain reads:

> But from that which things arise (But that from which things arise) also gives rise for their passing away according to what is necessary; for things render justice and pay penalty to one another for their injustice according to the ordinance of time.[11]

Heidegger's desire to retain this tainted section of the text as a secondary testimony is not difficult to appreciate. The matter under discussion in these phrases is τα οντα (*ta onta*). Literally translated, it means Being, the totality of Being, manifold Being in totality.[12] "According to its wording, the fragment speaks of

7 Heidegger, M., *Introduction to Metaphysics*, p. 150.
8 Heidegger, M., *Anaximanders Fragment*, in EGT, p. 40.
9 *Ibid.*, p. 3.
10 *Ibid.*, p. 30.
11 *Ibid.*, p. 20.
12 *Ibid.*, pp. 20–21.

onta, expressing what they involve, and how it is with them. Beings are spoken of in such a way that their being is expressed."[13]

Already Heidegger had glimpsed in the meaning of *ta onta*, a lightening flash of insight into "the Riddle of Being."[14] In seeking to penetrate this Riddle of Being which had come to language in the received fragment of Anaximander, Heidegger sought out a non-philosophical source. The source chosen was Homer's Illiad. "Thanks to him, we possess our reference in which the word appears as something more than a term in the lexicon. Rather it is a reference which poetically brings to language what *onta* means.[15] 'Him' here refers to Kalchas, a man well vast in the craft of the seer.[16] As the seer, he is the mantic man, the one who is beside himself, outside himself. It is the essence of the mantic that he is away from the self, "away from the sheer operations of what lies before us....Away to what is absent; and at the same time, away from what is presently present in so far as this is always only something that arrives in the cause of its coming and going."[17] The seer is the one who sees that which is in being, that which will be and that which once was. "The seer is outside himself in the solitary region, on the presencing of everything that in some way comes to presence."[18] The seer sees what becomes present in the lighting that penetrates its seeing. He sees the presently present. He sees the future and the past, i.e., that which is not presently present. The present does not simply mean something over against a subject. The present means something which comes alongside in unconcealment. The present means "an open expants of unconcealment into which and in which whatever comes alongside lingers."[19] "...*Eonta*...means for the Greece what is present in so far as it has arrived. To linger in expants of the unconcealment.[20]

What is presently present lingers in unconcealment. It has come from concealment, arrives in unconcealment and lingers a while. But in lingering, it "is

13 *Ibid.*, p. 22.

14 *Ibid.*, p. 33.

15 *Ibid.*, p. 32.

16 For nine days, the plague centered by Apollo has raged in the camp of the Greece. Finally Achilles the warrior commanded Kalchas the seer to interpret the anger of the gods. Cf. Illiad, bk.1, lines 68–72.

17 *Anaximanders Fragment*, p. 35. The mantic way of the seer does not demand that he rave, role his eyes or toss his limbs about. "...the simple tranquility of bodily composure may accompany the madness of vision."

18 *Ibid.*

19 *Ibid.* p. 34.

20 *Ibid.*

also already departing from unconcealment towards concealment."[21] The point that Heidegger was making was that what is presently present and lingering in unconcealment comes to presence out of absence. That which is presently lingering in unconcealment includes absence. For the seer to see that 'totality of what is present in its presencing' means that the present in its presencing brings the absence with it. "*Ta onta* in the pre-technical language of Homer meant that which is present and which comes into presence in the absence." Explicitly, the poetry of Homer says what is present. But implicitly and even ambiguously, the poetry of Homer allows the absent to come to language. Heidegger has already perceived in Homer the suggestion of "a long hidden fundamental trait…of being."[22] The outline of the ontological difference between Being and What-is is suggested in the Illiad of Homer. At the very source of human philosophy, the present and beyond that something concealed had come to language and addressed all thinking.

It was with Parmenides and Heraclitus that this suggestion latent in Homer's poetry explicitly came to language. Parmenides has transposed the one god of Xenophanes into one being of philosophy. Being is gathered in itself and is explicitly distinguished from the illusory world of what-is. There was in Parmenides a clearly stated insight into the ontological difference between Being and beings. For Heraclitus, one is all.[23] "All means here, all things that exists, the whole, the totality of beings…One means the One, the unique, the all uniting…Being gathers beings together, togethering together—Logos."[24]

Appropriation "appropriates man and Being in their essential togetherness."[25] This indispensible togetherness between Being and Dasein is understood by Heidegger with regard to the principle of identity: 'A' is 'A'. Talking about the Parmenedian logic of identity, Heidegger interprets this principle as 'A' is the same as (το αυτο) 'A'. The 'is' and the '*to auto*' in the principle of identity propose that each being is, in itself, identical with itself. That is to say, every being has an identity, namely, the union with itself that is born by Being. Therefore, "the principle of identity speaks of the Being of beings which holds beings in their unity and identity."[26] Having stated that the principle of identity refers to the Being of beings,

21 *Ibid.*, p. 37.
22 *Ibid.*, p. 36.
23 Heidegger, M., *Introduction to Metaphysics*, p. 136.
24 Heidegger, M., *What is Philosophy*, p. 49.
25 Heidegger, M., *ID*, p. 38.
26 *Ibid.*, p. 23–26. The usual formulation of the principle of identity reads: A=A. The principle of identity is considered the highest or ultimate principle of thought. The formula expresses the equality of A and A. According to Heidegger however, this is not what

Heidegger refers to the fragment of Parmenides.[27] The fragment states: "Το γαρ αυ το νοειν εστιν το κει ειναι" (*To gar auto noein estin to kei einai)* which is translated into English as "thinking and Being (*das Sein*) are the same".[28] Reminiscent of any other rendition of pre-Socratic terms and definitions, this translation is based on metaphysical categories, in which the original Parmenidian meaning is lost.[29]

Although Being and beings are so closely related to each other that we cannot think of one without the other, yet the relationship between Being and beings is one of difference.[30] The genitive 'of', taken subjectively and objectively, signifies a difference.[31] Heidegger refers to this as an ontological difference by which he means not mere rational distinction between Being and beings[32], but a difference as difference.[33]

---

the principle supposed to mean. In his understanding, that which is identical, the Latin "*idem*," is in Greek "το αυτο." The formula A=A speaks of equality. It does not define A as the same. So for Heidegger, the common formulation of the principle of identity conceals precisely what the principle is trying to say: A is A, that is, every A is itself the same.

27 Seidel, J. G., *Martin Heidegger and the Pre-Socrates*, Lincoln: University of Nebraska Press, 1976, p. 159.

28 Heidegger, M., *An Introduction to Metaphysics*, p. 101.

29 *Ibid.*, p. 101.

30 Heidegger, M., *ID*, p. 62.

31 Heidegger, M., *What is Called Thinking*, p. 144.

32 Heidegger, M., *ID*, p. 62.

33 In *Being and Time*, the term Ontological difference does not occur, although the issue to which the term refers is clearly present as we have seen in our examination of priority of Being over beings and entities. The expression first appeared in Heidegger's published works in *The Essence of Reason* (1929), *Kant and the Problem of Metaphysics* (1929) and he even already used the expression in a lecture course of 1927, namely, *The Basic Problems of Phenomenology*. For a detailed analysis and presentation of the problem of the ontological difference Cf. Kockelmans, J. J., *On the Truth of Being: Reflection on Heidegger's Later Philosophy*, Bloomington: Indiana University Press, 1984, especially chapter 4, pp. 73–93; "Ontological Difference, Hermeneutics, and Language." In *On Heidegger and Language*, ed. J. J. Kockelmans, Evanston: Northwestern University Press, 1972, pp. 195–234; Sallis, John, "*La différence ontologique et l'unité de la pensé de Heidegger.*" In Revue Philosophique de Louvain 65, (1967), pp. 192–206; Jean Greisch, "Identité et différence dans la pensé de Martin Heidegger. Le Chemin de l'Ereignis," in *Revue des sciences philosophiques et théologiques* 57, 1973, 71–111; Loy, M. Vail, *Heidegger and Ontological Difference*, University Park, Pa: The Pennsylvania State University Press, 1972. For a comparative analysis of this concept with the 'esse' and 'ens' in Thomas Aquinas, see Caputo, J. D., *Heidegger and Aquinas:*

Finally, Heidegger's preoccupation in much of his later writings is to work back through the history of philosophy so as to free us from taking for granted the categories of modern thought, and instead to tear open the possibility of other ways of thinking. This process, of course, takes him back eventually to the roots of Western philosophy in classical Greece.[34] This return to pre-Socratic philosophy was to discover and recover the ontological distinction which early Greek philosophy expressed in a distinction "between presencing and what is present." A failure to uncover the ontological difference is to distort the question of being. His realization produced a shift in emphasis from an analysis of the being of Dasein to an analysis of the event of being itself that occurs in the "there" (Da) of Dasein. This "Turn" in Heidegger's thinking in his later works is what we will examine in the following sections.

2.2. The 'Turn' (Kehre)

The *Kehre* or "turn," is definitely not an amendment in Heidegger's thinking. It is not an incident that could be dated in his philosophical career. Rather, it is one name among many topics of Heidegger's work. It is a "change in thinking", a shift in orientation from the transcendental-horizontal approach of 1928–29 to the *seinsgeschichtlich* (dispensing-of-being) approach of the rest of his career. In his letter to William J. Richardson (April 1962) Heidegger acknowledged "*eine Wendung in meinem Denken*," "a change in my thinking."[35] According to this letter, the change is neither a break in his philosophical expedition nor a desertion of one goal for another. As he explains:

> First and foremost the *Kehre* is not a process that took place in my thinking and questioning.
> It belongs, rather, to the very *issue* that is named by the titles "Being and Time/ "Time and Being."

An Essay on Overcoming Metaphysics, New York: Fordham University Press, 1982, pp. 147–184.

34 Rudd, A., *Expressing the World*, p. 204.

35 Richardson, W., *Heidegger, Through Phenomenology to Thought*, "Vorwort," xvi–xxiii; For the question of the problem of Heidegger I and II see especially, p. 242. Cf. also Heidegger's "eine Wandlung des Denkens": "*Vom Wesen der Wahrheit,*" Gesamtausgabe (GA) 9, 187.21–22; *Brief über den Humanismus* (Hereafter, BH), *Wegmarken*, (Weg), 3., durchgesehen Aufl. Frankfurt a. M: Vittorio Klostermann, 1996, (331–364) p. 325.

> (...) The turn operates within the *issue itself*. It is not something that I did, nor does it pertain to my thinking only.[36]

The question of Being which Heidegger tries to reveal, has to be shown as revealing itself. The change in question has not occurred suddenly; the call to his turn, according to Heidegger, was present from the preliminary steps of SZ.[37] Hence, even after the 'turn', SZ "remains even today, a necessary one, if our Dasein is to be stirred by the question of Being."[38] The turn purportedly accounts for the transition from SZ to Heidegger's later thought. In other words, "the notion of the turn is the major synthesizing link between early and later Heidegger."[39] One should distinguish between the notions of the turn as it occurs in Heidegger's writings, and the quite different concept of a change in his philosophy.[40] Heidegger developed

---

36 *Ibid.*, p. xix. 1–3 and 6–8 "Die Kehre ist in erster Linie nicht ein Vorgang im fragenden Denken; sie gehört in den durch die Titel 'Sein und Zeit', 'Zeit und Sein' gennanten Sachverhalt selbst.... Die Kehre spielt im Sachverhalt selbst. Sie ist weder von mir erfunden, noch betrifft sie nur mein Denken."

37 There is a little tension in the claim that is subject to questioning. In S.Z p. 333 Heidegger states that "we need an idea of Being in general, and this idea needs to have been adequately illumined in advance. So long as this idea is one at which we have not yet arrived, then the temporal analysis of Dasein, even if we *repeat it*, will remain incomplete and fraught with obscurities." Elsewhere in SZ. p. 436, he reiterated that "our aim is to work out the question of Being in general. The thematic analytic of existence, however, first needs the light of the idea of Being in general, which must be clarified beforehand." In both statements, he proposes to repeat the analysis of Dasein in the light of Being. Could these statements be understood that Dasein can adequately be seen only in the light of the truth of Being? Cf. *BT.*, pp. 382 & 487.

38 Cf. Author's preface to the seventh German edition of *Being and Time*, p. 17. Elsewhere in his writings Heidegger equally states: "Ich habe einen früheren Standpunkt verlassen, nicht um dagegen einen anderen einzutrauschen, sondern weil auch der vormalige Standort nu rein Aufenthalt war in einem Unterwegs. Das Bleibende im Denken ist der Weg. Und Denkwege bergen in sich das Geheimnisvolle, daß wie sie vorwärts und rückwärts gehen können, daß sogar der Weg zurück uns erst vorwärts führt." Heidegger, M., *US*, pp. 98–99.

39 Herman, P., *Heidegger's Philosophy of Being: A Critical Interpretation*, p. 233.

40 Grondin, J., *Le tournant dans la pensée de Martin Heidegger*, Paris: Presses Universitaires de France, 1987, pp. 12–13: "A la rigueur, il faudrait peut-être dissocier la problématique du tournant de la question de l' evolution de la pensée de Heidegger." There is a tendency to even discover and ascribe more turns in Heidegger's philosophical career as when John Caputo described three or even four turns. Cf. Caputo, J, D., "Heidegger and Theology" in Guignon, C. B., ed. *The Cambridge Companion to Heidegger*, Cambridge and New York: Cambridge University Press, 2000, pp. 270–288;

his later notion of the turn in *Beiträge zur Philosophie*, written in 1936–38 but published posthumously in 1989.[41] It is in his *Letter no Humanism* that Heidegger speaks of the incompletion of SZ and of the "turn" in his thought for the first time:

> The adequate execution and completion of this other thinking that abandons subjectivity is surely made more difficult by the fact that, in the publication of *Being and Time*, the third division of the first part "Time and Being" was held back. Here everything is reversed. The section in question was held back because thinking failed in the adequate saying of this turning (*Kehre*) and did not succeed with the help of the language of metaphysics [...]. This turning is not a change of stand point from Being and Time, but in it the thinking that was sought first arrives at the location of that dimension out of which Being and Time is experienced, that is to say, experienced from the fundamental experience of the oblivion of Being.[42]

The means to understanding Heidegger's *turn* (*Kehre*) is therefore through "the problematization of the unfinished project of *Being and Time* within the context of a promised continuation of his work on fundamental ontology vis-à-vis a reassessment of the whole project overlaid by subjectivist themes."[43] What this entails is that "in the pre-*kehre* period, Heidegger raises the question about the meaning of being and attempts to answer that question on the basis of the assumption that human understanding of being is determined by its transcendental structure. In the post-*kehre* period, Heidegger moves from the question about the meaning of being revealed to Dasein to the question about being in itself."[44] The latter is concerned with the problem of temporality of being which encourages a necessary contemplation of the finite mode of being-in-the-world. Hence, the "*Brief über den Humanismus*," the text "*Von Wesen der Wahrheit*" ("On the Essence of Truth"), a lecture Heidegger gave many times from 1930 on, provides a certain insight in the thinking of the turn, which is characterized as a turn from "*Sein*

See also Ott, Hugo, *Martin Heidegger: unterwegs zur seiner Biographie*, Frankfurt a. M. & New York: Campus Verlag, 1988, pp. 345–346.

41 Heidegger, M., *Gesamtausgabe*, Bd. 65, *Beiträge zur Philosophie* (*Von Ereignis*), 2., durchgesehen, Aufl., hrsg. v. Friedrich -Wilhelm von Hermann. Frankfurt a.M.: Vittorio Klostermnn, 1994, especially 8–11, 27, 91, 140–142, 190, 202, 226, 233, and 255.

42 Heidegger, M., "Letter on Humanism" in *Basic Writings*, Revised and Expanded Edition, ed. Davis Farrell Krell, New York: HarperCollins Publications, 1993, pp. 231–232 (BW); BH, *Wegmarken*, p. 325.

43 Rivas, V. The 'Turn' to Time and the Misscarriage of Being, in *Kritike*, 1(2), 2007, p. 68.

44 Partenie, C & Rockmore, T. (eds), *Heidegger and Plato: Toward Dialogue*, Evanston, Illinois: Northwestern University Press, 2005, (introduction), pp. xxiii–xxiv.

und Zeit" to "*Zeit und Sein*."[45] As this text states, "this turning is not a change of standpoint from *Being and Time*, but in it the thinking that was sought first arrives at the location of that dimension out of which *Being and Time* is experienced, that is to say, experienced from the fundamental experience of the oblivion of Being."[46]

The *turn* wanted to leave subjectivity behind alongside the task of submitting being to the ultimate project of comprehension. Jean Grondin recapitulates the progress of Heidegger's thought that reveals the expectation of the *turn*, quietly foreclosing the project of fundamental ontology in *Being and Time*:

> Being is that in which (or rather, by which since we have nothing to do with it) we are projected into this world and to which we could never claim to possess the key. This thinking of *Geworfenheit* is certainly already present in *Being and Time*, and in this respect, the work was already on the way to leaving the horizon of intelligibility of subjectivity behind. This same thinking will eventually present an obstacle to a project which is rigorously conceptual (or, if we may risk the term, strictly rational), a project which seeks to attain being from the horizon of Dasein. For Dasein proves to be too finite and too historically situated to obtain a perspective on being which would enable it to drive sub *specie aeternitas* the transcendental structure of its most fundamental being.[47]

The drastic reformulation of the question of the meaning of being would bring Heidegger into the questioning of the boundaries of traditional metaphysics, such as his examination of the definition of 'the thing':

> The unpretentious thing evades thought most stubbornly. Or can it be that this self-refusal of the mere thing, this self-contained independence, belongs precisely to the nature of the thing? Must not this strange and uncommunicative feature of the nature of the thing become intimately familiar to thought that tries to think the thing? If so, then we should not force our way to its thingly character.[48]

Heidegger's rebuttal to include the sovereign status of the thing to the subjective will of the agent exposes his break with traditional metaphysics with its emphasis on subjectivity especially at the dawn of modernity with its technological advancements. Heidegger's account of the turning, according to Julian Young, "is

---

45 Heidegger, M., *Letter on Humanism*, BW, p. 231. "Der Vortrag 'Vom Wesen der Wahrheit,' der 1930 gedacht und mitgeteilt, aber erst 1943 gedruck wurde, gibt einen gewissen Einblick in das Denken der Kehre von 'Sein und Zeit' zu 'Zeit und Sein.'

46 *Ibid.* The translation of "Vom Wesen der Wahrheit" into English is found in *Basic Writing*, entitled " On the Essence of Truth" pp. 111–138.

47 Grondin, J., *Sources of Hermeneutics*, Albany: University of New York Press, 1995, pp. 68–69.

48 Heidegger, M., *Poetry, Language, Thought*, trans. Albert Hofstadter, New York: Perennial Classics edition, 2001, p. 31. (Hereafter PLT)

profoundly critical with respect to modern technological practice."[49] Or as John Caputo opines, "Heidegger's turnabout on modernity and his privileging of a mythic age of early Greeks are central and defining features of the turn (Kehre) in his thought."[50] Such Greek influence of the later Heidegger means that the intrinsic power of the concepts in the earlier works still dominated his philosophical framework. We should not lose sight of that connection. As Jeff Malpas explains:

> The development of the later thinking is directly tied to the problems evident in the earlier, and so an adequate engagement with the earlier thinking must require an engagement with the later thinking also—and such an engagement demands a respect for the later thinking as well as an appreciation of the way in which it both breaks with and nevertheless also continues the project of which *Being and Time* is merely a part.[51]

In his increasing uncertainty even about the use of the word "being", Heidegger looked for new word that would mark the post-metaphysical era. David Halliburton observed that "Heidegger could find Being in existence, that is, in Dasein, because Being proved to be 'other' than Dasein, as it proved to be other than beings that are present-at-hand or ready-to-hand."[52] He chose the term "belongingness" or event of appropriation (*Ereignis*).[53] Thomas Sheehan has variously elaborated how Heidegger associated the *Kehre* with *Ereignis* "and specifically with the way *Ereignis* operates." He explains that "the turn is the inner movement of *Ereignis* whereby (a) finitude opens a clearing in human being (b) in which entities can

---

49 Young, J., *Heidegger's Later Philosophy*, Cambridge: Cambridge University Press, 2002, p. 84.

50 Caputo, J. D., *Demythologizing Heidegger*, p. 10 Here Caputo argues that from the 1930s, Heidegger undertook to write a history of the declining fortunes of Being from the splendor of the early Greeks to the dark days of subjectivistic modernity. He brought his metaphysical orientation in line with his social and political agenda, and in so doing, he was going against one of the basic tendencies of Being and Time. In other words, his turnabout on modernity was not a feature of Being and Time or of the works of the 1920s or even the inevitable outcome of that work.

51 Malpas, J., Heidegger and the Thinking of Place: Explorations in the Topology of Being, The MIT Press, 2012, p. 41.

52 Halliburton, D., *Poetic Thinking: An Approach to Heidegger*, Chicago: University of Chicago Press, 1981, p. 21.

53 Here I chose to avoid several recent controversies in Heideggerian scholarship, i.e., whether and where there was a turn or reversal (*Kehre*) in his thought and the haggling over translations of such key terms as *Ereignis* and even Dasein. Cf. Sheehan, T., "A Paradigm Shift in Heidegger Research" in *Continental Philosophy Review*, 34, 2001, pp. 183–202.

appear as this or that."[54] According to Sheehan, all said, the 'turn' should be seen as merely a shift in Heidegger's presentation, a change in his method, approach and orientation, but not specifically a change from his central topic, namely, the question of Being.[55] The 'turning' therefore symbolizes the point of unity in Heidegger's work. This point of unity shows how Dasein and being 'belong together' in 'the event' (*das Ereignis*). These later thoughts are what we will investigate in the subsequent sections of this chapter which will enable us to clearly understand the 'why' of Heidegger's teaching on technology per se in the second part of our study.

2.3. The Essential/Meditative Thinking of Being

Heidegger's critique of traditional metaphysics and the 'turn' in his philosophical undertaking gave rise to his dedicated emphases and development of the 'thinking of Being' (*Seinsdenken*). In *Being and Time*, Heidegger suggests that Dasein play a central part in the correlation between Being and man, but in the later works, it is obviously Being itself which holds the primacy in its personal self-disclosure. Again, whereas in *Being and Time*, the coming-to-pass of Being is said to take place in and through Dasein's care, in the later works, we find that it is Dasein's thinking which is to occupy this important place.[56]

As we stated in the beginning, Heidegger criticized the traditional Western thinking and declares in the early pages of his work on thinking that "the most thought-provoking is that we are still not thinking."[57] This means that our thinking, according to Heidegger, is still representational conceptual system of thinking; enmeshed in a chain of logical categories which leads to certain valid and practical conclusions, symptomatic of metaphysical and technological culture of the West. "Our own manner of thinking still feeds on the traditional nature of thinking, the forming of representational ideas."[58] While explaining what real representational thinking means, Heidegger declared:

> Is there anyone among us who does not know what it is to form an idea? When we form an idea of something—of a text if we are philologists, a work of art if we are art historians, a combustion process if we are chemists—we have a representational ideas of those

54 Sheehan, T., Kehre and Ereignis: A Prolegomenon to Introduction to Metaphysics. In Polt, R. & Fried, G., (eds.), *A Companion to Heidegger's Introduction to Metaphysics*, New Haven and London: Yale University Press, 2000, pp. 3–16.

55 *Ibid.*

56 Kockelmans, J. J. *On the Truth of Being*, p. 32.

57 WD, pp. 2, 3, 11; *What is Called Thinking*, pp. 4, 6, 14.

58 WD, p. 60; *WCT*, p. 45.

objects. Where do we have those idea? We have them in our head. We have them in our consciousness. We have them in our soul. We have the ideas inside ourselves, these ideas of objects.[59]

In this way, thinking is merely a reflection on thinking. Hence, Heidegger admonishes his students not to think about what thinking is. "Great thinkers, first Kant and then Hegel, have understood the fruitlessness of such reflection."[60] He however declares: "Let us learn thinking."[61] To think against "logic," according to Heidegger, "does not mean to break a lance for the illogical but simply to trace in thought the *logos* and its essence."[62] He designates this thinking of Being with many appellations: meditative thinking (*besinnliches Denken*),[63] essential thinking (*wesenliches Denken*),[64] primordial thinking or recollective thinking etc.

Essential thinking is not something that is pragmatic which man can perform when and as he wants. Man becomes an essential thinker only in so far as he stands in the 'lighting' of Being.[65] Hence, Heidegger asserts in a poem: "We never come to thoughts. They come to us."[66] Thinking establishes the relation between Being and man. It does not cause this relation to be. It simply gives this relation back to Being as something that was first handed over to it by Being. Thus, Being calls man to think. Man is a thinker only because he is called to think.

Thinking lets itself be claimed by Being in order to say the truth of Being. Thinking lets itself be claimed by Being and for Being. Or as Heidegger says it, using the French expression: "Thinking is *l' engagement par l'Etre pour l'Etre* (engagement by

59 Heidegger, M. *WCT*, p. 39.

60 WD, p. 9; *WCT*, p. 21 "Wir denken…nicht über das Denken". Thinking that is merely thinking on thinking would either lead to logic, as in the contemporary Anglo-Saxon philosophy, or like in the metaphysical tradition of the West, dominated by calculations and logic.

61 *WD*, p. 75; *WCT*, p. 87.

62 BH, *Wegmarken*, p. 345; "Letter on Humanism" in *BW*, p. 251.

63 Heidegger, M., *Gelassenheit*, 11. Aufl. Stuttgart: Günther Neske, 1999, p. 13 (Hereafter, GL).

64 Heidegger, M., "*Was is Metaphysik?*"(1929). In *Weg*, 3., durchgesehene Aufl. Frankfurt a. M.: Vittorio Klostermann, 1996, 103–122. (WM); Heidegger, M., "*What is Metaphysics?*", trans. David Farrell Krell, In *Basic Writing*, New York: HarperCollins Publishers, 1993.

65 Heidegger, M., *Der Satz vom Grund*, 5. Aufl. Pfullingen: Neske, 1978, p. 147. (Hereafter, SG)

66 Heidegger, M., *Aus der Erfahrung des Denken*, 4 Aufl. Pfullingen: Neske, 1977, p. 11. (Hereafter, ED) Cf. "The Thinker as a Poet" in *Poetry Language, Thought*, p. 6.

Being for Being).[67] Thinking is thus seen as both a call that *belongs* to Being and a responds that *listens* to Being.

> Thinking accomplishes the relation of Being to the essence of man. It does not make or cause the relation. Thinking brings this relation to Being solely as something handed over to it from Being. Such offering consists in the fact that in thinking Being comes to language. Language is the house of Being. In its home man dwells. Those who think and those who create with words are the guardians of this home [...] Thinking does not become action only because some effect issue from it or because it is applied. Thinking acts insofar as it thinks.[68]

Conventionally, we put action over thought, thinking that thought never accomplished anything. But Heidegger invites us to consider that thinking may be the most vital action of all. For it changes our correlation to Being. And because thinking brings Being to language, our thinking is not separate from the language we use. Although thinking and Being are essentially related, they are not identical: "Thinking does not arise. It exists insofar as Being becomes present."[69] Therefore, thinking is both other than and dependent upon Being.

In the history of Being, the truth of Being was forgotten; thinking paid attention then primarily on beings. At the same time, the essence of man remained obscured, and a metaphysical explanation of language covered up the essence of language. Thinking, in the early age of thinking warily says, according to Parmenides, "*esti gar einai*": "For there is Being," *Il y a l'être*, *Es gibt Sein*. Literally, the German expression say: "It gives Being." That which gives is again Being itself. "This 'there is/It gives' rules as the destiny of Being. Its history comes to language in the words of essential thinkers. Therefore, the thinking that thinks into the truth of Being is, as thinking historical."[70]

We are *still not* thinking—despite Parmenides' directive—because we have missed the object and source of thinking—Being. Thinking, hence, is the thinking of Being. That means first that thinking belongs to Being. Thinking is of Being, as long as thinking, belonging to Being, listens to Being. In the "Postscript" to *What is Metaphysics*? Heidegger talks of the call of Being. "Being is the silent voice which makes itself heard in Dasein through the attunement of anxiety. Through anxiety, Being lights up in man its own relation to Dasein's essence. The silent voice of Being is a call and a plea to Dasein to be the place where its truth can be preserved.

---

67 BH, *Wegmarken*, p. 311; "Letter on Humanism" in *BW*. P. 218.

68 "Letter on Humanism," in *BW*, p. 217.

69 Heidegger, M., *Holzweg*, 7. Durchgesehen Aufl. Frankfurt a. M.: Vittorio Klostermann, 1994, p. 325. (Hereafter, HW)

70 Letter on Humanism, *BW*, p. 238.

This call, according to Heidegger, is not from Dasein but something emanating from Being. Hence, the essential thinking is an occurrence of Being which comes from Being's initiative."[71] This aspect which best describes Heidegger's notion of Essential Thinking is clearly put forth in Kaufmann's *Existentialism*:

> The thought of Being seeks no hold in what-is. Essential Thinking looks for the slow signs of the incalculable and sees in this the unforeseeable coming of the incalculable. Such thinking is mindful of the truth of Being and thus helps the Being of truth to make a place for itself in man's history. This help effects no results because it has no need of effect. Essential Thinking helps as the simple inwardness of existence, insofar as this inwardness, although unable to exercise such thinking or only having theoretical knowledge of it, kindles its own kind.[72]

Such inward thinking is non-representational, rather it recollects. But Dasein must respond actively by concentrating upon the call of Being. According to Heidegger, this Dasein's response to the call of Being is a 'corresponding' to the call. Far from its normal usage, Heidegger used the word 'correspondence' (*Entsprechung*) in the sense of 'conformity' or 'agreement.'[73] To correspond to the call of Being is, therefore, to be determined (*be-stimmt, être disposé*), to attune (*abstimmt, accorder*) oneself to the call of Being. The call to thinking, on the part of Being and the parallel response, on the part of Dasein, involves a dual relationship: Being 'calls' and 'gives'; Dasein 're-calls' (re-collects) and 'thanks'. As Heidegger explains:

> The thanc means man's inmost mind, the heart, the heart's core, that innermost essence of man which reaches outward most fully and to the outermost limits, and so decisively that, rightly considered, the idea of an inner and an outer world does not arise. When we listen to the word thanc in its basic meaning, we hear at once the essence of the two words: thinking and memory, thinking and thanks, which readily suggest themselves in the verb "to think." The thanc, the heart's core, is the gathering of all that concerns us, all that we care for, all that touches us insofar as we are, as human beings.[74]

The 'giving' becomes complete only in a receiving or 'responding' - in a 'thanking' for the gift of 'giving'. From the side of man, therefore, thinking is 'thanking'. We have to respond to this gift of thinking, the highest and the most lasting gift

---

71 WM, pp. 46–48; Cf. Heidegger, M., "What is Metaphysics", 'Postscript' in *Existence and Being*, ed. Werner Brock, Indiana: Regnery/Gateway Inc., 1949 edition, pp. 380–392.

72 Kaufmann, W., *Existentialism from Dostoevsky to Sartre*, pp. 263–264.

73 Heidegger, M., Was ist das—die Philosophie? 5 Aufl. Pfullingen: Neske, 1972, p. 21, *What is Philosophy*? p. 69.

74 Heidegger, M., *WCT*, p. 144.

given to us,[75] not by any type of response, but by a 'corresponding' (*entsprechend*) response. Man corresponds to Being by responding to its invitation to think; and man's essential belonging to Being becomes more concretely realized through his corresponding to Being. Thus:

> To "belong" [...] means to be in the order of Being. But man's distinctive feature lies in this, that he, as the being who thinks, is open to Being, face to face with Being; thus man remains referred to Being and so answers to it. Man *is* essentially this relationship of responding to Being and he is only this. This "only" does not mean a limitation, but rather an excess.[76]

With his vast erudition in philology, and in a long citation in the Postscript to *What is Metaphysics* Heidegger lays bare the attributes of essential thinking, substantiating his subject by the etymological relatedness of the word '*Denken*' (thinking) with its related terms, *Gedächtnis* (memory), '*Gedachtes*' (thought) and '*Dank*' (thanks). According to him, essential thinking

> expands itself in being for the truth of being.... This sacrifice is the expense of our human for the preservation of the truth of Being in respect of what-is. In sacrifice there is expressed that hiddenthanking which alone does homage to the grace wherewith Being has endowed the nature of man, in order that he may take over in his relationship to Being the guardianship of Being. Original thinking is the echo of Being's favour wherein it clears a space for itself and causes the unique occurrence: that what-is is. This echo is man's answer to the Word of soundless voice of Being. The speechless answer of his thanking through sacrifice is the source of the human word, which is the prime cause of language as the enunciation of the Word in words. Were there not an occasional thanking in the heart of historical man he could never attain the thinking – assuming that there must be thinking (Denken) in all doubt (Bedenken) and memory (Andenken) – which originally thinks the thought of Being. But how else could humanity attain to original thanking unless Being's favour preserved for man, through his open relationship to this favour, the splendid poverty in which the freedom of sacrifice hides its own treasures? Sacrifice is the valediction to everything that "is" on the road to the preservation of the favour of Being.[77]

Dasein responds to the call of Being by re-calling the gift of Being. Thus, thinking is to be taken in its rich meaning, i.e., in relation to 'thanking' and 're-collecting.' Thinking is the ultimate thanking. "As we give thought to what is most

---

75 WD, p. 94; *WCT*, p. 142 "Die höchste und eigentlich währende Gabe an uns bleibt jedoch unser Wesen, mit dem wir so begabt sind, daß wir aus dieser Gabe erst die sind, die wir sind."

76 ID. p. 18; Identity and Difference, p. 31.

77 Heidegger, M. What is Metaphysics, Postcript, in *Existence and Being*, p. 389–390. See also *WD*, pp. 91–94; *WCT*, p. 138–143.

thought-provoking, we give thanks."[78] Memory and thanksgiving go together and belong together. That is why Heidegger pointed out in a 1955 memorial address that every memorial service is a thanksgiving service.[79]

But the question may arise: how do we thank by thinking? It is not by merely thinking on thinking or thinking thought, but thinking as thanking has to demonstrate itself concretely in man. It is to assert Being in his action. In thanking Being, Dasein thinks of Being, and in the thinking of Being, Dasein accepts the gift of existence as the thinker of Being. To think, then, means to move into the nearness of Being and to let oneself be addressed by Being itself. As Haynes observed, "Essential Thinking is holistic because it is able to be reflective of itself and of the Being by which it itself comes forth. That is, it is able to be both within itself as thinking and also outside itself – in a very important sense to be *other* than itself – in terms of assessing itself."[80] By responding appropriately to Being's address, man will be able to bring to light the truth of Being, discover the true essence of language, recognize the seduction of public opinion, and re-discover the genuine humanity of man.[81] In responding to the voice of Being, in thanking for the gift of thinking, man speaks, man thinks. "It is precisely in this careful thinking of Being's coming and what is necessary for it that the thinker, by way of essential thinking and saying, participates in the coming of Being."[82]

So our next inquiry will take us into the meaning of human dwelling which Heidegger sees as a staying with things and how this staying stands in relation to the fourfold gathering which is at the heart of our being able to dwell on earth as human beings. And such dwelling, Heidegger states, calls for essential/meditative thinking wich requires dwelling poetically on earth.

2.3.1. Dwelling in the Nearness of Being (Dwelling Thinking)

The most fundamental way of being free from representational thinking and being at home with essential/meditative thinking is through Dwelling. Being-in-the-world is not a simple property which man sometimes has and sometimes has

78 *WD*, p. 158; *WCT*, p. 146.

79 Cf. *GL*, pp. 9–26. Heidegger delivered this address to his kinsmen during commemoration of the 175th birthday of the German composer Konrad in Kreutzer on 30th Oct., 1955 in Messkirch. He said that commemoration of a person is a thoughtful and thankful remembrance of that person.

80 Haynes, J. D., *Perspectival Thinking For Inquiring Organizations*, Santa Rosa, California: Informing Science Press, 2007, p. 44.

81 *Letter on Humanism*, pp. 220–223; Cf. Kockelmanns, J. J.,*On the Truth of Being*, p. 35.

82 Mugerauer, R. *Heidegger and Homecoming: The Leitmotif in the Later Writings*, p. 162.

not. Being-in is not set up as the relation between two entities extended in space and regarding their location in that space as water is in the glass. Initially, being-in means to dwell, to reside along side, to be familiar with. It is "the mode of just tarring along side" as we saw earlier. But in 1951, this being-in received a more extensive and philosophical treatment. In a public lecture on dwelling ("*Building Dwelling Thinking*") Heidegger states: "The way in which you are and I am, the manner in which we humans are on earth, is *buan*, dwelling [...] we do not dwell because we have built, but we build and have built because we dwell, that is, because we are *dwellers*."[83] This statement shows us that 'Dwelling is more primordial than 'building'. But as the title of the lecture suggests, the main subject is as much about thinking as it is about dwelling and building. Julian Young tells us that "dwelling can plausibly be said to constitute the central topic of the thinking of late Heideggr."[84] And the question of nearness of Being, as we shall see below, has more to do with "our concernful dealings in the world," a matter of "the pre-occupying proximity of the object of our concern."[85] Dwelling, therefore, is the antithesis of our modern groundlessness and homelessness as put succinctly in Thompson and Steiner:

> For the viewer of contemporary culture each image relates to place without needing to become a place, and refers to the past, present, and future without relating to time. Yet people retain their need for substance to underlie and inform their experiences, and they desire an encoded environment they can understand, if only for the comfort of familiarity.[86]

Basically, the condition of Dasein which emanates from essential thinking is what Heidegger refers to as dwelling (*Wohnen*). He also calls it an "indwelling"[87]

---

83 Heidegger, M., "Building Dwelling Thinking," First published as "Bauen Wohnen Denken (1951) in Vorträge und Aufsätze, 8. Aufl. Stuttgart: Günter Neske, 1997, 139–156. (Hereafter, VA). Our English edition is from *Poetry, Langauge, Thought,* trans. Albert Hofstadter, Perennial Classic Edition, New York: Harper & Row, 2001, pp. 145–146 (*BDT* in *PLT*)

84 Young, J., "What is Dwelling? The homelessness of Modernity and the worlding of the world," In *Heidegger, Authenticity, and Modernity*, eds. Mark Wrathall & Jeff Malpas, Cambridge, MA: MIT Press, 2000, (pp. 187–203), p. 187–188.

85 Mitchell, A. J., Heidegger Among the Sculptors: Body, Space, and the Art of Dwelling, Stanford, California: Standford University Press, 2010, p. 5.

86 Mark, J., "Ecology and the Urban Aesthetic," in Ecological *Design and Planning,* George, F. Thompson & Frederick R. Steiner, eds., New York: John Wiley & Sons Inc., 1997, p. 183.

87 Heidegger, M., *Gelassenheit*, p. 70; Cf. Heidegger, M., *Discourse on Thinking*, trans., John M. Anderson and Hans E. Freund. London: Harper Colophon Edition, 1969, p. 81.

as well as "standing in the openness of Being".[88] It is the opening of Dasein for openness and the remaining of Dasein in his 'origins'.[89] In Heidegger's interpretation, the German lexis for dwelling (*Wohnen*) and building (*bauen*) are chiefly connected to one another. To dwell is to abide, to live, to build, to make a home and to protect and shelter. In fact, as Heidegger shows with an investigation of the etymology of the word "build", to build is already to dwell: "dwelling is the mode of human being, so human being is essentially a being in place, just as it is also a being in the world."[90] In Heidegger's opinion, "...building is not merely a means and a way towards dwelling—to build is in itself already to dwell."[91]

Etymologically speaking, '*Bauen*' is an equivalent of the term '*Wohnen*'. According to Heidegger, the Old English and High German word for building, *bauen*, means to dwell. This signifies: to remain, to stay in a place. Thus, '*bauen*' broadly speaking is suggestive of the way in which Dasein is on earth. "To be a human being means to be on the earth as mortal. It means to dwell."[92] Although the original meaning of the verb '*bauen*' is lost in the German usage, Heidegger reminds us that there is luckily a trace that still remains to this day, "preserved in the German word *Nachbar*, neighbour." As he states further "the neigbour is in Old English the *neahgebur*; *neah*, near, and *gebur*, dweller. The *Nachbar* is the *Nachgebur*, the *Nachgebauer* the near-dweller, he who dwells nearby."[93] Furthermore, the related verbs *büre, beuron, beuren* all means dwelling, the abode or place of dwelling (*Wohnstätte*). These root words all bear affinity to the German verb *bin*, which is a word for the verb 'to be', ("am"), viz '*Ich bin*', and '*du bist*', ('I am', you are'). Hence, the primordial saying of "I am", "you are" points to our dwelling, which means the way that we are on earth.

> What then does *ich bin* mean? The old word *bauen*, to which the bin belongs, answers: *Ich bin, du bist* means: I dwell, you dwell [...]. To be a human being means to be on the earth as a mortal. It means to dwell.[94]

---

88 Heidegger, M., "The Way Back to the Ground of Metaphysics" in *Existentialism: From Dostoevsky to Sartre*, p. 272.

89 *Gelassenheit*, pp. 60–61; *Discourse on Thinking*, pp. 82–83.

90 Malpas, J., *Heidegger and the Thinking of Place: Explorations in the Topology of Being*, p. 63.

91 *VA.*, p. 140; BDT, in *PLT*, p. 144.

92 *VA.*, p. 141; BDT, in *PLT*, p. 145.

93 *Ibid.*

94 *VA.*, p. 141; BDT, in *PLT*, p. 145. "Was heiß dann: ich bin? Das alte Wort bauen, zu dem das „bin" gehört, antwortet: „ich bin", „du bist" besagt: ich wohne, du wohnst...Mensch sein heißt: als Sterblicher auf der Erde sein, heißt: wohnen".

Likewise, the German "word for peace, '*Friede*', means 'the free', *das Frye*; and *fry*" has the distinction of being "preserved from harm and danger, preserved from something, safeguarded."[95] In other words, "To dwell is to experience oneself as safe in, cared-for by, the dwelling-place in a way one is not safe in or cared-for by foreign, and to dwell is to care for the place where one dwells in a way one does not care for the place where one does not."[96] Accordingly, to dwell means to be set at peace; to remain at peace within the free; the preserve; the free sphere that safeguards'. To be free according to Heidegger really means to spare. He emphasized the profound meaning of this sparing:

> Real sparing is something positive and takes place when we leave something before hand in its own nature, when we return it specifically to its being, when we "free" it in the real sense of the word into a preserve of peace.[97]

From the foregoing considerations of the primordial meaning of essential words, Heidegger is able to arrive at the fundamental insight: "*The fundamental character of dwelling is this sparing and preserving.* It pervades dwelling in its whole range. That range soon reveals itself to us as soon as we reflect that human beings consist in dwelling and, indeed, dwelling in the sense of the stay of mortals on the earth."[98] When man 'dwells', he allows the things to be in the essence and thus 'spares' everything within the openness of Being, the world, the 'fourfold'. Man spares by 'dwelling in the 'fourfold'. This means that by dwelling, we are spared from harm and thereby preserved.

Building, in the sense of sparing or dwelling involves the impression of achieving something by tough grind or doing something by work as in tilling the soil or cultivating the vine. In a brief summary of the meaning of the word *bauen*, Heidegger mentions two modes of building, viz:

1. Building is really dwelling.
2. Dwelling is the manner in which mortals are on the earth.
3. Building as dwelling unfolds into the building that cultivates growing things and the building that erects building.[99]

---

95 BDT in *BW*, p. 351; PLT, p. 147.

96 Young, J., *Heidegger's Later Philosophy*, p. 64.

97 VA., p. 143; BDT, in *PLT*, p. 147. „Das eigentliche Schonen ist etwas Positives und geschieht dann, wenn wir etwas zum voraus in seinem Wesen belassen, wenn wir etwas eigens in sein Wesen zurückbergen, es entsprechend dem Wort freien: einfrieden."

98 Ibid.

99 VA., 143: BDT, in *PLT*, p. 146.

To enable us get to the root the question of human dwelling then, we have to examine the nature of the 'thing' and the fourfold because as we shall see, is not possible for dwelling to take place except by way of staying with things and Heidegger clarifies the notion of the fourfold in relation to the thing.

2.3.1a. The 'Thing' and the 'Fourfold'

The Thing: From Heideggerian perspective, a thing must be understood in relation to its being. In the public lecture "*The Thing*", Heidegger begins with a reflection on what he described as the characteristics of technological age, that is, the shrinking of all distances in space and time, even though this elimination has brought us no genuine nearness. The first statement of his 1950 lecture on "*The Thing*" reads: "All distances in time and space are shrinking."[100] He wasted no time in giving clear-cut examples in the areas of transportation and communication. The inventions of the airplane and the radio may bring the distant corners of the world within reach in far less time than was formerly possible, but this does not mean that we have brought things to nearness. Modernity with its scientific and technological developments has to certain degree contributed to human progress, yet it is true that both are not without attendant depravities. So, while with our modern technology we are capable of shrinking distances in spatio-temporal world, nearness continues to elude us. This is because "the frantic abolition of all distances brings no nearness; for nearness does not consist in shortness of distance."[101]

Modern technology may remove distances (*Entfernungen*, *Abstände*), but it does not grasp the essence of the nearness (*Nähe*) and remoteness (*Ferne*) of things because the latter never arises as a question for it.[102] By explicating the difference between these two ways of relating to the thing, Heidegger raises not only the issue of the essence of the thing, but also the issue of the essence of technology.

Despite the allusions made by Heidegger in his various writings on the question of nearness, the centrality of the question is rarely overstressed. Already at the beginning of *Being and Time*, Heidegger speaks of Dasein as being ontically close to us and yet for just this reason, it is ontologically the farthest.[103] That is

100 Ding, in *VA*, p. 157; The Thing, in *PLT*, p. 163.

101 Ding, in *VA*, p. 157; The Thing, in *PLT*, p. 163. "Allein das hastige Beseitigen aller Entfernungen bringt keine Nähe; den Nähe besteht nicht im geringen Maß der Entfernung."

102 Ding, *VA*, p. 167–168; The Thing, in *PLT*, p. 165–166.

103 *SZ*., p. 15; *BT*, p. 36.

perhaps why Emil Kettering considers the issue of nearness, in its ontological profundity, as "the key of all keys" to Heidegger's understanding of Being.[104]

Heidegger attempts to say something about nearness by concentrating on what is near to us, namely, things. He builds up this idea with something that is recognizable to us, namely a jug, and in so doing, he sharply distinguishes a scientific notion of the jug-thing with the thing as it emerges in attentive reflection.

A thing, fundamentally, is not something that is; but rather something that 'things' (*dingt*). This constitutes one of the most important theoretical strengths of the concept of the thing: *connectedness*. The old High German word '*Ding*' (thing) means 'gathering' (*Versammlung*). This term points to an aspect of the thing, which is emphasized by many theorists: the thing as something that 'draws other things together'. A thing partakes in giving; for instance, water or wine is given over to a jug, and the jug, in turn, pours out the liquid as a gift. What belongs to the jug, and what the jug belongs to, are gathered in this saying. Each thing may have its own giving and gathering.

The Fourfold: According to Heidegger, in 'thinging,' the thing have the capacity to bring together what he calls in his poetic language the "fourfold" of mortals, gods, Earth and sky,' all which are referred to as *das Geviert*. "The thing stays—gathers and unites—the fourfold. The thing things world. Each thing stays the fourfold into a happening of the simple onehood of world."[105] Is good to point out from the onset that the 'fourfold' does not refer to any particular kinds of objects.

> The four terms in the fourfold cannot be taken literally to mean: (a) things down on the ground, (b) things up high in the sky, (c) deities, and (d) people. There is only one case where Heidegger seems to veer slightly from this tacit principle: in his treatment of mortals, which at times he does seem to identify literally with human dieties.[106]

In an elaborate passage, Heidegger reflects on what comes to the jug and how it pours it out to express the intersection between the elements of the fourfold:

> The giving of the outpouring can be a drink. The outpouring gives water, it gives wine to drink.The spring stays on in the water of the gift. In the spring the rock dwells, and in the rock dwells the dark slumber of the earth, which receives the rain and dew of the sky. In the water of the spring dwells the marriage of sky and earth. It stays in the wine given

104 Kettering Emil, *NÄHE: Das Denken Martin Heidegger*, Pfullingen: Günther Neske, 1987, p. 18. „Die Erfahrung der NÄHE ist nicht nur eine Grunderfahrung Heidegger unter anderen, sondern sie ist in einem noch genauer zu bestimmenden Sinne die Grunderfahrung, der Schlüssel (aller Schlüssel) zu seinem Seinsverständnis."

105 "TheThing", in *PLT*, p. 178.

106 Harman, G., *The Quadruple Object*, Alresford, UK: Zero Books, 2011, pp. 85–86.

> by the fruit of the vine, the fruit in which the earth's nourishment and the sky's sun are betrothed to one another. In the gift of water, in the gift of wine, sky and earth dwell. But the gift of the outpouring is what makes the jug a jug. In the jugness of the jug, sky and earth dwell.
>
> The gift of the pouring out is drink for mortals. It quenches their thirst. It refreshes their leisure. It enlivens their conviviality. But the jug's gift is at times also given for consecration. If the pouring is for consecration, then it does not still a thirst. It stills and elevates the celebration of the celebration of the feast. The gift of the pouring now is neither given in an inn nor is the poured gift a drink for mortals. The outpouring is the libation poured out for the immortal gods.
>
> The gift of the outpouring as libation is the authentic gift. The consecrated libation is what our word for a strong outpouring flow, "gush," really designates: gift and sacrifice. "*Gush*," Middle English *guschen, gosshen*—cf. German *Guss, Giessen*—is the Greek *cheein*, the Indoeuropean *ghu*. It means to offer in sacrifice. To pour a gush, when it is achieved in its essence, thought through with sufficient generosity, and genuinely uttered, is to donate, to offer in sacrifice, and hence, to give [....] In the gift of the outpouring that is drink, mortals stay in their own way. In the gift of the outpouring that is a libation, the divinities stay in their own way, they who receive back the gift of giving as the gift of the donation.[107]

In the jugness of the jug, the jug gathers earth and the sky, mortals and divinities, in its giving. This fourfold which thing gathers is called 'world' by Heidegger. Thus things give and gathers world. With 'the fourfold,' Heidegger brings to an end the metamorphosis of the subject as Dasein by rationalizing the world in a different way from its appearance in *Being and Time*, i.e., wordly existence is at this moment understood, neither in terms of 'things' which are *ready-to-hand* nor *present-to-hand*, but as confluence of four elements, namely mortal, earth, sky and divinities, each of which are 'connections' onto the world. For that reason, the four could be seen as "an incorporation of the two in the four in the sense of the sky and earth being symbolic of divinities and mortals, and the earth and mortals relating to *dwelling*."[108]

---

107 "The Thing", in *PLT*, p. 170–71. In A Companion to Heidegger, James Edwards said that we should visualize the fourfold as the intersection of two axes. "At the head of each of the four semi-axes is one of "the four": *earth; sky; mortals; divinities.* One axis is formed at either end by earth and sky; the other is formed at either end by divinities and morntals. At the center, at the intersection of the axes, sits *the thing*. As that which sits at the center in this way, the thing, as Heidegger says, "gathers the fourfold." Cf. Edwards J. C. The Thinging of the Thing, in *A Companion to Heidegger*, Dryfus, H. L & Wrathall, M. A. (eds), p. 457.

108 Willems, B., *Hopkins and Heidegger*, London and New York: Continuum International Publishing Group, 2009, p. 67.

In *Being and Time*, Heidegger had described man as that being whose essence is to be in the world, and that man therefore is to be characterized as that being that stands out into the openness of Being which is the world. Vycinas is of the opinion that now in his later writings; Heidegger came to conceive of dwelling as the distinctively human way of being.

> Dwelling is disclosed there as the preservation of the foursome, as letting earth and sky, mortals and gods bring up the structural world in which things can become what they are and in which man can live his life as placed in his history in the sense of befalling. Dwelling is specifically human way of being. An animal does not dwell, but merely live biologically. To dwell means to be in the world as world.[109]

Man *is* insofar as he dwells in the fourfold. We are only insofar as we dwell here. This means that we *are* only insofar as we "save the earth". Thus:

> Mortals dwell in that they save the earth—taking the world in the old sense still known to Lessing. Saving does not only snatch something from a danger. To save really means to set something from into its own presencing. To save the earth does not master the earth and does not subjugate it, which is merely one step from spoliation.[110]

Human beings dwell on earth and under the sky as mortals in the presence of the divinities. The four synchronize in a primordial unity. Man, as the 'dweller' and 'builder', has an exceptional function to play in bringing the four into a unity, into a thing. This is an effort to disclose a more genuinely human 'dwelling'. As John Caputo writes of Heidegger's thought, "the analysis of the fourfold is a protest against the dehumanization of the earth, against rendering it inhumane and unlivable."[111] Dwelling refers to way of being that has to do with a vigilant and guarded attitude. The main feature of dwelling is to preserve and care for, to allow thing to exist in their essence. What has to be nurtured and preserved is the dweller's relationship with the *das Geviert*: Talking about the eath is already thinking of the other there along with it, according to Heidegger. "Earth is the building bearer." "The sky is the sun's path, the course of the moon…the year seasons, the light and dusk of day, the gloom and glow of night"; the "divinities are the beckoning messengers of the godhead"; and "out of the hidden sway of the divinities the

109 Vycinas V., *Earth and Gods: An Introduction to the Philosophy of Martin Heidegger*, p. 115.

110 BDT, in *PLT*, p. 148.

111 Caputo, J. D., *The Mystical Element in Heidegger's Thought*, New York: Fordham University Press, 1986, p. 243.

god emerges as what he is." And the people "are called mortals because they can die;" they are "capable of death as death.'[112]

In *Being and Time*, Heidegger presented death as one's "ownmost possibility," a possibility that, when faced, can bring one's life into sharp relief, forcing one to take it seriously. Death is the condition of possibility for an authentic existence. In "*What are Poets For*?" Heidegger states that "Death and the realm of the dead belong to the whole of beings as its other side."[113] So he was quick to point out that "the self-assertion of technological objectification is the constant negation of death."[114] What Heidegger wants to underscore in this passage is the ways that the technological domination over nature leads to a denial of death, as well as our place in the fourfold. Acknowledging our mortality means recognizing that death is the ultimate limit to technological mastery over nature. In other words, the person who "dwells" is someone who is open to these fundamental dimensions of "being."

Furthermore, since 'sparing' is the basic character of dwelling, mortals 'dwell' insofar as they 'spare' or preserve the presencing of the fourfold. As opposed to the effort to bend nature to the will of the human being, exemplified in Sir Francis Bacon's philosophy, Heidegger's post-metaphysical thought aims "to befriend nature…to dwell with it, and to find in it a hidden address of Being."[115] Conversely, the fourfold has today become materialized to the extent that the gods have commonly withdrawn. The mortals have demanded all their privileges for themselves and death is something they would not like to be reminded of. The negation of essential thinking and preference for calculative thinking, as the native trait of modern era, has led to a universal technologization of human life. Consequently, humans are seen not as mortals but as raw materials to meet the consumerism of technological process. The heaven and the sky has completely disappeared and demystified. Finally, the earth, the place of abode for the mortals, is so radically abused that it has become endangered and on the brink of 'fading away'.[116]

However for Heidegger, true Being means to be open to the fourfold, to tend the fourfold in its essence. But that is just what is absent in our present situation. Modernity is pigeonholed by forgetfulness of Being: people no longer grasp "Being"; they are not open to the fourfold. What reigns is this instrumentalist outlook based on considerations of expediency and efficiency, from which discretion and cherishing are far removed.

---

112 The Thing, in *PLT*, p. 176.

113 "What Are Poets For?" in *PLT*, p. 122.

114 *Ibid.*

115 Caputo, J. D., *The Mystical Elements in Heidegger's Thoughts*, p. 243.

116 Cf. Kockelmans, J. J., *On the Truth of Being*, p. 95.

Dwelling Thinking: Having examined the nature of the thing, let us consider a little the nature of building in relation to dwelling, the thing and the fourfold. "Building" means to make a place out of undifferentiated space, where the earth appears as earth, the heaven as heaven, the divinities as divine and mortals as mortal. So, in essence, building has nothing to do with the human activity of fashioning or erecting the concrete structure as in our ordinary understanding. Heidegger states clearly:

> ...the nature of the erecting of buildings cannot be understood adequately in terms either of architecture or of engineering construction, nor in terms of a mere combination of the two. The erecting of buildings would not be suitably defined *even if* we were to think of it in the sense of the original Greek *technē* as *solely* a letting-appear, which brings something made, as something present, among the things that are already present.[117]

The essential nature of building, then, is *letting dwell (letting-be)* or bringing forth by which Dasein brings forth things as things. Here we can see the essential relations between the fourfold, things and human dwelling, which Heidegger detailed as follows:

> Dwelling preserves the fourfold by bringing the presencing of the fourfold into things. But things themselves secure the fourfold *only when* they themselves as things are let be in their presencing. How is this done? In this way, that mortals nurse and nurture the things that grow, and specially construct things that do not grow. Cultivating and construction are building in the narrower sense. *Dwelling*, insofar as it keeps or secures the fourfold in things, is, as this keeping, *a building*.[118]

We can see from the above the inseparability of dwelling and things. This lies in the fact that dwelling cannot be accomplished without a staying with things. Things can only be things in their essential unfolding when human beings are able to dwell. Hence, both dwelling and things preserve the fourfold. As an example of authentic dwelling, Heidegger refers to the two-hundred- year-old-peasant-farmhouse in the Black Forest. The farmhouse is placed in the wind-sheltered mountain slope looking south. Its overhanging single roof bears up the burden of snow and shields the chambers from storms. Inside, the alter corner behind the community table is not forgotten, and there are hallowed places for the childbed and the laying of the dead and is designed for the different generations under one roof. The farmhouse not only provides shelter, it offers a dwelling for man as well

117 *VA*, p. 155; BDT, in *PLT*, p. 157.
118 *VA*., p. 145–146; BDT, in *PLT*, p. 149.

as God, and nature. It thus assembles the fourfold and bears witness to an earlier, authentic mode of dwelling. As Richardson comments:

> ...it (Dasein) can let things shine forth in their own "place," occupying their own "space," because from the very beginning its openness to Being is an openness to all possible "space," [...] its ontological dimension is a constitutional nearness to things. But only when this ontological dimension is articulated on the ontic level in the things among which There-being sojourns, does There-being find itself genuinely "at home" in its nearness to things.[119]

Now, man must learn to think for the sake of dwelling. Thinking which enters into the mode of dwelling stands out into *phusis* (world/earth) and lets beings be. To let beings be, to engage oneself with the open region in which beings are made to stand, is a mode of dwelling-thinking. But such a letting be of beings in the open region is a concealment of being itself. It does not think Being. Dwelling thinking is always a staying with things and letting things be in their essence. It lets beings be. According to Heidegger,

> Such thinking is neither theoretical nor practical [...] such thinking has no result. It has no effect. It satisfies its essence in that it is. Historically, only one saying belongs to the matter of thinking, the one that is in each case appropriate to its matter. Its material relevance is essentially higher than the validity of the sciences, because it is freer. For it lets Being—be.[120]

Thus in dwelling thinking, man's thinking becomes thoughtless when his thinking is merely thinking on thinking. "Only that can lie fallow which in itself is a ground for growth such as a field. An expressway, where nothing grows, cannot be a fallow field."[121]

Dwelling thinking then is not the kind of thinking which contemplates Being, rather it thinks beings. Dwelling thinking is the resolute openness that does not yet open the concealed being. But being-in-the-world for Heidegger is neither a theological nor a moral statement. It is a phenomenological and ontological assertion. It affirms that man, by virtue of his existence is stood out into the world and is able to tarry alongside things. It is a being near to things, so that the being of such entities may be able to receive the manifestation. Being-in-the-world is that ontological *a priori* which makes the existence of Dasein to be basically existent, into the openness of beings. But to think Being, there is required the event of *Called Thinking* 'through which being claims man for the truth of being.'

---

119 Richardson, W. J. *Heidegger Through Phenomenology to Thought*, p. 586.
120 Letter on Humanism, in *BW*, p. 359.
121 Heidegger, M., "Memorial Address", *Discourse on Thinking*, p. 45.

2.3.1b. Poetic Dwelling

Still on his 1950 lectures, in another essay, "*Poetically Man Dwells*" Heidegger refines the theme of dwelling by linking it with the poetic. The 'dwelling' that 'spares' the fourfold is a poetic dwelling. He makes a parallel between authentic dwelling, preserving the fourfold and poetry. The 'poetic' does not belong to the realm of phantasm, frivolous rhapsodizing and disappearing into the unknown. Poetry thus understood is not some fanciful flight of imagination or escapism. On the contrary, "Poetry is what first brings man onto the earth (under the sky, and before the divinities), making him belong to it, and thus brings him into dwelling."[122] It is the illuminating projection which lets the open happen, so that now, in the midst of beings, the open can bring the beings to shine or to ring out.[123] The poet is the person who attends to language and speaks impersonally, or allows language to speak through him or her. Poetry in this sense expresses the culture, with its history, whose language it is, but also the world in which and out of which that culture has grown and lives.

Heidegger also characterizes poetry as taking measure. "Poetry is a measuring."[124] This measuring has nothing to do with scientific activity, which relates to a very accurate dimension. The measuring here is not precise or purposive in any ego-centric, avaricious way. The taking of our measure is "a taking which at no time clutches at the standard but rather takes it in a concentrated perception, a gathered taking-in that remains a listening."[125] Such listening or concentration does not elucidate things or solve any problems (although it might make them seem no longer important). The poet, after all, takes measure of the "between" that brings together heaven and earth, divinities and mortals. It is a question of measuring in the stringent sense of the word: the measuring that indicates a measure for the scope of "being", the measuring that extends to the inauguration of the fourfold. The world and our place in it, and thus our essence, remain mysterious. Indeed, "poetry, as the gauging of that strange measure, becomes ever more mysterious. And so it must doubtless remain, if we are really prepared to make our stay in the domain of poetry's being."[126] The spirit of modern technology is quite opposed to the spirit of such poetry.

---

122 *VA.*, p. 186; "...Poetically Man Dwells..." in *PLT*, p. 216.

123 Kockelmans, J.J. *On the Truth of Being*, p. 196.

124 *VA.*, p. 187; "...Poetically Man Dwells..." in *PLT*, p. 219.

125 *VA.*, p. 189; "...Poetically Man Dwells..." in *PLT*, p. 221.

126 *VA.*, p. 190; "...Poetically Man Dwells..." in PLT, p. 222.

Heidegger's phenomenology challenges us to contemplate and face the mystery of Being- in- the-world (Dasein) as an assemblage of truths that surrounds human existence. To this end, he invites us to think poetically. Philosophical thinking, like other forms of discursive reasoning, relies on representation as we discussed in the section on Essential thinking. Philosophical thinking can suggest "mystery" (what truths conceal) but it remains, necessarily, horizon bound. Poetic thinking is instinctive and direct, bringing mystery to presence. Heidegger comments during a stopover in Provence: "These days in Cézanne's homeland are worth more than a whole library of philosophy books. If only one could think as directly as Cézanne painted."[127] Heidegger seeks to conquer metaphysics by appealing to a poetic sensibility that transcends ordinary language.

Poetry breaks open prospect of disclosure, as "it says, means, more than can even be captured in words"[128] George Pattison reverberates these feelings in summing up Heidegger's later work: "There is certain plausibility in seeing the later Heidegger as essentially a poetic thinker."[129] This is also reiterated by Richard Rorty who notes that, for Heidegger, "philosophical truth depends upon the very choice of *phonemes*, on the very *sound* of words."[130]

To think poetically, Dasein must open himself to the poetic presencing of Being by poetic dwelling, so as to dwell in the neighbourhood of Being. Dwelling in the fourfold is a dwelling in the neighbourhood of Being and this involves a 'homecoming' (*Heimkommen*) of the poet as a return (*Rückkehr*) to the source or origin (*Ursprung*).[131] This question of homecoming implies a 'homelessness' (*Heimatlosigkeit*) to which Heidegger variously referred in his writings. However, homelessness here has nothing to do with housing-shortage in its social understanding. Heidegger himself explains:

> However hard and bitter, however hampering and threatening the lack of houses remains, the real plight of dwelling (die eigentliche Not des Wohnens) does not lie merely

127 Young, J., *Heidegger's Later Philosophy*, p. 19. For a more detailed discussion of Heidegger's relationship to Cézanne, see also Young, Julian, *Heidegger's Philosophy of Art,* Cambridge: Cambridge University Press, 2001, especially chapter 4 sections 19–20.

128 Young, J., "What is Dwelling? The Homelessness of Modernity and the Worlding of the World" in *Heidegger, Authenticity and Modernity*, p. 196.

129 Pattison, G., *The Late Heidegger*, London: Routledge, 2000, p. 190.

130 Rorty, R., *Contingency, Irony and Solidarity*, Cambridge, UK: Cambridge University Press, 1989, p. 114.

131 Heidegger, M., *Erläuterungen zu Hölderlins Dichtung*, 5. Aufl. Frankfurt am Main: Vittorio Klostermann, 1981, p. 23 (Hereafter, HD); *Existence and Being*, p. 259.

> in a lack of houses.The real plight of dwelling is indeed older than the world wars with their destruction, older also than the increase of the earth's population and the condition of the industrial workers. The real dwelling plight lies in this, that mortals ever search anew for the nature of dwelling, that they *must ever learn to dwell.*[132]

Heidegger was not being insensitive to millions of people who were left homeless after the Second World War. 'Homelessness' for him consists in being 'strangers' in one's own homeland (*Heimat*), and being a neighbour to the world dominated by technology. Contemporary man is "homeless". Men escape from their homeland and suppose themselves at home in the skyscrapers of the big cities, where they live in the neighbourhood of an invented world, sufficiently supplied by the mass-media: television, radio, films, newspapers etc.[133] The most concrete confirmation of this fact, according to Heidegger is the behaviour of his fellow countrymen.

> Many Germans have lost their homeland, have had to leave their villages and towns, have been driven from their native soil. Countless others, whose homeland was saved, have yet wandered off. They have been caught up in the turmoil of the cities, and have resettled in the waste lands of industrial districts. They are strangers now in their former homeland.[134]

The ultimate cause of this alarming development is attributed to industrialization and urbanization. But there is much more than that when one considered those Germans who remained in their native villages. Heidegger laments:

> And those who have stayed on in their homeland? Often they are more homeless than those who have been driven from their homeland. Hourly and daily they are chained to radio and television. Week after week the movies carry them off into uncommon, but merely common realms of the imagination, and give the illusion of a world that is no world. Picture magazines are everywhere available. All that with which modern techniques of communication stimulate, assails, and drive man—all that is already much closer to man than his fields around his farmstead, closer than the sky over the earth, closer than the change from night to day, closer than the conventions and customs of his village, than the tradition of his native world.[135]

---

132 VA., p. 156; Building Dwelling Thinking, in *PLT*, p. 159. "So hart und bitter, so hemmend und bedrohlich der Mangel an Wohnungen bleibt, *die eigentliche Not des Wohnens* besteht nicht erst im Gehlen von Wohnungen. Die eigentliche Wohnungsnot ist auch älter als die Weltkriege und die Zerstörungen, älter als die Lage des Industrie-Arbeiters. Die eigentliche Not des Wohnens berht darin, daß die Sterblichen das Wesen des Wohnens immer erst wieder suchen, *daß sie das Wohnen erst lernen müssen.*"

133 Cf. *GL*, p. 15; *Discourse on Thinking*, p. 48.

134 "Memorial Address" *Discourse on Thinking*, p. 48.

135 *Ibid.*

While the appeal and enticement of the metropolis reduces the contemporary metropolitans homeless, the country dweller are forcefully dislodged by technological appliances and print media (transistors, television sets, cinemas, newspapers and magazines).

In *Letter on Humanism*, Heidegger argues that homelessness "consists in the abandonment of Being by beings. Homelessness is the symptom of oblivion of Being. Because of it the truth of Being remains unthought. The oblivion of Being makes itself known indirectly through the fact that man always observes and handles only beings."[136] Not only is modern homelessness a "symptom" of the forgetfulness of Being, it is also a sign that prevents man from addressing the root cause of his homelessness. As such, homelessness is the consequence of the history of Being and therefore its destiny:

> Homelessness is coming to be the destiny of the world. Hence it is necessary to think that destiny in terms of the history of Being. What Marx recognized in an essential and significant sense, though derived from Hegel, as the estrangement of man has its roots in the homelessness of modern man. This homelessness is specifically evoked in the form of metaphysics, and through metaphysics is simultaneously entrenched and covered up as such.[137]

Also homelessness is loss of man's rootedness (*Bodenlosigkeit*) in Being. "The most proper and the best feature of the homeland consists in its being in the nearness to the Origin,—and nothing but this."[138] The rootedness of contemporary humanity is endangered at its core. Even more: the loss of rootedness is not something caused by mere outside circumstances or destiny. Nor is it due exclusively to the laziness and shallow lifestyle or superficiality of human beings. The loss of rootedness stems from the spirit of the age into which we all were born...the atomic age.[139] So the spirit of the age is technological: "The power concealed in modern technology determines the relation of man to that which exists."[140] Hence, modern homelessness is a manifestation of the technological spirit of the modern age.

---

136 Letter on Humanism, in *BW*, p. 242.

137 *Ibid.*, p. 243.

138 *HD*, p. 23. "...Das Eigenste und das Beste der Heimat ruht darin, einzig diese Nähe zum Ursprung zu sein—und nichts anderes außerdem...."Cf. William, J. Richardson, *Heidegger: Through Phenomenology to Thought*, p. 445.

139 Heideger, M., Reden und andere Zeugnisse eines Lebensweges": 1940–1976. Ed. Hermann Heidegger, Frankfurt: Klostermann, 2000, *Gesamtausgabe,* Vol. 16, pp. 521–22.

140 "Memorial Address," in *Discourse on Thinking*, pp. 49–50.

Dwelling cannot properly happen if we are forgetful of our relation to our world. Construction alone cannot suffice; we must take into consideration the spaces in which we build: "the relationship between man and space is none other than dwelling, strictly thought and spoken."[141] Dwelling is our thought about our relation to space, and about our stay on earth among things.

Arriving in one's homeland, returning to the 'fatherland', meeting with the countrymen there, those who are dwelling on the same native soil, should be handled with "care". Heidegger insisted on the proximity between the poetic notion of the word "dwelling" and the word "care" as referred to the "care of homecoming". He affirmed that dwelling and building are not the same" "Letting-dwell" means getting closer to the nature of something, with respect to the nature of the word "home": the existence of human beings subsists in the nature of dwelling. To dwell does not mean to reside in a house or a lodge. Dwelling implies the mortality of human beings living temporally in the world to reveal the immortality of sky and earth. Authentic dwelling therefore is a poetic dwelling, a dwelling in the neighbourhood of Being, a dwelling that spares and preserves the play of the fourfold.[142]

Dasein's homecoming is dwelling in its homeland and nearness to Being. He is summoned to the homeland by a primordial poetic presencing to which he responds by poetic dwelling. But according to Heidegger, Dasein is on the way towards home, he has not yet reached it. The people and things at home seem pleasantly familiar, yet they are not really so. Commenting on Hölderlin's poetry on "*Remembrance of the Poet*", Heidegger writes:

> Even with his arrival, the returning one has not yet reached home [...]. Therefore the newcomer still remains in search of it. Only what he seeks is already coming to meet him. It is near. But what is sought is not yet found, if "find" means to receive what is found as one's own, to be able to dwell in it as a possession.[143]

The joy of return cannot be accomplished simply by the arrival on the shores of one's ancestral birth. A real homecoming should be echoed by the immeasurable shifts that the place had experienced. In his "*Homecoming/to Kindred Ones*"

---

141 BDT in *PLT*, p. 155.

142 Puthenpurackal, J.J., *Heidegger Through Authentic Totality to Total Authenticity*, p. 159.

143 Heidegger, M., "Remembrance of the Poet" trans. Douglas Scott, in *Existence and Being*, p. 263. "Mit der Ankunft hat der Heimkehrende die Heimat noch nicht erreicht. Also ist sie 'schwer' zu gewinnen...'...Deshalb bleibt auch der Ankommende noch ein Suchender. Allein das Gesuchte begegnet ihm schon. Es ist nahe. Aber das Gesuchte ist noch nicht gefunden, wenn 'finden' heißt, den Fund zu eigen bekommen, um in ihm als dem Eigentum zu wohnen." HD, pp. 13–14.

Hölderlin reflects upon the burden of homecoming: "*What you seek is near and already coming to meet you.*"[144] But one's returning home has not yet been reached merely by arriving there, therefore, man "must ever learn to dwell."

The essay on poetry end with the question: "Do we dwell poetically? Presumably we dwell altogether unpoetically".[145] Heidegger suggests that this unpoetic dwelling results from our inability to take measure, from our being cursed with a calculating measuring that does not suffice for an authentic dwelling. Authentic dwelling is nonetheless inseparably linked with the poetic: "The poetic is the basic capacity for human dwelling [...]. When the poetic appropriately come to light, then man dwells humanly on the earth, and, then—as Hölderlin says in his last poem—"the life of man" is a 'dwelling life'"[146] How does these thinking and dwelling poetically answer the question of the truth of Being? This is what will will answer the the sections that follows.

2.4. Seeing the Truth of Being

The discussion on Being and thinking led to the question of truth. "Heidegger turned the question of the meaning of being in the direction of the question of the truth of being."[147] Truth involves judgement. Judging is a way of being towards the thing itself. Judgement involves an assertion in which an entity is pointed out as the very entity which one had in mind in the assertion. To judge is to assert; to assert is to point out; to point out forces the thing to show itself and uncovers the entity. Heidegger attempted a statement on truth on the basis of these statements. "To say that an assertion "is true" signifies that it uncovers the entity as it is in itself. Such an assertion asserts, points out, 'lets' the entity 'be seen' (αποφανσις) in its uncoveredness."[148] He declares that assertions are not representations; they are tools for calling attention to pertinent aspects of a situation occurring within an already disclosed world. To say that an assertion is true implies that it uncovers the thing it is about in the "how" of its being, relative to a disclosure. The entities that are uncovered are true "in a second sense." The being that does the uncovering, Dasein, is true in a more fundamental sense.

---

144 "Remembrance of the Poet", in *Existence and Being*, p. 259.

145 *VA.*, p. 192; "...Poetically Man Dwells..." in *PLT, p.* 225.

146 *VA*,. P. 194; "...Poetically Man Dwells..." in *PLT*, p. 227.

147 James, Risser, ed. *Heidegger Toward the Turn, Essays on the Work of the 1930s,* New York: State University of New York Press, 1999, p. 4.

148 *SZ.*, p. 218; *BT.*, p. 261.

> Uncovering is a way of being for being-in-the-world. Circumspective concern, or even that concern which we tarry and look at something, uncovers entities within the world. These entities become that which has been uncovered. They are "true" in a second sense. What is primarily 'true'—that is, uncovering—is Dasein.[149]

Overall Dasein's activities disclose the worldliness of the world. That is, they set up a world that allows entities within the world to show up and hence be discovered. Only through Dasein's activities can truth get a grip on the world.

In this part of our study, we shall examine how Dasein can see the truth of Being by moving away from representational and logic-dominated thinking to the realm of *Ereignis* in which Dasein and Being are naturally appropriated to each other in their essential nature.

2.4.1. Dasein and Aletheia

Heidegger has *affection* for a kind of primitivism which makes him look into the dark reaches of past time in order to find the more "primordial" insights that have been subsequently covered up.[150] Shifting from an epistemological perspective to an ontological approach, Heidegger discusses truth in terms of it Greek origin, *aletheia*.[151] According to him, the most fundamental form of being was αληθεια, which means 'un-concealment' (*Un-verborgenheit*). "The prefix "un-" corresponds to the Greek 'α-' which grammar calls "α privativum,"[152] which means undoing the

---

149 SZ., p. 220; *BT.*, p. 263.

150 Ruud, A., *Expressing the World: Skepticism, Wittgenstein, and Heidegger*, p. 204. According to Ruud, Heidegger locates his Golden Age in Pre-Socratic Greece, where poets and thinker lived in openness to Being that was covered up when thinking started to take a wrong turn with Plato.

151 Heidegger refused to accept the metaphysical rendering of *aletheia* as "truth". In *Sein und Zeit,* he translated it as "uncoveredness" and as "Being-uncovering." Cf. *SZ*, p. 220; *BT.*, p. 263. He accepts the notion of truth as correspondence to reality as obviously right, as far as it goes, but he is interested in the more fundamental question of what makes it possible for there to be a correspondence between a statement or judgment, and a state of affairs in the world.

152 Heidegger, M., *Parmenides*, trans. André Schuwer and Richard Rojcewicz, Bloomington, IN. Indiana University Press, 1998, p. 14. Published in German in *Gesamtausgabe 54* as *Parmenides* by Vittorio Klostermann, Frankfurt an Main, 1982. In this book, Heidegger rendered Aletheia in four different senses. The first two corresponds to the concealing and revealing aspects of aletheia, based on the emphasis given either to the 'a' or to the 'lether'. While the former points to the revealing, the later stresses the concealing. (Cf. pp. 14–16). The third way of rendering *aletheia* refers to the metaphysical understanding of the term, in which it comes to mean truth,

concealing that is there. This term can therefore mean to the ancient as well as the contemporary Greek "honest," "blunt," "genuine," or "real." Heidegger extends the meaning to describe a situation or process which unexpectedly or progressively shows itself, and can be concretely appropriated.

The etymological meaning of this word was inherently linked to a form of unconcealing, revealing, and disclosing. As his thinking advances, we notice that Heidegger did not use the term *aletheia* in its Greek equivalent to 'truth' (*Wahrheit*), (correspondence or correctness) i.e., conformity of intellect and thing, making truth reside ultimately in the judgment or as Heidegger defines it, "correspondence, grounded in correctness, between proposition and thing."[153] However, he did not continue to consider truth as aletheia. He simply focuses on aletheia as aletheia. According to him, "to raise the question of aletheia, of unconcealment as such, is not the same as raising the question of truth. For this reason, it was inadequate and misleading to call aletheia in the sense of opening, truth."[154] Thus, the essence of aletheia as 'unconcealment' is openness, which is unconcealed in aletheia. Truth was synonymous to 'bringing things out of concealment' into the open, into disclosedness. So for Heidegger, Aletheia was the primordial truth which has to be disclosed first of all in being:

> Dasein, as constituted by disclosedness, is essentially in the truth. Disclosedness is a kind of Being which is essential to Dasein. 'There is' truth only in so far as Dasein is and solong as Dasein is […]. Before there was any Dasein, there was no truth; nor will there be any after Dasein is no more.[155]

What this means is that the phenomenological analysis of Dasein as being-in-the-world and existence appears to lead of necessity to the essence of truth, and in this essence man's own being finds its radical explanation. In order words, the "essence of truth is the history of man's essence."[156]

Truth as *aletheia*, according to Heidegger, is an "event" which happens when something is presented to us as it really is, without any concealment or distortion. Truth originally is Dasein's disclosedness, and to this, belong the disclosure of

certitude and correctness as opposed to falsity, uncertitude and incorrectness. (Cf. pp. 17–20). The fourth interpretation of aletheia means the unconcealment or the clearing of Being. (Cf. pp. 131ff) Cf. also VENSUS, G., *The Experience Being as Goal of Human Existence, The Heideggerian Approach*, p. 182.

153 HEIDEGGER, M., *The Essence of Truth: On Plato's Cave Allegory and Theaetetus*, trans. T. SADLER, Athlone: Continuum International Publishing Group, 2002, p. 1.

154 HEIDEGGER, M. *On Time and Being*, p. 70.

155 *SZ*, p. 226; *BT.*, p. 269.

156 HEIDEGGER, M., *The Essence of Truth: On Plato's Cave Allegory and Theaetetus*, p. 81.

entities. There is a genuine freedom associated with this disclosedness. Heidegger stated also that "the essence of truth is freedom."[157] The freedom in question is not a mere lack of constraint. Freedom is primordially not a characteristic of human activity, but, as being-in-the-world, man is openness, he transcends being-necessitated and has the positive possibility to transcend and to project. Primordially, therefore, freedom indicates the being of Dasein on the proper level of his Daseinhood. "Freedom for what is opened up in an open region lets beings be the beings they are. Freedom now reveals itself as letting beings be."[158] This letting-be is freedom which consists in the disclosure of being as such.

To let beings be is in reality to let things flare up into self-sufficient emergence. But everything that emerges does so in some light. Only in virtue of the light can what emerges show itself and be seen. Heidegger calls this *Lichtung* which means "clearing" or open space".[159] The openness is the 'play-ground' (*Spielraum*) and is the lightening or clearing. This opening which is also called '*lighting*' is compared to a clearing in the forest. "The forest clearing (opening) is experienced in contrast to dense forest [...]. The substantive "opening" goes back to the verb "to open." The adjective *licht* "open" is the same word as "light." To open something means: To make something light, free and open, e.g., to make the forest free of trees at one place."[160] This 'lighting' or clearing denotes the way truth sheds light on things, brings them out of obscurity. As Heidegger states it further in another text, *The Origin of the Work of Art*:

> In the midst of beings as a whole an open place occurs. There is a clearing, a lighting [...]. That which is can only be, as a being, if it stands within and stands out within what is lighted in this clearing. Only this clearing grants and guarantees to us humans a passage to those beings that we ourselves are not, and access to the being that we ourselves are. Thanks to this clearing, beings are unconcealed in certain changing degrees. And yet a being can be concealed, too, only within the sphere of what is lighted. Each being we encounter and which encounters us keeps to this curious opposition of presence in that it always withholds itself at the same time in a concealedness.[161]

These texts give us a clearer understanding of the latest thinking of Heidegger on the question of the truth of Being. In the first instance, this clearing is contrasted with a dense forest, and to open it is to make a clearing. The German word

---

157 On the Essence of Truth in *BW.*, p. 123.

158 *Ibid.*, p. 125.

159 *SZ*, p. 133; *BT*, p. 171. (see also footnote)

160 Heidegger, M., *Zur Sache des Denkens,* 2. Auflage, Tübingen: Max Niemeyer, 1976, p. 72. (Hereafter, SD); *On Time and Being,* p. 65.

161 Heidegger, M., The Origin of the Work of Art, in *PLT*, pp. 51–52.

"opening," (*Lichtung*) is a borrowed translation of the French *clairiére*.[162] When the clearing is made, light can stream into the opening and struggle with the darkness. This "open" is for Heidegger the world as the necessary sphere within whose limits every tangible being can be truly brought to light by Dasein. It is in this opening that things are able to radiate, to appear and to be seen. It is this opening which first grants the possibility of truth. "...aletheia, unconcealment in the sense of the opening may not be equated with truth. Rather, aletheia, unconcealment thought as opening, first grants the possibility of truth. For truth itself, just as Being and thinking, can only be what it is in the element of the opening."[163]

Through this image, Heidegger understands aletheia as the clearing. There is a clearing within which an understanding of being or essence can prevail when irreconcilable possibilities of being are concealed or held back. However, according to Wrathall, "'unconcealment,' when understood as the clearing, does not name a thing, or a property or characteristic of things, or a kind of action we perform on things, or even the being of things. It names, instead, a domain or structure which allows there to be things with properties and characteristics, or modes of being."[164]

The life of Dasein is a continuous struggle to unconceal truth that is dwelling within its presence and acknowledging the hiddenness of the truth of being. To Heidegger, truth is personal and historical as well as ontological. Wrathall sums up this hiddenness as "a non-assertoric dealing with the world in the sense that, in such pre-predicative comportments, the world is experienced in a way that lacks determinacy, that is, propositional articulation."[165] In this propositional concealment, however, the Dasein who would be authentic has to overcome the 'They', (inauthentic collectivity of the '*das man*'), show a 'difference' from entities which persists in dwelling in a space and time stuffed with only partial insight. "The understanding, dispositions, and skills that Dasein has in the first instant are the banalized understandings, dispositions, and skills of the one (*Das Man*)."[166]

162 *SD*, pp. 72–72; *On Time and Being*, p. 65.

163 *SD*, p. 78; *On Time and Being*, p. 69.

164 Wrathall, M. A., "Unconcealment," in *A Companion to Heidegger*, H. L. Dreyfus & M. A. Wrathall, (eds.), Oxford, England: Blackwell Publishing Ltd., 2007 edition, p. 340.

165 Wrathall, M., *Heidegger and Concealment. Truth, Language and History*, New York: Cambridge University Press, 2011, p. 19.

166 *Ibid.*, p. 24.

In *Identity and Difference,* Heidegger sees the 'difference' and the 'clearing' together.[167] Thus, Being as the 'difference' between Being and entities, and truth as the 'clearing,' for concealing and revealing belong together. The unity and the difference in this dualistic nature of truth, and thus, of Being which is the spine of Heideggerian thinking calls for more attention as it is affixed in the belonging-together of Being and Dasein.

> Being presences in the mode of concealing and it is in the unity of the revealing-concealing by holding itself away from man; but in this distancing it is nevertheless there. Being presences in its withdrawing itself, but in the withdrawal it is present.[168]

To sum up our study of the essence of truth as clearing, it is necessary to stress that the clearing should be understood as something like a room of possibilities. As Heidegger points out; "The clearing grants first of all the possibility of the path of presence, and grants the possible presencing of that presence itself. We must think aletheia, unconcealment, as the clearing that first grants Being and thinking and their presencing to and for each other."[169] This means that "the clearing makes it possible for a certain understanding of being—a particular mode of presence—to come to prevail among entities."[170]

---

167 *ID,* p. 57; *Identity and Difference,* p. 65. " Die Differenz von Sein und Seiendem ist als der Unter-Scheid von Überkommnis und Ankunft der entbergend-bergende Austrag beider. Im Austrag waltet Lichtung des sich verhüllend Verschließenden, welches Walten das Aus- und Zueinander von Überkommnis und Ankunft vergibt.

168 Bretschneider Willy, *Sein und Wahrheit: Über die Zusammengehörigkeit von Sein und Wahrheit im Denken Martin Heideggers,* Meisenheim am Glan: Verlag Anton Hain, 1965, p. 172. "Das Sein west in der Weise des Verbergens und ist in der Einheit des entbergenden Verbergens, indem es sich als es selber vom Menschen fernhält, aber im Fernhalten gleichwohl da ist. Das Sein west als sich entziehendes, ist aber im Entzug anwesend."

169 "The End of Philosophy and the Task of Thinking" in *BW.*, p. 445.

170 Wrathall, M. A., *Unconcealment,* in *A Companion to Heidegger,* p. 356. Concealment does not allow aletheia to have complete exposure of beings; Concealment preserves what is most proper to aletheia as its own." This concealment does not occur simply because the "knowledge of beings is always fragmentary as unconcealment is older than every openedness of this or that being" and "letting-be itself, which in disclosing, already holds concealed and comports itself towards concealing." In order for being to unconceal beings, there must be concealment, it must preexist any particular unconcealing of beings that lets them be as they are. That which is 'conserved'/'preserved' is exactly that which is concealed—"being as such." Since every comportment is a comportment of beings, "beings as such" are "what are most proper to aletheia as its own." Hence, untruth/unconcealment is also that which is

The essence of aletheia as 'unconcealment' is openness, which is unconcealed in aletheia. Herein lies the hub of Heidegger's thinking of the truth of Being, namely, this "letting be" takes things from concealedness, it brings them to light and makes them participate in the truth of being. We can then assert that aletheia, as the unconcealment of Being, takes place only in relation to the reciprocal gaze of Being and Dasein, that is, in the belonging together. Besides his openness to *aletheia*, Dasein can attain the truth of Being by his openness to language, the house of Being. "It is the light of language that illuminates the clearing with the revelation of truth. The world is lit up with the conception of words as the focal points of recognition."[171]

2.4.2. Language: The house of Being

Language, says Heidegger, is something more than an activity of man, more than expression or a means to it, more than a representation of the actual or the imaginary. Language does not refer to the metaphysico-technological language which is subjectivistic instrument for communication, clarification and dominion.[172] Language is not only and primarily a phonetic and written expression of that which is to be communicated. We must, in Heidegger's view, first overcome this current conception of language as communication. According to the classical view of language as laid down for posterity by Aristotle (in his *peri hermēneias* or *de interpretation*):

> Spoken words are the signs of the soul's experiences, and written words are the signs of spoken words. Just as all men have not the same writing, so all men have not the same speech sounds; but the soul's experiences, which they immediately signify, are the same for all, as also are those things which our experiences are the images.[173]

Here, the essence of language is not to be sought by focusing on voice and vocalization, or on sign and signification. True, signification and expression are indeed characteristics of language, but they do not reach into the domain of what, as

---

proper to aletheia as its own. This is "the one mystery" that "holds sway throughout man's Dasein." See also Wrathall, M., *Heidegger and Concealment. Truth, Language and History*, p. 34.

171 Oberst, J. *Heidegger on Language and Death: The Intrinsic Connection in Human Existence*, London and NewYork: Continuum International Publishing Group, 2009, p. 80.

172 Puthenpurackal, J. J, 1987, p. 214.

173 Aristotle, *Peri Hermēneias* c. 1. 16a3–8. Cited in Kockelmans, J. J. *On Heidegger and Language*, Evanston, IL.: Northwestern Univ. Press, 1980, p. 4.

far as language itself is concerned, is primordial and essential.[174] Understanding language in this way does not take us to the essence of language. To reach the essence of language, we must consider "the being of language: the language of being."[175] This phrase suggests that the essence of language can be reached, only by considering the 'Being' (*Wesen*) of language, i.e., language in its coming-to-be.

> Language is the primal dimension within which man's essence is first able to correspond at all to Being and its claim, and in corresponding, to belong to Being. This primal corresponding, expressly carried out, is thinking. Through thinking, we first learn to dwell in the realm in which there comes to pass the restorative surmounting of Enframing (the destining of Being).[176]

The disjunction, or alienation, stemming from man's homelessness, his *ek-sistence* (thrownness), in regard to Being can, it seems, only be overcome through the medium and experience of language (the house of Being) and thinking. The thinking of Being (*Das Denken des Seins*) becomes a thinking of language. "The way is a way of thinking. All ways of thinking, more or less perceptibly, lead through language in a manner that is extraordinary."[177] Thus, instead of speaking about language, we shall let language to speak to us. Only by letting the language speak within itself, can we bring language, as language into language, i.e., into its being.

> Language is not only and not primarily an audible and written expression of what is to be communicated. It not only puts forth in words and statements what is overtly or covertly intended to be communicated; language alone brings what is, as something that is, into the Open for the first time.[178]

In Heidegger's view, one way to get to the essence of language consists in a careful listening to the original meaning of the Greek words λογος (*logos*) and λεγειν (*legein*). Logos here "speaks simultaneously as the name for Being (*Sein*) and for saying (language)."[179] The verb *legein*, equivalent to the Latin '*legere*' (lessen, to read) and German '*legen*' (to lay) has the nuance of 'collecting or together', i.e., a laying which gathers. "To lay means to bring to lie. Thus, to lay is at the same time to place one thing beside another, to lay them together. To lay is to gather

174 Kockelmans, J. J, *On the Truth of Being*, p. 150.
175 *US*, pp. 200–201; *On the Way to Language*, p. 94.
176 *QCT*, p. 41.
177 *QCT.*, p. 3.
178 Heidegger, M. "The Origin of the Work of Art" in *PLT.*, p. 71.
179 *US.*, p. 185; *On the Way to Language*, p. 80.

(*lesen*)."[180] As a laying that gathers, 'λεγειν' keeps the gathered in the open region, in unconcealment. To allow something lie in unconcealment means to say (*sagen*). So, *legein* unquestionably means "to say" and "to speak," for saying consists in the letting-lie-together, as gathered, before that which gathers.[181] Hence, the essence of language in its Greek profundity is "the gathering letting-lie-before of what is present in its presencing."[182] The unconcealing of the concealed into non-concealment is the very coming-to-pressence of what is present, and that we call the Being of beings. As Biemel states: "In naming a being one first makes it appear. Where there is no naming, there is no openness. Therefore, Heidegger equates saying with the project of the clearing; through saying, unconcealment comes into being."[183] What this suggests is that, for Heidegger, *logos* primarily let language, as language, speak of itself from within itself. i.e., in its being, as a saying that shows something or that lets something emerge.

The things which language calls and names, gather to themselves the fourfold—heaven and earth, mortals and gods, and from the beginning, these four are united in being primarily toward one another; they make up the original fourfold. When things are named, they are called into their 'thinging' as things, through which they unfold the world in which they themselves can then dwell and abide.

> Saying, (Die Sage) as the way-making movement of the world's fourfold, gathers all things up into the nearness of face-to-face encounter, and does so soundlessly as quietly as time times, space spaces, as quietly as the play of time time-space is enacted. The soundless gathering call, by which the Saying moves the world-relation on its way, we call the ringing of stillness. It is: the language of being.[184]

This means that it is language alone that genuinely speaks. But its speaking is not audible. "*Language speaks as the peal of stillness.*"[185] Language is the house of Being as saying, that is, it shows the world in its time-space-lay and let happen the dif-ference (*Unter-Schied*) for world and things by worlding the world in the fourfold. In his lecture "*Die Sprache*", Heidegger hints his main train of thought on language:

---

180 *VA.*, p. 201; *EGT*, p. 61. „Legen heißt: zum Liegen bringen. Legen ist dabai zugleich: eines zum anderen-ist zusammenlegen. Legen ist leden."

181 *VA.*, p. 205; *EGT.*, p. 64. "Sagen ist gesammelt-sammelndes beisammen-vor-liegen-Lassen."

182 *VA.*, p. 220; *EGT.*, p. 77.

183 Biemel, W., "Poetry and Language in Heidegger," in Kockelmans, J. J., *On Heidegger and Language*, p. 76.

184 *US.*, p. 215; *On the Way to Language*, p. 108.

185 *US.* p. 30; *PLT.*, p. 205 "Die Sprache spricht als das Geläut der Stille".

> Language speaks. Its speaking bids the dif-ference to come which expropriates world and things into the simple onefold of their intimacy. Language Speaks. Man speaks in that he responds to language. This responding is a hearing. It hears because it listens to the command of stillness.[186]

This sounds like the résumé of Heidegger's teaching on language. Man plays a huge role in the linguistic appearance of Being. Man is neither the originator nor the speaker of language. He is merely the 'co-speaker' and is capable of passing on the speaking of Being. The speaker can speak only because he listens to language, and he is able to hear only because he belongs to language. As Charles Taylor explains this;

> So language, through its telos, dictates a certain mode of expression, a way of formulating matters which can help to restore thingness. It tells us what to say, dictates the poetic, or thinkerly words... We can go on talking, mindful only of our purposes, unaware that there is anything else to take notice of. But if we stop to attend to language, it will dictate a certain way of talking. Or otherwise put, the entities will demand that we use the language which can disclose them as things.[187]

This aspect of language is further referred to by Heidegger in a series of imaginary conversations in the winter of 1944/45 in which the interlocutors discussed wide variety of issues ranging from thinking to truth, language to technology. As regards the role of language in conversation Heidegger says: "[...] it seems to me as though in a proper conversation an event takes place wherein something comes to language."[188] Instead of the interlocutors initiating, the most appropriate comportment in the conversation should be waiting. He states further:

> Yet a conversation first waits upon reaching that of which it speaks. And the speakers of a conversation can speak in its sense only if they are prepared for something to befall them in the conversation which transforms their own essence.[189]

In the "*Letter on Humanism*" (1947), Heidegger describes once more how the relation between Being and man is brought to completion by means of thought. The interrelation between language, the essence of man, Being and thinking is strongly

---

186 *US.*, p. 30; *PLT.*, p. 207. Cf. HEIDEGGER, M., *Phenomenologie und Theologie*, Frankfurt am Main: Vittorio Klostermann, 1970, p. 41. „Die Sprache Spricht. Der Mensch spricht nur, indem er der Sprache entspricht."

187 TAYLOR, C., Heidegger on Language, in *A Companion to Heidegger*, H. L. DREYFUS & M. A. WRATHALL, (eds.), Malden, MA, USA: Blackwell Publishing Ltd., 2005, p. 451.

188 HEIDEGGER, M., *Country Path Conversations*, trans. BRET W. DAVIS, Bloomington, IN: Indiana University Press, 2010, p. 36.

189 Ibid, p. 37.

affirmed in the opening passage, emphasizing another highly important aspect of his "*Weltanschauung*": the process of giving:

> Thinking accomplishes the relation of Being to the essence of man. It does not make or cause the relation. Thinking brings this relation to Being solely as something handed over to it from Being. Such offering consists in the fact that in thinking Being comes to language. *Language is the house of Being.* In its home man dwells.[190]

In other words, language is the house that gathers everything in it, so that in this house they find their essence, their name and their being. Even in his premier lecture course in Marburg (1923–1924), Heidegger astutely declared that language is "the being and becoming of human being himself."[191] According to him, language is "a specific way of being human being, of being in the world."[192] Therefore, "language is of an essential quality of the highest ontological order. Through language Dasein remains in touch with being itself, even if it is not aware of this."[193] Language is an event that has Being as its ultimate origin, a house that is arranged according to a pattern inscribed and prescribed by it. According to Wrathall, "Language is the house of being means, then, that a world is kept and preserved by a consolidation of the relationships that determine a thing as the thing it is."[194] In that case, Being makes manifest the presence of beings through language. Being therefore reveals the truth of beings through language. It is in language that the truth of Being is guarded. "It is the keeper of being present, in that its coming to light remains entrusted to the appropriating show of saying. Furthermore, language is the house of Being because language, as saying, is the mode of Appropriation."[195] When seen in terms of the event of appropriation (*Ereignis*), language is not inaccessible to man. Man essentially dwells in the house of Being not as its creator, but as its shepherd. This brings us to another trait of Dasein in quest for the truth of Being.

---

190 *BH*, *Wegmarken*, p. 311; "Letter on Humanism" in *BW.* p. 217. (Italics ours)

191 Heidegger, M., *Einführung in die phänomenologische Forschung*, GA, 17, Frankfurt am Main: Klostermann, 1994, p. 16; Heidegger, M. *Introduction to Phenomenological Reasearch*, trans. Daniel, O. Dahlstrom, Bloomington, Indiana: Indiana University Press, 2005, p. 12.

192 Ibid., p. 317: p. 240.

193 Oberst, J. L., *Heidegger on Language and Death: The Intrinsic Connection in Human Existence*, p. 82.

194 Wrathall, M., *Heidegger and Concealment. Truth, Language and History*, p. 152.

195 *US.*, p. 267; *On the Way to Language*, p. 135.

2.4.3. *Man: The Shepherd of Being*

Still in his poetic style of expression, Heidegger opines that man's essence consists in his being the friend and neighbour of Being. Man dwells in the neighbourhood of Being by preserving and guarding Being. Man has the exclusive privilege to play the role of guardianship of Being, not by domineering over it, but by merely attending on it. Hence, man is not primarily the overlord of that-which-is, as pragmatism and the technical mentality maintains. He is first of all the shepherd of being:

> Man is not the lord of beings. Man is the shepherd of Being. Man loses nothing this "less"; rather, he gains in that he attains the truth of Being. He gains the essential poverty of the shepherd, whose dignity consists in being called by Being itself into the preservation of Being's truth [...]. In his essential unfolding within the history of Being, man is the being whose Being as ek-sistence consists in his dwelling in the nearness of Being. Man is the neighbour of Being.[196]

Heidegger is saying here that man is "thrown" from Being itself into the truth of Being so that ek-sisting in this fashion he might "guard the truth of Being, in order that beings might appear in the light of Being as the being they are."[197] This means that to be a guard is to allow things to appear in the light of being. The house becomes a house only through dwelling. Man in his genuine speaking, thinking and poetizing, dwells in this 'house of Being'. "Those who think and those who create with words are the guardians of this home. Their guardianship accomplishes the manifestation of Being insofar as they bring the manifestation to language and maintain it in language through their speech."[198] By dwelling in the house, through '*saving the earth*', '*receiving the sky*', '*awaiting the divinities*' and '*initiating mortals*', man is able to preserve the fourfold: "To spear and preserve means: to take under our care, to look after the fourfold in its essence. What we take under our care must be kept safe."[199] In keeping safe, "Being is shepherded in Dasein's own being."[200] Heidegger explains further:

> Since Being is never the merely precisely actual, to guard Being can never be equated with the task of a guard who protects from burglars a treasure stored in a building. Guardianship of Being is not fixated upon something existent. The existing thing, taken for itself,

196 "Letter on Humanism" in *BW.*, p. 245.

197 *Ibid.*, 234.

198 *BH*, *Wegmarken*, p. 311; "Letter on Humanism," in *BW.*, p. 217.

199 BWT in *PLT*, p. 149.

200 Vensus, A. George, *From Being-in-the-World to Being -towards- Being: Martin Heidegger's Philosophy of the Authentic Human Person*, Nagpur: SAC Publications, 1996, p. 154.

never contains an appeal of Being. Guardianship is vigilance, watchfulness for the has-been and coming destiny of Being, a vigilance that issues from a long and ever-renewed thoughtful deliberateness, which heeds the directive that lies in the manner in which Being makes its appeal. In the destiny of Being there is never a mere sequence of things one after another; now frame, then world and thing; rather, there is always a passing by and simultaneity of the early and late.[201]

Dasein is not a stranger to the neighbourhood of Being, but "he at first fails to recognize the nearest and attaches himself to the next nearest. He even thinks that this is the nearest. But nearer than the nearest and at the same time for ordinary thinking farther than the farthest is nearness itself: the truth of Being."[202] Only by dwelling there does Dasein become a shepherd. Modern technological man lacks a neighbourhood since he is not 'at home' anywhere. He is estranged from the world of the really nearest, i.e., being but enchanted by the apparently nearest, i.e., entities. But only a dweller can be a shepherd of Being.

The shepherd, in shepherding the truth of Being, stands in the light of Being's presence and thereby sees the truth of Being. Hence, "the shepherd of Being is the seer of Being: the shepherd that 'shepherds' and the seer that 'sees' Being as a revealing-concealing process [...]. "A shepherd who attends on, watches over... sees the revealing-concealing play of Being"[203] in history. So, seeing is determined not by the eye, but by the lighting of Being that has been given to him already or as Richardson puts it; "He sees because the lighting-up of Being has been visited upon him."[204]

However, Dasein can only realize this at the realm of Ereignis. The whole process of seeing the truth of Being belongs to the realm of Ereignis, one further step, in which Dasein and Being own each other. So we have to take a closer look at this word that has been re-occurring in our text of late.

201 The Thing in *PLT*, pp. 182–183.

202 *BH. Wegmarken*, p. 348; "Letter on Humanism"in *BW.*, p. 235.

203 Puthenpurackal, J., J. *Heidegger Through Authentic Totality to Total Authenticity*, p. 202.

204 Richardson, W. J., *Heidegger Through Phenomenology to Thought*, p. 525.

2.4.4. *Ereignis: The Event of Appropriation*

The word "*Ereignis*[205]," along with the image of Dasein being appropriated by the absent, emerges in Heidegger's thought only in the 1930s.[206] However, this future word reverberates what Heidegger had previously called Dasein's thrownness, that is, the idea that Dasein is plunge into possibilities, forestalls its self-absence, and so is "previously" involved in the world-disclosure. As Thomas Sheehan observes, "In the language of Heidegger I, man is *In-der-Welt-sein*, a thrown-projective engegment-with-meaning. In that sense, the appropriation of man of meaning in Heidegger II is the same as the thrownness of man into meaning in Heidegger I."[207] This '*Grundstimmung*', this 'fundamental mood' are not 'inner feelings' tacked on to cognitive experience as causal byproducts. They are, rather, the way 'beings as a whole' are disclosed.[208]

In ordinary German, Ereignis means an "event" or a happening. But Heidegger employed this word as the central concept in his post-1938 thinking[209] not as 'event' but as a precondition for the emergence of all events.[210] Playing on

205 This word has been given many interpretations by many authors. We prefer to render it here as Ereignis. Translations such as 'event', 'event of appropriation', 'appropriation', 'happening' or 'enowning', although they each have their merits, overly limits the scope of the word and fail to capture the yet unforeseen mystery behind the word. Depending on the context, the word should invoke such meanings as happening, owning and seeing, to various degrees and in various combinations.

206 Recent studies have taken into account the appearance of the word in his very early 1919 lecture course and in scattered places in his work of the 1920s. See essay by Polt, R., "Ereignis," in *A Companion to Heidegger*, 2007, pp. 375–391.

207 Sheehan, T. Facticity and Ereignis., in *Interpreting Heidegger: Critical Essays*, Dahlstrom, D.O, editor, Cambridge: Cambridge University Press, 2011, p. 53

208 Julian, Y., *Heidegger's Philosophy of Art*, Cambridge: Cambridge University Press, 2001, p. 105.

209 This does not deny the fact that the single, defining concern of his path of thinkng right to the end of his life was about the originary, fundamental, unfying meaning of Being which he called Being itself (dasSein Selbst), Being as such and Being as Being (dasSein als Sein) Cf. Capobianco, R., "The Fate of Being in Heidegger's Four Seminars 1966–1973," *Existentia: An International Journal of Philosophy*, XV 3–4, 2005, p. 163.

210 Heidegger, M., *Contributions to Philosophy (From Enowning)*, trans. Parvis Emad & Kenneth Maly, Bloomington: Indiana University Press, 1999, p. xxi. Here Heidegger shows that 'event' cannot live up to the demands put on it by Ereignis because 'event' emerges from within 'time-space'and as such is itself enowned by Ereignis. "*Ereignis*...cannot be thought as an event *in* the course of serial time, since it is the *source* of any kind of time as well as of everything else." Cf. Stambaugh, J., *Th*

the adjective *eigen* ("one's own"), he creates the word *Ereignung*: movement as the process of being drawn into what is one's own. This being-drawn into one's own absence, is what Heidegger calls "appropriation" or "event of appropriation" and he gives it a superlative pride of place in his thinking.[211] Hence, he speaks of its etymological affinity with two root words: '*er-eigen*' (*eigen* = own) means, to come into one's own, to come to where one belongs etc., and '*er-äugnen* (*Auge* = eye) means, to catch sight of, to see directly, to see face to face, etc.[212] When these two are put together, we see that *Ereignis* has the nuance of being "abstract" as well as "concrete", i.e., it is something far from us, and yet remains close. Thus Heidegger explains the word in candid terms: "'Appropriate' does not mean here 'opportune'; instead, it signifies entrance into the eventuation which assigns all beings to the inceptuality of the beginning, such that they essentially occur no longer in their 'examination' for the human being but, rather, out of their 'absence' (i.e., here, departure) towards the event."[213]

Finitude of Being, Albany: State University of New York Press, 1992, p. 154. In French, one could speak of the évenir at the source of every événement. Cf. François Fédier, Traduire les Beiträge zur Philosophie (Vom Ereignis), in *Heidegger Studies*, no. 9, 1993, p. 32. The challenge then is to investigate a "happening" or "doing" deeper that what we ordinarily call events.

211 A footnote in *On the Way to Language* reads: "Today, when so much thoughtless and half-thought matter is rushed into print any which way, it may seem incredible to many of my readers that I have used the word 'appropriation' (Ereignis) in my manuscripts for more than twenty-five years to indicate what is here in my thoughts. The matter, while simple in itself, still remains difficult to think, because thinking must first overcome the habit of yielding to the view that we are thinking here of "Being" as appropriation. But appropriation is different in nature, because it is richer than any conceivable difinition of Being. Being, however, in respect of its essential origin, can be thought of in terms of appropriation." p. 129. *US*, p. 260.

212 *ID*, p. 24–25; *Identity and Difference*, footnote, p. 14. Cf. Also *Deutsches Wörterbuch von Jacob Grimm und Wilheim Grimm*, Leipzig: S. Hirzel, 1854–1954, s. vv. *eräugen, ereigen, ereignen, ereignis*. Some authors like Richard Polt have argued that *Ereignis* has no etymological connection to *eigen* (own, proper, particular) and its congnates. Instead it is related to *Augen*, „eyes", and originally means something like „coming into view". (Cf. Richard Polt, Ereignis, in *A Companion to Heidegger*, p. 378. See also Sheehan, T., A Paradigm Shift in Heidegger research. *Continental Philosophy Review*, no. 34, 2001, pp. 196–98.

213 Heidegger, M., *The Event*, trans. R. Rojcewicz, Bloomington, Indiana: Indiana University Press, 2013, p. 140. Published in German as Heidegger, M., *Gesamtausgabe* 71: *Das Ereignis*. Edited by Friedrich-Wilhelm v. Herrman, Frankfurt am Main: Vittorio Klostermann, 2009.

Consequently Ereignis, 'event of appropriation' is an enigma shrouded in ambiguity. "It is itself the most inconspicuous of inconspicuous phenomena, the simplest of simplicities, the nearest of the near, and the farthest of the far in which we mortals spend our lives."[214] The event of appropriation is equally " that realm, vibrating within itself, through which man and Being reach each other in their nature, achieve their active nature…"[215] and in doing so, they appropriate or own each other. Accordingly, "the appropriation appropriates man and Being in their essential togetherness;"[216] that is to say, it lets them belong together. "The outcome of thrownness/appropriation is the togetherness or bond (*Zusammengehörigkeit*) of man and meaning, the state of affairs that is itself meaning-giving."[217]

The encounter of man and Being in 'this mutual appropriation,' within the open of the clearing and outside the grasp of calculating thinking, is the new kind of thinking, made accessible through the discipline of phenomenology. As a kind of thinking that does not rely on 'the outline' it "neither offers a doctrine nor brings about a moral action".[218] The 'event of appropriation' (*Ereignis*), into which man and Being plunge in *Der Sprung*, is therefore 'a key term in the service of thinking'. Heidegger continues:

> As such a key term, it can no more be translated than the Greek λογοσ or the Chinese *Tao.* The term event of appropriation (*Ereignis*) here no longer means what we would other-wise call a happening, an occurrence. It now is used as a *singular tantum.* What it indicates happens only in the singular, no, not in any number, but uniquely.[219]

---

214 *US.* p. 259; *On the Way to Language*, p. 128. „Das Ereignis ist das Unscheinbarste des Unscheinbaren, das Einfachste des Einfachen, das Nächste des Nahen und das Fernste des Fernen, darin wir Sterblichen uns zeitlebens aufhalten."

215 *ID*, p. 26; *Identity and Difference*, p. 37. „Das Er-eignis ist der in sich schwingende Bereich, durch den Mensch und Sein einander in ihrem Wesen erreichen, ihr Wesendes gewinnen…"

216 *ID*, p. 27; *Identity and Difference*, p. 38. „ Das Ereignis vereignet Mensch und Sein in ihr wesenhaftes Zusammen."

217 Sheehan, T., *Facticity and Ereignis*, p. 53.

218 Heidegger, M. *Contributions to Philosophy*, p. 171.„…das weder eine Lehre gibt noch ein 'moralisches' Handeln veranlaßt." *Gesamtausgabe* 65, *Beiträge zur Philosophie, (Vom Ereignis)* ed. Friedrich-Wilhelm von Herrmann, 2nd, rev. ed. Frankfurt am Main: Vittorio Klostermann, 1994, p. 242.

219 *ID*, p. 101; *Identity and Difference*, p. 36. " Als so gedachtes Leitwort läßt es sich sowenig übersetzen wie das griechische Leitwort λογοσ und das chiesische *Tao.* Das Wort Ereignis meint hier nicht mehr das, was wir sonst irgendein Geschehnis, ein Vorkommnis nennen. Das Wort ist jetzt als Singulare tantum gebraucht. Was

It does not any more mean a happening or occurrence in the normal sense, but it does retain the sense that what happens and does so uniquely in the singular, according to which the singular occurrence constantly and promptly disappear from view, holds back itself in what it grants. Because this withholding can in effect be experienced as a lack or deficiency, Heidegger portrays technology as the ceaseless organization of this lack—where technology camouflages the barrenness of being underneath an unrelenting functionality. It is as though the whole of the lecture *Der Satz Der Identität* has been pressing towards the "mere dominance of the frame to turn it into a more original appropriating (*Das Ereignis*)"[220] representing, as it does, the final resolution of the problematic of technology and post-modern ontology.

2.4.4.a. Features of Ereignis

Taken cognizance of the above clarifications of the term, we can then say that Ereignis, as the realm of Dasein's experiencing of Being, first of all, makes the relationship of belonging-together (*Zusammengehörigkeit*) between Being and Dasein possible. Talking of this reciprocal appropriation of Being and Dasein in the realm of Ereignis, Heidegger says that because Dasein listens to Being and is appropriated to being, he belongs to Being.

> To "belong" here still means to be in the order of Being. But man's distinctive feature lies in this, that he, as the being who thinks, is open to Being, face to face with Being; thus man remains referred to Being and so answers to it. Man is essentially this relationship of res-ponding to Being, and he is only this. This "only" does not mean a limitation but rather an excess. A belonging to Being prevails within man, a belonging which listens to Being because it is appropriated to Being.[221]

It is on the basis of the event or appropriation that Dasein experiences Being in beings and materialization of Being in history. To think in tune with the history of Being is not to tell a story about the ways Being has been conceived, but to grasp how Being itself takes place. History here must be understood not merely in terms of change but in terms of how we belong or fail to belong to a unique dispensation of meaning.[222] Accordingly, history (*Geschichte*) is the happening (*Geschehen*) in

---

es nennt, ereignet sich nur in der Einzahl, nein, nich einmal mehr in einer Zahl, sondern einzig.

220 *ID*, p. 101; *Identity and Difference*, p. 37 "....Verwindung des Ge-Stells aus dem Ereignis..."

221 *ID*, p. 18; *Identity and Difference*, p. 31.

222 Cf. Guignon, C., The History of Being, in *A Companion to Heidegger*, ed. Dreyfus and Wrathall, pp. 398–406.

which our fate and destiny (*Schicksal and Geschick*) are drawn into the sending (*Schicken*) of the Being of beings to us.[223]

Furthermore, out of this 'belonging-together', Being shows itself as the difference as such between Being and entities. "We think of Being rigorously only when we think of it in its difference with beings, and of beings in their difference with Being."[224] Therefore, Ereignis is the domain in which Dasein, in his belonging-together to Being, comes upon the correct understanding of Being and its correlation with beings as difference.

Again, Ereignis is the sphere in which the connection of Being to Dasein and entities is manifested in the spatio-temporal history. Temporality which defines time in its concreteness and process in past, present and future is defined by Heidegger as the Dasein outside-of-itself, in, and for-itself. These three aspects of *Zukunft* (future), *Gewesenheit* (past), and *Gegenwart* (present) are the ecstasies of time.[225] In various texts, Heidegger explains this aspect of Ereignis as Being's space-time *continuum* to Dasein and entities as presencing. In his letter to Richardson Heidegger states: "The disquieting, ever watchful question about Being under the guise of presence (Present) developed into the question about Being in terms of its time-character."[226] The presencing of 'having-been', i.e., the past and the 'not-yet' i.e., the future is in the mode of absencing, whereas the presencing of the present is in the mode of presencing (*Anwesen*).[227] The presencing of Being, when thought with regard to what is present, shows itself as a 'letting-presence' (*Anwesenlassen*). In *On Time and Being*, Heidegger explains: "Letting shows its character in bringing into unconcealment. To let presence means: to unconceal, to bring to openness. In unconcealing prevails a giving, the giving that gives presencing, that is, Being, in letting-presence."[228]

The different ways of obscuring or withdrawing speak of the various ways in which Being has condescended to presence itself. "As it reveals itself in beings,

---

223 *SD*, pp. 8–9; *On Time and Being*, pp. 8–9. 'Geschick' and 'Geschehen' are closely related in their meaning: 'Schicken' means, 'machen daß etwas geschieht.' Cf. *Grimm Wörterbuch*, s.v. 'Schicken'.

224 *ID*, p. 53; *Identity and Difference*, p. 62.

225 Iroegbu, P. O., *Kpim of Time, Eternity*, p. 64.

226 Heidegger, M., "Letter to Richardson," in Richardson, W. J., *Heidegger Through Phenomenology to Thought*, p. xii. „Die beunruhigende, ständig wache Frage nach dem Sein als Anwesenheit (Gegenwart) entfaltete sich zur Frage nach dem Sein hinsichtlich seines Zeitcharakters." p. xiii.

227 *ID*, pp. 55–56; *Identity and Difference*, pp. 64–65.

228 *SD*, p. 5; *On Time and Being*, p. 5.

Being withdraws."[229] This aspect of withdrawal belongs essentially to Being as sending. Thus, "in sending itself, Being withdraws; in giving itself, Being withholds; in presencing itself, Being absences; in revealing itself, Being conceals."[230] In "*Der Spruch des Anaximander,*" Heidegger states unequivocally:

> What is past and what is to come also become present, namely as outside the expanse of unconcealment. What presents itself as non-present is what is absent. As such it remains essentially related to what is presently present, in as much as it either comes forward into the expanse of unconcealment or withdrawal from it. Even what is absent is something present for as absent from the expanse, it presents itself in unconcelament.[231]

Equally Heidegger gives us a clearer indication as to how to think Being and Time together in his letter to Richardson: „[The process of] Presencing (Being) is inherent in the lighting-up of self-concealment (time). [The] lighting-up of self-concealment (time) brings forth the process of presencing (Being)."[232] Ereignis is not an entity, but what enables beings to appear.[233] Therefore *Ereignis* is the happening of being, the unfolding of being, the temporalizing of being, and the turning of thought.

Another attribute of *Ereignis* is that the realm of *Ereignis* which opens to Dasein is nothing but that of '*Aletheia*' (truth) of Being. The question of Being thus considers the origin or ground of the truth of beings as such. The question of Being, then, is to some extent the question of truth: how are beings given, that is, how are they reachable and graspable? How are beings given as beings rather than nonbeing—as other than nothing? In this sense, Being is the basis of the distinction between beings and beingness.[234] Beings must continue to *be*; but they are to

---

229 *HW*, p. 311; *EGT*, p. 26. „Das Sein entzieht sich, indem es sich in das Seiende entbirgt."

230 Puthenpurackal, J. J., *Heidegger Through Authentic Totality to Total Authenticity,* p. 180.

231 *HW*, p. 320; *EGT*, p. 34–35. "Auch das Vergangene und Zukünftige ist Anwesendes, nämlich außerhalb der Gegend der Unverborgenheit. Das ungegenwärtig Anwesende ist das Ab-wesende. Als dieses bleibt es wesensmäßig auf das gegenwärtig Anwesende bezogen, insofern es entweder in die Gegend der Unverborgenheit hervorkommt oder aus ihr weggeht. Auch das Abwesende ist Anwesendes und, *als* Abwesendes aus ihr, in die Unverborgenheit anwesend."

232 Richardson, J. W., *Heidegger, Through Phenomenology to Thought,* p. xx „Anwesen (Sein) gehört in die Lichtung des Sichverbergens (Zeit). Lichtung des Sichverbergens (Zeit) erbringt Anwesen (Sein). p. xxi.

233 Polt, R., *The Emergency of Being, On Heidegger's Contributions to Philosophy*, Ithaca and London: Cornell University Press, 2006, p. 76.

234 Polt, R., p. 59–60.

specify the event of Being. So, truth basically happens only "where it establishes itself in a being."[235] Therefore, taking shelter of Being in aletheia, which entails reciprocal looking at each other by Dasein and Being, is exposed exclusively in the realm of *Ereignis*. When sheltering takes place, the clearing is grounded in the beings that lie open within it and this must also include self-concealment. In his vast erudition in philology, Heidegger semantically echoes this in his thoughts: *Bergung* (sheltering) involves *Verbergung* (concealment) as well as *Unverborgenheit*; unconcealment is primal truth (*Wahrheit*), so in order for truth to happen it must be guarded and preserved (*bewahrt, verwahrt*) within beings.[236]

On another note, *Ereignis* entails a *Schritt Zurück* ('Step-back') from metaphysical thinking. This is made manifest in Heidegger's critique of Western metaphysical tradition as we shall see in the next section. This 'step-back' is nothing but a 'leap' (*Sprung*) into the 'ground' or essence of metaphysics and technology. It is neither a historical return to the earliest Western philosophy nor a mere change of attitude, but an attempt to overcome metaphysical thinking and take a 'leap' into the unthought source and region of the genuine thinking of Being (*Seinsdenken*). As J. L. Mehta vividly puts it:

> This 'region of all regions,' suspended in itself, is itself above all regional loyalties and the Babel of conflicting tongues. It is the realm of that universality and simplicity of primordial truth, the happening of aletheia, of overtness, in the belongingness of man and Being in the Self-same, where alone divergent traditions, disfranchised of their exclusive claims and yet without losing their own identity, can meet together as one, as belonging-together in the Self-same.[237]

This leap into the primordial realm of the 'Self-same' or the 'event of appropriation' (*Ereignis*), Heidegger asserts, is only by attempting a 'step-back' from the metaphysical thinking wherein the experiencing of Being as Being is made possible. In view of that, the thinking of *Ereignis* is the thinking of how Being can be revealed by unhiddenness or unconcealed by unconcealment when this unconcealment cannot be in or even by Being and Being itself cannot be. This is the genesis of the concept and its necessity: turning away from Dasein and turning towards unconcealment itself, *Ereignis* becomes crucial for Heidegger's philosophy

In summary, *Ereignis* is the realm in which the truth of Being is made manifest. Thinking of Being reaches its purity and excellence when it is thought from the

---

235 "The Origin of the Work of Art" in *BW*, p. 162.

236 *Contributions to Philosophy*, p. 244; GA 65, p. 349.

237 Mehta, J. Lal, *Martin Heidegger: The Way and the Vision*, Honolulu: University of Hawaii Press, 1976, pp. 463–464.

realm of *Ereignis*. It is not accessible to the representative-calculative thinking and to individual experiences of men. It is, rather, given to the essential thinker, the poetic dweller, the seer and the shepherd, in his realization of his belonging-together with Being. Thus in *Ereignis*, the philosophical question of the 'I' (Dasein) and the 'world' has been absolutely transformed through modification from the beginning of Heidegger's career. "It is not a question of how one thing called 'I' is connected up to another thing called 'world'. It is rather a question of experiencing the way in which this world takes place as an event through which all meaningfulness and valuation takes place. It is an event that is our own and to which we reciprocally belong."[238] In the event, the ontic is made to transcend into the ontological by which it is appropriated.

2.5. Levinas' Critique of Heidegger's Project: Ontology vs Ethics

Emmanuel Levinas (1906–1995) is one of the most fascinating and important philosophers of the twentieth century. The hub of his teaching is that *ethics is fundamental to philosophy and to human existence*. This is expressed, for instance by Edith Wyschogrod that: "The prime objective of Levinas' work has been to develop a metaphysics upon ethical foundations by showing man's being in the world to be moral being."[239] He overcame the alleged indifference and ethical apathy of Heidegger's ontology in favour of ethics, and split with Heidegge's entire supposition of social being in favour of an unalterable and always "unbounded" responsibility.

The Lithuanian-born French Philosopher of Jewish origin was a disciple of Edmund Husserl. Levinas was present at Husserl's concluding lectures of 1928–29 in Freiburg and became influenced by Husserl's *Logical Investigation*. Nonetheless he promptly became an admirer of Heidegger's *Being and Time*, which was to have a profound effect on his thinking. This is manifested as Levinas recalled in a well-known phrase: "I went to Freiburg because of Husserl, but discovered Heidegger."[240] Levinas became influential in France for his translations of Husserl and Heidegger into French as he began an exploration of Heidegger's thought. Many French students were then attracted to Levinas by the urge to hear him

238 Greaves, T. *Starting with Heidegger*, p. 49.

239 Wyschogrod, E., *Emmanuel Levinas: The Problem of Ethical Metaphysics*, New York: Fordham University Press, 2000, p. 228.

240 Françios, Poirié, *Emmanuel Levinas, Qui êtes-vous?*, Lyons: La Manufacture, 1987, p. 97.

transpose *Sein und Zeit* into their own language.[241] In fact, many of them first heard the words *"Dasein"* as *"être-la"* and *"Sorge"* as *"souci"* from the lips of Levinas. This is obvious from the statement of Maurice de Gandillac, the French literary critic, in his memoir as he describes his first encounter with Levinas:

> Student at Strasbourg, Emmanuel Levinas, a young Lithuanian, my exact contemporary, perfectly managed the French that he had chosen as his language. There was no better aid besides him for penetrating into the thoughts of the master, whose instruction he was then following at Marburg. How can one forget that beautiful afternoon when for some assembled Frenchmen he translated and commented upon several pages of *Sein und Zeit*? The sun was bit by bit melting away the mark of snow on the spot where Emmanuel was sitting, dressed in street clothes, heeled dance-hall shoes protected in rubber galoshes. When Levinas rose, we remarked that, like the biblical Job, but without the urge to interrogate his God, he had just spoken to us of "being-there" and of "care" upon a pile of dung.[242]

The above painted image might sound derogatory, but the underlying messages is indicative that Levinas was at the beginning an enthusiastic adherent of Heideggerian phenomenology. But the upswing of Nazi fascism and Heidegger's alliance with it led him to somber and enduring doubts about the practicality of Heidegger's philosophy. By the late 1940s these doubts were articulated in two short books and subsequently in essays and writings that showed the appearance of his personal response to Husserl and Heidegger and the contemporary crisis of Western civilization. Like Husserl, Levinas commenced his own deliberate push away from Heidegger and en route for the ethical thinking of *alterity* for which he is at present largely known. These reflections culminated in his first major work *Totality and Infinity*, published in 1960. The thoughts examined in this book continued to be magnified, ramified, and enhanced in a collection of essays, articles, commentaries, dialogues, and subsequent books in the years that followed until his death in 1995.

---

241 On Levinas' role in the early reception of Heidegger in France, see Janicaud, D., *Heidegger in France*, Paris: Albin Michel, 2000, especially pp. 31–36. A prominent history of twentieth-century French thought present him—in its last chapter—as the cutting edge. See Gary, Gutting, *French Philosophy in the Twentieth Century*, Cambridge: Cambridge University Press, 2001, chapt. 12.

242 Maurice de Gandillac, *Le siècle traverse: Souvenirs de neuf décennies*, Paris: Albin Michel, 1998, pp. 132–33. Cited from Peter, Eli Gordon, "Fidelity as Heresy: Levinas, Heidegger, and the Crisis of the Transcendental Ego," in *Heidegger's Jewish Followers*, edited by Samuel Fleischacker, Pittsburgh, Pennsylvania: Duquesne University Press, 2008, p. 191.

In this part of our study, we shall see how Levinas' critique of Heidegger's fundamental ontology, namely, his replacement of the emphasis on *Being* to emphasis on alterity, the *Other* led to an ethical responsibility which was more or less lacking in Heidegger with its attendant impact on his reactionary interpretation and solution to the impact of modern technology.

2.5.1. Phenomenology

Phenomenology, according to Cohen's description, is "the broadest, deepest, most flexible conception of science ever to be conceived."[243] As a phenomenologist, Levinas' prologue is the philosophies of Husserl and Heidegger from out of which he built up an exceptionally unique 'Post-Modern' ethics. The phenomenology of Edmund Husserl "sought merely to describe what appears to consciousness in the most unprejudiced, naïve way possible."[244] If we learn to look at things in the right way as they are, we realize that they present themselves to consciousness directly and unequivocally. Hence, "the motto of phenomenology is, 'Back to the things themselves! (*Zu den Sachen selbst*!). And these things are the *phenomena*."[245]

This is because, "phenomenological description requires a 'philosophical reduction,' a bracketing of everything one knows—every theory, system, category, and name—in order to reflect on the originating sources of cognition and experience."[246] We have to go back to the phenomena, according to Husserl, if we are "to know how things *do* and *must* appear."[247] The phenomenological procedure unveils the structure of conscious experience, the way phenomena emerge in consciousness via intention and reduction, or *epoche*.[248] Franz Brentano states:

> Every mental phenomenon is characterized by what the scholastics of the Middle Ages called the intentional (or mental) inexistence of an object, and what we might call, though not wholly unambiguously, references to a content, direction towards an object

---

243 Cohen, R., *Ethics, Exegesis, and Philosophy: Interpretation after Levinas*, Cambridge: Cambridge University Press, 2001, p. 12.

244 Craig, M., *Levinas and James: Towards a Pragmatic Phenomenology*, Bloomington, Indiana: Indiana University Press, 2010, p. 34.

245 Lewis, M. & Steahler, T., *Phenomenology: An Introduction*, London: Continnum International Publishing Group, 2010, p. 5.

246 Craig, M., *Op. Cit.*

247 Lewis, M. & Steahler, T. *Op. Cit.*

248 "The *epoche* is a suspension of judgement; the Greek *epechein* means to suspend, refrain, bracket. Specifically, the phenomenological epoche means a suspension of judgement regarding the being of the world which is neither affirmed nor denied." Ibid., p. 14.

(which is not to be understood here as meaning a thing), or immanent objectivity. Every mental phenomenon includes something as object within itself...[249]

According to Husserl's description, each intentional action has an *intentional* object, an object that the action is about, but they definitely need not all have a *real* object.[250] Instead of being perceived as plain *cogito,* the human knowledge and experience can now appear to persistently lead in the direction of something in the concrete world. As such, for Husserl, phenomenology ought to begin by bracketing the question of the existence of the objects of intention. Phenomenology should concentrate exclusively on issues around which we can attain certainty—the immediate prevailing intuitions and intentions, their fundamental qualities, and the mode wherein intuitions and intentions are constituted by the mental and rational procedures located in pure subjectivity. "Intentionality for Husserl," according to Michael Hammond, Jane Howarth, and Russell Keat in their text *Understanding Phenomenology*, means that "consciousness is intentional or possesses intentionality." Intentionality does not mean that the observer's "conscious acts...are intended or deliberate acts, but that they point to, or reach out towards, objects." Intentionality is both a "fundamental characteristic of 'psychic phenomena' and presents "the

249 Brentano, F., *Psychology from an Empirical Standpoint,* trans. A. C. Rancurello, D. B. Terrell, & L. McAlister, London: Routledge, 1995, p. 88. Ordinarily, at the philosophical level, it was Franz Brentano of Marienberg, Germany (1838–1917), teacher of Husserl and founder of descriptive phenomenology of the operations of the human mind, who first thematized the datum of intentionality. For Brentano intentionality is the psychological relation of the conscience to a given mental object including images, hopes, fears and desire. Husserl carries his master's meaning of intentionality to a higher level: the transcendental. It is a transcendental relation of the mind to any given object. That mind, in creating the meaning of the object, creates itself. Intentionality, not only brings out relationships, it does the higher function of constructing and constituting the objects of intention in their meaningfulness. Intentionality gives meaning to objects. Cf. Iroegbu P. O., *Metaphysics: The Kpim of Philosophy*, p. 207. A further point is that Husserl himself received the term "phenomenology" from Hegel. As Lyotard notes, "it is from Hegel that the term phenomenology received its full and singular meaning with the publication of *Die Phenomenologie des Geistes*. Phenomenology is "science of consciousness", in that consciousness is, in general, knowledge of an object, either exterior or interior. Lyotard, J-F, *Phenomenology,* trans. Brian Beakley, Albany, USA: State University of New York Press, 1991, p. 65. See also Husserl, Edmund, *Cartesian Meditations: An Introduction to Phenomenology*, Dordrecht: Kluwer Academic Publishers, 1995.

250 Husserl, E., *Logical Investigations*, Ed. Dermot Moran, 2nd ed. 2 vols, London: Routledge, 2001 [1900/1901], p. 127.

method for a descriptive transcendental-philosophical theory of consciousness."[251] As Levinas in the same way explains: "Intentionality means that all consciousness is consciousness of something, but above all that *every object calls forth and as it were gives rise to the consciousness through which its being shines, and in doing so, appears.*"[252] Nevertheless, the object which is being observed or experienced and the observer or experiencer of the object is not connected by intentionality. Mental phenomena to Husserl "contain an object intentionally within themselves".[253] In other words, mental consciousness does not exist independent of the object, but the object is a part of the mental phenomenon, rather than a part of the world outside the observer's consciousness. In that case, it should be said that "intentionality designates the realization that consciousness is never empty, but is always consciousness of something. …consciousness is directed towards its object in a dynamic fashion, in the sense that it wants to get an ever closer grasp of this object."[254]

Husserlian phenomenology is not concerned with details, beings, events, hypotheses, whether real or ideal, imagined or confirmed; it is concerned with the conditions for the experience of "realities, values, theories, facts" and all thinkable objectivities. Whether such realities, hypotheses, objectivities exist or not, are for Husserl ontological and at the end of the day metaphysical questions. On the whole, Husserl was bothered with the dilemma of how objective reality could be constituted in and through subjective acts of consciousness, what he called "the enigma of subjectivity."[255]

Heidegger expands the concept of intentionality, arguing that it missed the mark to describe what the most fundamental form of intentionality is. "Dismantling Husserl's ban against any investigation of an "outside" beyond the intentional sphere, Heidegger assaulted the transcendental ego as a metaphysical dogma, and called instead for a phenomenology of the so-called "natural attitude," an inquiry into human existence as a worldly and practical event."[256] He argues:

---

251 Hammond, M., Howarth, J. & Keat, R., *Understanding Phenomenology*, Oxford: Basil Backwell, 1991, p. 62.

252 Levinas, E., *Discovering Existence with Husserl*, trans. and eds. Richard A. Cohen and Michael B. Smith, Evanston: Northwestern University Press, 1998, p. 119.

253 Hammond, M., Howarth, J. & Keat, R., *Understanding Phenomenology, Op. Cit.*

254 Lewis, M. & Steahler, T., *Phenomenology: An Introduction*, p. 22.

255 Moran, D. & Mooney, T., (eds.), *The Phenomenology Reader*, London: Routledge, 2002, p. 61.

256 Peter Eli Gordon, "Fidelity as Heresy: Levinas, Heidegger, and the Crisis of the Transcendental Ego," in *Heidegger's Jewish Followers*, pp. 198–199.

> The usual conception of intentionality [...] misconstrues the structure of the self-directedness-towards the intention [...]. An ego or subject is supposed, to whose so-called sphere intentional experiences are then supposed to belong. The ego here is something with a sphere in which its intentional experiences are, as it were encapsulated [...]. (T)he mode of being of our self, the Dasein, is essentially such that this being, so far as it is, is always already dwelling with the extant. The idea of a subject which has intentional experiences merely inside its own sphere and is not yet outside it but encapsulated within itself is an absurdity which misconstrues the basic ontological structure of the being that we ourselves are.[257]

Levinas admitted that his interpretation of Husserl was profoundly inspired by Heidegger and depicts Husserl as underpinning the work of Heidegger. For that reason, he explored for those clues in Husserl's works that foretold Heidegger's overhauling of phenomenology. In his words, "*Being and Time*, which is much more significant and profound than any of Heidegger's later works, represents the fruition and flowering of Husserlian phenomenology."[258]

In Levinas' *Dialogue with Richard Kearney*, he disclosed that the phenomenology proposed by Husserl was a methodological revelation of exactly how we discover and comprehend meaning in the world of experience and in what way this meaning appears in our sensible, conscious and intentional affiliation with the world. In this case, "intentionality would be a means of explaining the way consciousness relates to the world, a way of answering the question: "How does a subject reach a transcendent object?"[259] But then, in his understanding and interpretation, "Intentionality is not the way in which a subject tries to make contact with an object that exists besides it. Intentionality is what makes up the very subjectivity of subjects. The very reality of subjects consists in their transcending themselves."[260] In fact, Levinas phrases his central claim or critique of Husserl in this way:

> The intentionality of consciousness allows one to distinguish the ego from things, but it does not make solipsism disappear: its element—light—renders us master of the exterior world but is incapable of discovering a peer for us there.[261]

257 Heidegger, M., *The Basic Problems of Phenomenology*, pp. 63–64.

258 Levinas, E.& Kearney, R., "Dialogue with Emmanuel Levinas", in *Face to Face with Levinas*, Cohen, A. Richard, (ed.), New York: State University of New York Press, 1986, p. 15.

259 Levinas, E., *The Theory of Intuition in Husserl's Phenomonology*, second edition, trans. André Orianne, Evanston, Illinois: Northwestern University Press, 1995 p. 41.

260 Levinas, E., The *Theory of Intuition in Husserl's Phenomenology*, p. 41.

261 Levinas, E., *Time and the Other*, trans Richard, Cohen, Pitsburgh, PA: Duquesne Unversity Press, 1987, p. 65. (Hereafter, TO)

In this interpretation, the Husserlian theory of intentionality is essentially unable to account for the presence of other as others, and as companions. Whereas Husserl understands the social relation as a unique sort of objectification, Levinas, conceives of it as an undiminished *ethical* relation.

Levinas recognizes his indebtedness to the works of Husserl and Heidegger even though he was critical of both. In his 1930 thesis, *La théorie de l'intuition dans la phénoménologie de Husserl*, Levinas turned against the idealism of Husserlian phenomenology i.e., to bracket any deliberations concerning the contingent and accidental, and focus on (intuit) the essential natures or essences of the objects and acts of consciousness[262] and moved in the direction of the uniqueness, ingenuity and autonomy of conscious existence. He joined Heidegger in discarding Husserl's phenomenology as being too hypothetical and abstract without any foundation in reality. As he puts it:

> Husserl conceives philosophy as a universally valid science in the manner of geometry and the sciences of nature, as a science which is developed through the efforts of generations of scientists, each continuing the work of others [...]. *In this conception, philosophy seems as independent of the historical situation of man as any theory that tries to consider everything sub specie aeternitatis* [...]. The historical structure of consciousness, which occupies a very important place in the thought of someone like Heidegger [...] has not been studied by Husserl, at least in the works published so far. He never discusses the relation between the historicity of consciousness and its intentionality, its personality, its social character.[263]

---

262 Husserl, E., *Ideas Pertaining to a Pure Phenomenology and a Phenomenological Philosophy*, trans. F. Kersten, The Hague: Nijhoff, 1982 (1913), sec. 2. See also, Zahavi Das, *Husserl's Phenomenology*, Stanford: Stanford University Press, 2003, pp. 38–39.

263 Levinas, E., *Theory of Intuition in Husserl's Phenomenology*, 2nd Edition, Evanson, IL, USA: Northwestern University Press, 1995, pp. 155–156. In *Otherwise Than Being*, Levinas defines intentionality as "an aspiration to be filled and fulfilled, the centripetal movement of a consciousness that coincides with itself, recovers, and rediscovers itself without ageing, rests in self-certainty, confirms itself, doubles itself up, consolidates itself, thickens into substance." p. 48. Rather than starting with subjectivity as intentionality, founded on auto-affection, Levinas approaches the subject in terms of the passivity of time, a lapse of time, irrecuperable and outside all will, the exact contrary of intentionality. The Other appears to the subject, not as phenomena or an apparition in the full light, but as a trace and an enigma that disturb phenomena. Cf. Levinas, E., *Totality and Infinity: An Essay on Exteriority*, trans. Alphonso Lingis, Pittsburgh: Duquesne University Press, 1969, pp. 50–51, 66–67, (Hereafter TI); Levinas, E., *Otherwise Than Being, Otherwise than Being or Beyond Essence*, trans. Alphonso Lingis, Pittsburg: Duquesne University Press, 2000, pp. 47–48, 53, 96 (Hereafter, OB); See also Levinas E., "Phenomenlogy and Enigma," in *Collected*

Towards the dawn of 1950s and beginning of 1960s, Levinas started to articulate his personal philosophy that turn out to be more and more disapprobation of Heidegger's philosophy and with emphasis on the superiority and importance of the ethical relationship with the Other.

As we have seen already, Heidegger's fundamental ontology dwelt on the revival of the meaning of Being. In the treatise "Is Ontology Fundamental?" Levinas outlines his critique of Heidegger's scheme of fundamental ontology and presents in basic expressions the notion of the Other. His aspiration in his critical assessment of Heidegger's venture is to demonstrate a relation that surpasses the limits of ontology or the science of being. A key trait of this perception of being is that it is not merely intellectual or abstract, but somewhat consists in all facet of human activity. Specifically, Ontology is not basically about knowing, it is about being pragmatic. Such a distinction presents a primary step for Levinas' shift towards an ethics, a march that first of all, requires a repudiation of Heidegger's ontology.

2.5.2. Levinas' Critique of Heidegger

While Levinas persistently acknowledged the extreme importance of Heidegger's place in the philosophical tradition of the twentieth century, his critique of Heidegger's work became increasingly explicit. Levinas had trouble disentangling Heidegger's later thinking from his affiliation to National Socialism and Nazi party as well as his submission to Hitler's leadership. In 1933, he became the Rector of Freiburg University and delivered his Rectoral inaugural speech approving the Nazi regime and policy which according to Zimmerman, he "never unequivocally apologized for such support and never publicly expressed regret about the victims of Nazism."[264] Thus it was difficult for Levinas to forgive Heidegger for this lack of regret. As a result, he advanced a moral ethical principle which is "postmodern, post-phenomenological and post-Heideggerian."[265] His principal critique of Heidegger stem from his appraisal of Heidegger's phenomenology as curbing the possibility of dealing with the questions of ethics, an issue which according to him, Heidegger left at the level of ontology, as a simple avenue to more profound

---

Philosophical Papers, trans. Lingis, A., Dordrecht, The Netherland: Martinus Nijhoff, 1987, p. 70.

264 Zimmerman, E. M., *Contesting Earth's Future, Radical Ecology and Postmodernism,* Berkeley and Los Angeles: University of California Press, 1994, p. 114 See also Critchley, S., Bernasconi, R., (eds.), *The Cambridge Companion to Levinas*, Cambridge: Cambridge University Press, 2002, p. 8.

265 Colin, Davis, *Levinas: An Introduction*, Notre Dame: University of Notre Dame Press, 1996, p. 8.

understanding of Being. In a statement insidiously directed at Heidegger, Levinas states: "Every civilization that accepts being—with the tragic despair it contains and the crime it justifies—merits the name 'barbarism'"[266] Levinas, therefore, stressed the need to advance a phenomenology that provides an ethical metaphysics that goes beyond being with emphasis on responsibility for the 'Other'. Here, we shall examine his critique of Heidegger from these two perspectives; the question of Being and ontology and the question of ethical responsibility.

2.5.2.a. On the Question of Being and Ontology

Although Levinas calls for a profound need to leave the climate of Heideggerian philosophy, he acknowledges his debt to Heidegger's ontology. "At the beginning, our reflections are in large measure inspired by the philosophy of Martin Heidegger where we find the concept of ontology and of the relationship which man sustains with Being."[267] Thus, Taminiaux avers that "Levinas agree with Heidegger as far as the concept of ontology is concerned." According to him, "the task of ontology is not the task of a superscience defining the ultimate properties of all beings and characterizing their relations. The task is to ask what it means to be, a question that only makes sense for the human beings, a question which points to a relation between a being, or an existent, and its Being or its existence."[268] While Levinas agrees with Heidegger on this formal concept, Taminiaux observes that he did not "agree with Heidegger's definition of the relation existent-existence in ekstatic terms," and his *De l'existence à l'existant* expresses his disagreement.[269]

So the most significant critique of Heidegger's project by Levinas lies in a rejection of the neutrality of Being. This neutrality originated from his understanding of Heidegger's undertaking as a fundamental ontology. "According to Levinas, ontology reduces all relationship with beings to knowledge; it conquers beings and never lets them out of its iron grip of understanding."[270]

As noted above, it is Heidegger's own critique of Husserlian intentionality which Levinas eventually applied in a modified form to Heidegger himself. In

---

266 Levinas, E., *On Escape De l'invasion*, trans. Berttina Bergo, Stanford, Calif: Stanford University Press, 2003, p. 73.

267 Levinas, E. *Existence and Existents*, p. 19.

268 Taminiaux, J. The Presence of *Being and Time* in *Totality and Infinity*. In Hansel J., editor. *Levinas in Jerusalem: Phenmenology, Ethics, Politics, Aesthetics,* (Amsterdam Studies in Jewish Philosophy), Springer, 2009, p. 6.

269 *Ibid.* pp. 6–7.

270 Rychter, E., *(Un)Saying the Other Allegory and Irony in Emmanuel Levinas's Ethical Language*, New York: Peter lang Publishing, 2004, p. 22.

so doing, however, he relegates his appraisal of the ontological horizon to *Being and Time* (and other texts) where the phenomenological hermeneutic of Dasein has a different legacy in Husserlian intentionality. In fact, it has been stated that "Heidegger's criticism of the epistemological tradition—and ultimately of the entire philosophical tradition—is that it is guilty of over-intellectualism, that it has failed to question the primacy of theory. And Levinas' criticism of Heidegger is that he too is guilty of over-intellectualism; that fundamental ontology also remains under the sway of theory, which only fundamental ethics can escape."[271] This critique comes to culmination in *Totality and Infinity*, where Heidegger is accused of replicating the standard ontological gesture of subordinating ethics to ontology. Levinas states:

> To affirm the priority of *Being* over *existents* is to already decide the essence of philosophy; it is to subordinate the relation with *someone*, who is an existent (the ethical relation), to a relation to the *Being of existents*, which, impersonal, permits the apprehension, the domination of existents (a relationship of knowing), subordinates justice of freedom.[272]

For Levinas, Heidegger's pessimistic notion of the social relation rationally derived from the idea that Heidegger follows the examination of being in terms of isolation, namely, in terms of individual Dasein. If the social relation is understood in terms of individual Dasein, it follows that the self in its pre-theoretical understanding that being-in-the-world is at all times a being-with automatically subordinates the social relation of the self's pre-theoretical understanding of the Other. In this way, the self always already understands the Other before it *relates* to the Other. This self understanding accordingly comprises a preliminary act of possession in which the Other is condensed to the "same" in its relationship to the Other, an Other which cannot be reduced to the same, cannot be listed under the same, and which incessantly eludes any attempt to grasp it.

Although it may emerge at first that *Being and Time*, because of its idea of *mit-sein*, is concerned with an examination of existence from the point of view of the social relation, Heidegger's main concern is unquestionably individual Dasein. For Levinas, on the contrary, the main concern is specifically on the social relationship;

---

271 Frank, P., "Ontology and Ethics: Questioning First Philosophy in Levinas, Heidegger, and Fichte," in *Heidegger's Jewish Followers*, p. 178.

272 Levinas, E., *TI*, p. 45. Levinas agrees with Heidegger that the fundamental question of philosophy concerns the meaning of being. For an excellent discussion of the differing ways in which Heidegger and Levinas approach this fundamental philosophical issue, see Manning, J. S. Robert, *Interpreting Otherwise Than Heidegger: Emmanuel Levinas's Ethics as First Philosophy*, Pittsburgh: Duquense University Press, 1993.

it is the distinctive *otherness* of the Other that is to be conserved. In contrast to Heidegger, the relationship to the Other is not subordinated to an understanding of Being, but constitutes the very significance of Being, which subverts the solitude of the self. Thus, the "face to face" relationship is, as Levinas develops it in *Totality and Infinity*, the archetype of the ethical relationship. It characterizes the claim of the other who interrupts the spontaneity of my enjoyment and "brings me back" to the earnestness that the ethical relationship requires of me.

In his criticism, in which Levinas tried to show that metaphysics precedes ontology, he states: "The metaphysical desire tends toward *something else entirely, toward the absolutely other*", whereas ontology, of which Western philosophy has always been, involves "a reduction of the other to the same by interposition of a middle and neutral term that ensures the comprehension of being."[273] Dalton puts it in a classy manner that "metaphysical desire can be read as awakening a subject from the slumber of ontological actuality into the true life of ethical potentiality."[274] Such ethical potentialities go beyond the stagnant personal existence of the self to the world which only the other can neutralize.

Levinas' first serious attempt to depart from the climate of this Heideggerian thought occurs in his 1951 essay *"Is Ontology Fundamental?"*[275] For according to Critchley, "'Is Ontology Fundamental' demonstrates for the first time in Levinas' work the ethical of his critique of Heidegger."[276] In this essay, he argues that even though Heideggerian ontology was able to go beyond the intellectualism leveled against Husserl, it was unable to explain the interpersonal relation because in any given situation a particular being is always already understood within the horizon of being as such. This ontological outlook forms the definitive ground for the appearance of any particular being and is ultimately constrained by the Platonic tradition which makes claims to comprehension through allusion to a universal *eidos*. "Understanding is the very event that existence articulates. All incomprehension is only a deficient mode of comprehension. It turns out that the analysis

---

273 Levinas, E., *TI*, p. 33 and p. 43. (Italics in the original)

274 Dalton, D., *Longing for the Other: Levinas and Metaphysical Desire*, Pittsburg: Duquesne University Press, 2009, p. 42.

275 Levinas, E., "Is Ontology Fundamental?" *Emmanuel Levinas: Basic Philosophical Writings*. Eds. Peperzak, A.T., Critchley, S., & Bernasconi, R., Bloomington: Indiana University Press, 1996, pp. 1–10.

276 Critchley, S., Leaving the Climate of Heidegger's Thinking. In Hansel J., editor. *Levinas in Jerusalem: Phenmenology, Ethics, Politics, Aesthetics,* (Amsterdam Studies in Jewish Philosophy), Springer, 2009, p. 49.

of existence and of what is called its *haecceity* (*Da*) is only the description of the essence of truth, of the condition of the very understanding of being."[277]

On another note, even though Levinas' distinctive notion of being is in contrast to Heidegger's understanding, he originates it from Heidegger's conception of *Geworfenheit* (thrownness or facticity). Levinas endeavours to formulate a notion of Being that runs counter to the phenomenological ontology of Heidegger, as he declares:

> One must understand *Geworfenheit* as the 'fact-of-being-thrown-in' [...] existence. It is as if the existent appeared only in an existence that precedes it, as though existence were independent of the existent and the existent that finds itself thrown there could never become master of existence. It is precisely because of this that there is desertion and abandonment. Thus dawns the idea of an existing that occurs without us, without a subject, an existing without existents...[278]

This existing without existents is what Levinas refers to as the *il y a* (*there is*), it is the nameless being that exists prior to the subjects, that is, "a neutral, continuing existence without existents (and so never capitalized)."[279] The *il y a* is to be contrasted with Heidegger's *es gibt*, jointly of which can be translated by the phrase *there is*. Levinas identifies that "the most profound thing about Being and Time... is this Heideggerian distinction [of existence and the existent]. But in Heidegger there is a distinction, not a separation. Existing is always grasped in the existent... existing is always possessed by someone."[280] Heidegger, with his emphasis upon finitude, would not articulate a notion of existence (Being) that was separate from existents (beings). For his Being is known only by and through beings. As Levinas recognized, on the other hand, the idea of *thrownness* implies that there is definitely an existence that precedes the existent, if not, into what else would the existent be "thrown?" The *il y a* of Levinas, according to Dalton "is the sign of a rupture within the continuity of being, the separation of a particular being from the totality of being in general."[281] Levinas therefore uses Heidegger's concept as his point of exit from which to articulate an explanation of the meaning of both Being and beings that is in contrast to Heidegger.

Where Heidegger interprets being-with (*Mitsein*) as an ontological category that is ethically neutral, Levinas sees it as the very basis of ethics itself. He reiterates

---

277 *Ibid.*, p. 5.

278 Levinas, E., *TO*, p. 47.

279 Bernasconi, R. & Critchley, S., (eds.), Re-reading Levinas Bloomington, Indiana: Indiana Unviersity Press, 1991, p. 214.

280 Levinas, E. *TO*, pp. 44–45.

281 Dalton, D., *Longing for the Other: Levinas and Metaphysical Desire*, p. 96.

"metaphysics of the present, of materiality, and of desire, and counter-poses an ethics of social solidarity: a temporal ethics in which presence and *Mitsein* are no longer associated with a state of being 'fallen', but are held to be a fundamental pre-condition for an ethically good life."[282] Such an ethical life is what Levinas will describe as "the infinitude of the ethical demand that comes to me from outside, 'bringing me more than I contain', and remaining beyond understanding."[283]

2.5.2.b. On the Question of Ethics and Responsibility

Levinas did not delay in expressing his doubts about the objectivity of Heidegger's omission of ethical issues from his fundamental ontology. Heidegger's Dasein is locked on itself to the point that it is not linked to infinity or connected to the Other. As we have seen, "Heideggerian ontology subordinates the relation with the other to the relation with the Neuter, Being, and it thus continues to exalt the will to power, whose legitimacy the Other (*Autrui*) alone can unsettle, troubling good conscience."[284] It's deficiency or inauthenticity has to consist in some relation to itself. The result is a crumpled and defective ethics. This is not a private ethics, nor one that gets to the Other, but somewhat an aspect of Dasein's relation to Being in the emergence of the truth of Being for Dasein. As Peperzak eloquently expressed further:

> The idea of a debt or guilt toward others than the self is excluded from this thought. By the absence of a true alterity that could question and accuse *Dasein's* freedom, that is, by the absence of an ethical "principle," the Heideggerian perspective belongs to a tradition the barbarous depths of which were shown by Nazism. When Heidegger criticizes the essence of technology, he forgets that the source of modern evil, such as it was manifested in Nazism, is found at a depth that lies deeper than the realm of technology. Alluding to certain expressions found in Heidegger's later works, Levinas sketches the portrait of a pagan existence rooted in mother earth and prone to exploitation—very different from the sober existence of availability for the needs of others. The individuals are immersed in the *physis* that encompasses them like elements of it unfolding."[285]

---

282 Cf. Benso, S., "The Time of the Feminine: For a Politics of Material Corporeality," in *Continental Philosophy Review*, 36, 2003, pp. 195–202; Chanter, T., *Time, Death, and the Feminine: Levinas with Heidegger*, Stanford: Stanford University Press, 2001.

283 Strhan, A., *Levinas, Subjectivity, Education: Towards an Ethics of Radical Responsibility*, Chichester, UK: Blackwell Publishing Limited, 2012, p. 9.

284 Peperzak, A.T., *To the Other: An Introduction to the Philosophy of Emmanuel Levinas*, West Lafayette, Indiana: Purdue University Press, 1993, p. 103.

285 *Ibid.*, pp. 54–55.

As a result of Heidegger's association with Nazism whose cruelty and callous insensitivity seem to be devoid of any moral code, Peperzak was quick to conclude that "for Levinas, Heidegger's philosophy is the clear expression of a paganism that ignores the essential demands of morality and does not resist Nazi ruthlessness but rather is prone to collaborating with and even promoting it."[286] Rather than merely postponing questions of ethics, Levinas argues that this exclusion renders ethics a minor, insignificant, non-essential concern of philosophy.

Levinas looks at Heidegger's attitude as being indicative of a way of thinking that relegates moral considerations to a secondary and separate discipline. He emphasizes that the exhibition of the ethical is in effect a constitutive element of first philosophy; that every philosophy proper to experience must be ethical from the beginning. As John Caputo states; "the ethical is there from the start and does not require either ontological preparation (the Heidegger of *Being and Time*) or a deontological foundation (value theory, criticized by Heidegger). The ethical does not wait and does not need to have a space prepared for it."[287] It is therefore impracticable to think critically and essentially if moral perspective is set aside or bracketed.

In an essay entitled *The Search for a Heideggerian Ethics*, Michael Zimmerman asserts that "the burden of Levinas' critique of Heidegger" dwells in noticing how "Heidegger's fascination with the Greeks led him to discount Jewish and Christian insistence on the importance of personal responsibility for the concrete other."[288] Levinas' argument, against Heidegger and others, on the question of concreteness not only of day by day human existence per se but of the particular individual as such is manifest in his renowned comment that "Dasein in Heidegger is never hungry,"[289] and in the extensive examination of *jouissance* (enjoyment), the framework in which this remark become obvious, a concept which, Levinas claims, Heidegger does not take into account. For him, "Heidegger's project… does not assume a merely intellectual attitude, but rather the rich variety of intentional

---

286 Peperzak, T. A., *Beyond: The Philosophy of Emmanuel Levinas*, Evanston, Illinois: Northwestern University Press, 1997, p. 205.

287 Caputo, J., *Demythologizing Heidegger*, p. 197.

288 Zimmermann, M., "The Search for Heideggerian Ethics," in *Ethics and Responsibility in the Phenomenological Tradition*, p. 81

289 Levinas, E., *TI*, p. 134.

life – emotional, practical and theoretical – through which we relate to the Being of various beings."[290]

Meanwhile, in another essay called *The Pack*, Levinas, like Hiedegger, characterizes the ills of our society:

> [Ours is a society] whose boundaries have become, in a sense, planetary: a society, in which due to the ease of modern communications and transport, and the world wide scale of its industrial economy, each person feels simultaneously that he is related to humanity as a whole, and equally that he is alone and lost. With each radio broadcast and each day's papers one may well feel caught up in the most distant events, and connected to mankind everywhere: but one also understands that one's personal destiny, freedom or happiness is subject to causes which operate with inhumane force. One understands that the very progress of technology—and here I am taking up a common place—which relates everyone in the world to everyone else, is inseparable from a necessity which leaves all men anonymous. Impersonal forms of relationship come to replace the more direct forms, the 'short connections' as Ricoeur calls them, in an excessively programmed world.[291]

Today, according to above statement, ours is a world in which "men find themselves side by side rather than face to face."[292] This individualistic living in isolation is suggestive of the freedom and liberty being promoted. But for Levinas, freedom is not the ultimate price. Such a subjective freedom is overshadowed in its responsibility and accountability for the Other:

> It is not that I wish to preserve, over and against the structuralist critique, the idea of a subject who would be a substantial or mastering center of meaning, an idealist, self-sufficient *cogito*. These traditional ontological versions of subjectivity have nothing to do with the meontological version of subjectivity that I put forward in *Otherwise Than Being*. Ethical subjectivity dispenses with the idealizing subjectivity of ontology, which reduces everything to itself. The ethical "I" is precisely insofar as it kneels before the other, sacrificing its own liberty to the more primordial call of the other. For me, the freedom of the subject is not the highest or primary value.[293]

Finally, although Levinas asked for a profoundly challenging concept of responsibility, the prices to be paid by the individual seem to be phenomenal and tough. In

---

290 Peperzak, A.T., Critchley, S., & Bernasconi, R., (eds.), "Is Ontology Fundamental?" in *Basic Philosophical Writings*, p. 1; Critchley, S., & Bernasconi R., (eds.), *The Cambridge Companion to Levinas*, p. 9.

291 Levinas, E., *The Levinas Reader*, Sean Hand, (ed.) Cambridge, Mass: Basil Blackwell, 1989, p. 212.

292 *Ibid.*

293 Levinas, E., & Kearney, R., "Dialogue with Emmanuel Levinas" in *Face to Face with Levinas*, p. 27.

taking stock of the difficulties in Levinasian ethical discourse, we will seek to open a link, a rupture and point of departure which will be traced to the work of Hans Jonas later in our chapter five in order to rethink responsibility in a manner that is demanding as is located within the terrain of our shared modern existence. But at the moment, let us examine closely Levinas' tenets of subjectivity and ethical responsibility, against the backdrop of which he criticized Heidegger.

2.5.3. *Ontological Priority of Ethics*

In his major philosophical texts from 1930 to 1987, Levinas argued for the ontological priority of ethical responsibility. He challenged Heidegger's ontological priority, and therefore the appropriateness of "his phenomenological interpretation of subjectivity, by arguing that the relationship with the Other precedes experiencing oneself as an assimilating consciousness that internalizes a 'horizon' of objects toward which one 'moves'"[294] The question of Being for Heidegger is not a question of the Other and it excludes the ethical. "Being's interest takes dramatic form in egoisms struggling with one another, each against all, in the multiplicity of allergic egoisms which are at war with one another and are thus together. War is the deed or the drama of the essence's interest."[295] In his *Existence and Existents*, Levinas states categorically:

> If at the beginning our reflections are in large measure inspired by the philosophy of Martin Heidegger, where we find the concept of ontology and of the relationship which man sustains with Being, they are also governed by a profound need to leave the climate of that philosophy, and by the conviction that we cannot leave it for a philosophy that would be pre-Heideggerian.[296]

Indeed, in the Forward to *Otherwise Than Being or Beyond Essence,* Richard Cohen remarks that this title "alert us to the priority Levinas gives to his ongoing contestation of Heideggerian thinking, otherwise than Heideggerian Being; beyond Heideggerian essence."[297] In view of this, metaphysics is no longer the first philosophy; instead ethics has taken its place for as Levinas emphatically asserts: "Morality is not a branch of philosophy, but first philosophy."[298] This *firstness* of

---

294 Donald, R. Wehrs & David, P. Haney, (eds.), (Introduction), *Levinas and Nineteenth-Century Literature: Ethics and Otherness from Romanticism through Realism*, University of Delaware Press, 2009, p. 17.

295 Levinas, E., *OB*, p. 4.

296 Levinas, E., *Existence and Existents*, p. 19.

297 Levinas, E., *OB*, Preface, p. xiii.

298 Levinas, E., *TI*, p. 304.

first philosophy is not really in the order of grounds but in the order of *putting into question*, as Levinas explains further:

> But theory understood as respect for exteriority delineates another structure essential for metaphysic. In its comprehension of being (or ontology) it is concerned with critique. It discovers the dogmatism and naïve arbitrariness of its spontaneity, and calls into question the freedom of the exercise of ontology; it then seeks to exercise this freedom in such a way as to turn back at every moment to the origin of the arbitrary dogmatism of this free exercise. This would lead to an infinite regression if this return itself remained an ontological movement, an exercise of freedom, a theory. Its critical intention then leads it beyond theory and ontology, but all calls into question the exercise of the same. A calling into question of the same—which cannot occur within the egoist spontaneity of the same—is brought about by the other. We name this calling into question of my spontaneity by the presence of the Other ethics [...]. And as critique precedes dogmatism, metaphysics precedes ontology.[299]

Ethics for Levinas, therefore, "constitutes the moment when the arbitrary freedom of the individual egoistic subject is curbed, and when it learns to recognize its responsibility to others instead of merely using or assimilating alterity to serve its own egoist economy."[300] And as John Llewelyn observes: "Levinas's ontology calls into question the fundamentality of the 'ontological difference,' the distinction between being and beings, between the ontological and the ontic upon which [Heidegger's] *Being and Time* takes its stand....Levinas's ontology stands for the ontological significance of concrete empirical, hence ontic experience."[301] In other words, Levinas strives to advance an ethical ground for metaphysics by presenting Dasein's being-in-the-world as a moral being.

In this section of our study, we shall see that for Levinas, "to be" is not enough; "to be or not to be" is not the question.[302] It is rather the event of being in relation with the other that Levinas calls ethical, which replaces Heidegger's institution of ontology as the fitting sphere of phenomenology. This leads to a new understanding of subjectivity.

---

299 *Ibid.*, p. 43.

300 Hofmeyr, B., (ed.), *Radical Passivity: Rethinking Ethical Agency in Levinas*, Springer, 2009, p. 19.

301 LLEWELYN, J., *Emmanuel Levinas: The Genealogy of Ethics*, New York: Routledge, 1995, p. 108.

302 Levinas, E., *OB*, p. 3.

2.5.3.a. *Ethical 'Metaphysics': The Other*

In three primary theses, *Totality*, *Infinity*, and *Exteriority*,[303] Levinas developed what we shall call an ethical-metaphysics. His "metaphysics" is basically the thesis of the primacy of the infinite, the source of subjectivity and the imperative of the respect of the other-person (*d'autrui*). His ethical metaphysics "is essentially a meditation on what makes ethical agency possible—that which enables us to act in the interest of another, to put the well-being of another before our own."[304] This ethical undertone, according to Maesschalck is a break from the Heideggerian path.[305]

In *The Wisdom of Love in the Service of Love*, Roger Burggrave brings to light the fundamentals in Levinas's thought. Levinas portrays being (ego) both as the struggle for life and self-becoming. Being as the struggle for life can be seen on the level of the atom where the atom resists splitting. It can also be seen in plants that will push others aside in order to survive.[306] In the human level, Levinas catalogs a number of social consequences of the ego struggling for life:

> Violence occurs when one ego meets another, considers it as a threat, and therefore needs to eliminate it. Murder occurs when the ego raises itself to the level of unlimited power. Therefore it wants to reduce the other to sheer nothingness; it wants to annihilate the Other. Hatred is worse than murder because hatred annihilates the other by humiliating and causing suffering without doing away with him once and for all. War is power against power: two freedoms denying each other freedom.[307]

Having highlighted such a negative ontological depiction of the human person, Levinas' fundamental concern becomes ethical. What justifies ethics? What

---

303 LEVINAS, E., *Totalité et infini, essai sur l' extériorité*, Paris: Le Livre de Poche/Kluwer, 1994.

304 HOFMEYR, B., *Radical Passivity: Rethinking Ethical Agency in Levinas*, (Preface), p. v

305 MAESSCHALCK, M., *raison et pouvoir: les impasses de la pensée politique postmoderne*, Bruxelles: Facultés universitaires Saint-Louis, 1992, p. 67. "Levinas à particulierement développé la leçon éthique…assure de soi dans sa "correlation" aux autres. Le visage de l'Autre a attaint une telle impotance, pour lui, qu'il est venu rompre jusqu'au rythem patient de la question du Meme, qui isolait le chemin heideggérien."

306 BURGGRAEVE, R., *The Wisdom of Love in the Service of Love*, trans. Jeffery Bloechl, Milwaukee, WI: Marquette Univ. Press, 2002, p. 44.

307 *Ibid.*, pp. 58–67. As Andrew Shepherd noted, for Levinas, "'metaphysics' is the relationship with the Infinite Other that overcomes the totalizing violence of ontology. In place of an ontology of sameness—a totality—Levinas offers a metaphysics of otherness and difference—an alternative ontology of infinity." Cf. SHEPHERD, A., *The Gift of the Other: Levinas, Derrida, and a Theology of Hospitality*, Cambridge, UK: James Clerke & Co, 2014, p. 92.

grounds value and ethical imperative. To find a response to these questions, Levinas cannot resort again to any ontological foundation; he somewhat grounds it in the human face. The face is not an association with some empirical item with discernible qualities. When one stumbles upon the face of the other, one does not initially think about empirical features as one would when looking at a statue – the physical observable features or qualities. Because of the exceptional power of the Other's face, one does not notice the colour of the Other's eyes, the shape of the nose or mouth nor is it essential in this encounter. As Simon Glendinning remarked; "The face, for Levinas, is not the visual countenance or visible 'facial expression' of a human being. Rather, he uses the term 'face' to capture the sense that one's own relation to the Other is not just an attentiveness to a certain body in the world but to *this particular person*."[308] Further still, François Sebbah indicates that if the Other is the Other, or better still, if he is human, is because he does 'face'. He goes on to say that "the face is not a sign referring to pre-existing meanings."[309]

When the Other faces me, he or she inadvertently takes a position. With his or her face and language, the Other calls me out, challenging my tranquility and enjoyment in nature or in the elemental.[310] In a section suggestively entitled *The I of enjoyment is neither biological nor sociological,* Levinas states:

---

308 Glendinning, S., *In the Name of Phenomenology*, London and New York: Routledge, 2007, p. 154.

309 Sebbah François-David, *Lévinas et le Contemporain*, Besançon: Les Solitaires Intempestifs, Éditions, 2009, pp. 159–169. "Si autrui est autrui, ou bien encore s'il est humain, c'est parce qu' il fait visage, parce qu' il apparait comme ce bouleversement de toute apparition don't nous venons de rappeler les traits principaux. *Ce n'est pas, ce n'est jamais, à l' inverse, parce qu'il est "humain" ou d' "autui" selon des definitions déjà constituées de l' humain ou d' autrui, qu'un visage est visage: le visage n'est pas un signe renvoyant à des significations préexistantes.*"(Emphasis in the text)

310 The elemental, which Levinas sometimes refers to as "nature" or as "earth," should not be understood naturalistically as designating the totality of natural entities and systems (Cf. *Totality and Infinity*, pp. 140–142). Rather, as a phenomenological notion, which is to be seen as a formal concept of the natural environment that includes its relationship to human subjects, problematic as that relationship may be? In fact, it is within the context of his account of the genesis of subjectivity that Levinas introduces the elemental. In this context, the elemental (in contrast to "earth" or "nature") is literally environmental: it *surrounds* the subject. The subject develops itself, according to Levinas as "interiority,", that is, through a "centripetal" movement of "separation" from a primordial immersion in an elemental medium into its own interior, it is necessarily also treating the rest of the elemental medium as exterior. The subject thus develops itself as a hub around which a region of the outside is appropriated and oriented—"coiled," as Levinas says (Cf. *Totality and Infinity*, p. 175).

> In order that a pluralism in itself [...] be realized there must be produced in depth the movement from me to the other, an attitude of an I with regard to the Other [...], that would not be a species of relationship in general; this means that the movement from me to the other could not present itself as a theme to an objective gaze freed from this confrontation with the other, to a reflection. Pluralism implies a radical alterity of the other, whom I do not simply conceive by relation to myself, but *confront* out of my egoism. The alterity of the Other is in him and is not relative to me; it *reveals* itself. But I have access to it proceeding from myself and not through a comparison of myself with the other. I have access to the alterity and the Other from the society I maintain with him, and not by quitting this relation in order to reflect of its terms [....] The Other... (is) an alterity that, belonging to the essence of the other, is nevertheless visible only from an I.[311]

Levinas is saying here that it is basically in approaching the Other that I can address myself; it is simply in being conscientious and attentive to the call of the Other that I can become conscious of myself. Commanded by the Other, called to respond, I become cognizant of myself as subject to the Other. In speaking to the Other, in responding to the challenge of the Other, I express myself, reveal myself, and in that way become aware of myself. Self-conscious subjectivity can only arise when the privacy of alienated being is wrecked by the impetuous Other.

According to Levinas, this "relation with the Other alone introduces a dimension of transcendence..."[312] and as such, the face consists in a manifestation of God. Along these lines, "the epiphany of the face is ethical."[313] The reason why the human face has this absolute quality about it is because it reveals God who is the Infinite Other. The human face is the epiphany or manifestation of God in the world. It is a trace of the Absolute, of Infinite Godness who is unadulterated by being, by self-interest, and by egocentric predicament. God, according to Levinas, is beyond being and can only be encountered in the world in the human face of the other person. God in the human face of the Other assigns us to stand up for the weak, the defenseless, the suffering. God, in a word, is the Infinte Godness beyond egotistic being.[314] So, the definitive ground of ethical behaviour is a transcendent Other beyond being.

John Caputo in *Being and the Mystery of the Person* alludes to this interpretation when he asserts that something more reverberates through the human face than human language. He declares: "The face [...] is a place of transcendence by which I mean a place in which the transcendent breaks through, and in which we are initiated into deeper things, drawn into the mystery of self, of world, of

---

311 Levinas E., *TI.*, p. 121.

312 *Ibid.,* p. 193.

313 Levinas, E., *TI* p. 199.

314 Burggraeve, R., *The Wisdom of Love in the Service of Love*, pp. 116–121.

God."[315] By this he means that the face is the revelation of being. It both reveals and conceals the mysterious depths of the human person. But the other, for Heidegger, is devoid of this transcendence since his understanding of *Mitsein*, being-with, has its aim at the Being of beings, Being in its universality.

Owing to the unique relation with the face of the Other and oneself, the ontological subject comes after Levinas' ethical subject. From the phenomenological progression, while we place Heidegger's emphasis on the primacy of ontology over epistemological claims, in which Dasein is a break from subjectivity of the subject inaugurated by Descartes' cogito, we can place Levinas' tenets of alterity, the Other within a post-modern philosophical outlook. The modifications of Heidegger's primal ontico-ontological constitution of Being revealed to *Dasein* are obscured and overshadowed by Levinas' ethical imperatives that precede every ontological constitution, not excluding Heidegger's outline of Being as fundamental ontology.

2.5.3.b. Elemental Objects as "Subordinated" to Enjoyment

Levinas' account of the elements is outlined in a portrayal of the amorphous affective components of sensation. "Hence, we can say that the element comes to us from nowhere; the side it presents does not determine the object, remains entirely anonymous. It is wind, earth, sea, sky, air."[316] Lingis, in the translator's introduction to Levinas' *Existence and Existents*, gives a concise synopsis of the significance of Levinas' doctrine of the elements. Rather than beginning with the perception of objects, Levinas begins with the elements that support "being in the world," which is the foundation of Heidegger's web of *Zeug* (equipment):

> If the world is a field of things, there is then something else in subjectivity besides being in the world; there is a relationship with the terrestrial, with the light—and with the sensuous element, which, before being taken as so much data for cognition, is savoured, is assimilated, nourishes and contents life. There is the elemental; and an existence finds itself and rests in the elemental, and thus finds itself, prior to awakening to the world.[317]

He distinguishes the elements from things and proposes "to analyze more closely the way the things we enjoy come to us. Enjoyment precisely does not reach them qua things. Things come to representation from a background from which

315 Caputo, J., "Being and the Mystery of Person," in *The Universe as Journey: Conversation with W. Norris Clark, S. J.*, Bronx, N.Y: Fordham Univ. Press, 1988, p. 94.

316 Levinas, E., *TI.*, p. 132. "Wind" designates neither a singular object nor a plurality; the partitive construction renders this in the French: "C'est du vent, de la terre, de la mer, du ciel, de l'air." (cf. Footnote in the text)

317 Lingis, A., Translator's Introduction *Existence and Existent*, pp. 10–11.

they emerge and to which they return in the enjoyment we can have of them."[318] Consequently, for Levinas, our use of things, tools or implements takes for granted our enjoyment of them.

> The world of things calls for art, in which the intellectual accession of being moves into enjoyment [...]. Tools and implements, which themselves presupposes enjoyment, offer themselves to enjoyment in their turn. They are playthings [*jouets*]; the fine cigarette lighter, the fine car. They are adorned by the decorative arts; they are immersed in the beautiful, where every going beyond enjoyment reverts to enjoyment.[319]

In his clarification of enjoyment, Levinas talks of this reduction of our relationship with objects or equipment to our enjoyment. "This relationship with an object can be characterized by enjoyment (*jouissance*). All enjoyment is a way of being, but also a sensation—that is, light and knowledge. It is absorption of the object, but also distance with regard to it."[320] It is this notion of nourishment which is at the heart of his critique of Heidegger regarding enjoyment. For according to Levinas;

> Nourishment, as a means of invigoration, is the transmutation of the other into the same, which is the essence of enjoyment; an energy that is other, recognized as other, recognized [...] as sustaining the very act that is directed upon it becomes, in enjoyment, my own energy, my strength, me.[321]

In Heidegger's study, as we have already seen in chapter one, a piece of equipment (*Zeug*) reveals its referential relation to further pieces of equipment, because there is certainly not such thing as "*an* equipment" ("ein *Zeug*"). A piece of equipment immediately reveals a totality of equipment whose "for the sake of which" is Dasein. But for Levinas, every object, together with Heidegger's equipment refers to and takes for granted our enjoyment (*jouissance*) of them.

> Every object offers itself to enjoyment, a universal category of the empirical - even if I lay hold of an object-implement, if I handle it as a *Zeug*. The handling and utilization of tools, the recourse to all the instrumental gear of life, whether to fabricate other tools or to render things accessible, concludes in enjoyment."[322]

---

318 Levinas, E., *TI*, p. 130.

319 *Ibid.*, p. 140.

320 Levinas, E., *TO*, pp. 63. Levinas also develops the idea of enjoyment in *Existence and Existent*, pp. 37–45; *TI*, pp. 127–139, 143–151; and *Otherwise than Being*, pp. 72–74

321 Levinas, E., *TI*, p. 111. Levinas argues that "prior to being a system of tools, the world is an ensemble of nourishments. Human life in the world does not go beyond the objects that fulfill it....These are the nourishments characteristic of our existence in the world. It is an ecstatic existence—being outside oneself—but limited by the object" Cf. *TO*, p. 63.

322 Levinas, E., *TI*, pp. 132–133.

Enjoyment here is a process wherein things go back to their elemental standing in our nourishment and contentment with them. The food nourishes me and saturates my hunger on the elemental level as a necessity of life and its sustenance. Enjoyment is, accordingly, not at the outset the activity of an interested agent, but affectivity, namely, an intentionality somehow motivated by what is enjoyed.

To recap, ethics, for Levinas, begins in exteriority; one becomes an ethical subject in contact with the exterior commands of alterity, the other. The power of the face of the Other ask for a response and in this response, responsibility and subjectivity originates. Responsibility, according to Levinas, is here described as an "essential, primary, and fundamental structure of subjectivity [...]; the very node of the subjective is knotted in ethics and understood as responsibility."[323] Let us then examine closely what Levinas means by subjectivity and responsibility.

2.5.4. Ethical Subjectivity (Ipseity)

From what we have seen so far, the ethical subject is the hallmark of Levinas' teaching on responsibility. Simon Critchley observes that "the key concept in Levinas' work is ethical subjectivity. The precondition for the ethical relation to the other is found in Levinas' picture of the ethical subject. It is because of a disposition towards alterity at the heart of the subject that relatedness to the other is possible."[324]

While *Totality and Infinity* concentrates on ethical alterity, *Otherwise Than Being* pays special attention to ethical subjectivity. Expressing the structure of the book, Levinas explains:

> It aims to disengage the subjectivity of the subject from reflections on truth, time and being in the amphibology of being and entities which is borne by the said; it will then present the subject in saying, as a sensibility from the first animated by responsibilities. Then it will set out to show proximity to be the sense of the sensibility, substitution as the *otherwise than being* at the basis of proximity, and as a relationship between a subject and infinity, in which infinity *comes to pass*.[325]

Levinas starts by demonstrating that the subject, the "I," obtains its character as subject by first untying or isolating itself from what is not itself. This is accomplished in the process of gratifying desires, or the process of enjoyment, in which one becomes conscious of one's personal pleasure and sorrow, hence conscious

323 Levinas, E., *Ethics and Infinity: Coversations with Philippe Nemo,* trans. Richard A. Cohen, Pittsburg: Duquesne University Press, 1985, p. 95.

324 Critchley, S., *Infinitely Demanding: Ethics of Commitment, Politics of Resistance,* New York: Verso Press, 2007, p. 62.

325 Levinas, E., *OB*, p. 19.

of one's own ipseity, or one's selfhood. Ipseity involves uniqueness, or "unicity" as Levinas puts it. Therefore to be conscious of one's own ipseity, one's selfhood, is to be conscious of one's uniqueness, one's unicity. Enjoyment brings about this since in itself, enjoyment is isolation and isolation is the structure of the unicity of the "I". To be conscious of one's own ipseity, or unicity, is the primordial step toward the consciousness of subjectivity.

Levinas base his argument first and foremost on the contention that Western ontology as a whole, and particularly Heidegger's ontology, is in the long run, unavoidably egocentric in its notion of the subject as a being who is interested only in its own being. However, Levinas overturns the conventional "no other-than-self without a self" for a claim to "no self without another who summons it to responsibility."[326] The subject's relation to the Other is not mutual but is a relation of irrevocable force to which I am subjected. As David Fryer explains: "A theory of ethical subjectivity posits two things: (1) that there is an intimate connection between a person's subjectivity and her being ethical, that is, the structure of the human subject as an ethical structure, and (2) that an ethics is best conceived when one first understands that the structure of the human subject is ethical."[327] In order words, in studying both ethics and subjectivity, our center of attention is on the ways wherein we, as subjects, are ethical; in the sense of being accountable for others and accountable to ourselves; in the sense that, at our very interior, we, as subjects, have obligations that we need to be conscious of, that we have to foster and uphold, and that it is these obligations that make us who and what we are. Fabio Ciaramelli gives an elucidation of the interlocking of subject formation and ethical obligation in Levinas' works that can help us understand the distinctive solitary demand placed upon the subject. He states:

> I am obliged to the other, without being ontologically *compelled* to do so (since I can still refuse myself to the other). Levinas stresses that I am the only one who can respond to an appeal so personal and so direct that it arises immediately from my orientation toward the other, from my position in the relation [...]. Hence, in the same event which displays my own ipseity prior to any intelligible ontological identity whose form I might share with others, an absolute obligation arises concerning that which is irreducibly mine, not only now, in the contingency of his event, but always. The authority of this obligation, upon which the meaning of my subjectivity depends, is grounded not in the universality

---

326 Ricœur, P., cited in KEMP PETER, "Ricœur Between Heidegger and Levinas: Original Affirmation between Ontological Contestation and Ethical Injunction," in Richard Kearney, ed., *Paul Ricœur: The Hermeneutics of Action*, London: Sage, 1996, p. 46.

327 FRYER, R. D., *The Intervention of the Other: Ethical Subjectivity in Levinas and Lacan*, New York: Other Press LLC, 2004, p. 18.

of the *logos*, but in the immediacy of the transcendence of the other who places me under obligation. In my position as a subject, responsible for another, I am affected by an infinite transcendence that I am unable to comprehend through the (*arche* of the) *logos*.[328]

Levinas' subject is not a Hegelian self-consciousness because, "the ego ought to be considered essentially as a *subjectivity* that gives rise to difference, but only to the extent that one does not consider the I or the Other as a *monad*, or as purely exterior entity."[329] As Hegel states:

Self-consciousness exists in *itself* and for *itself*, in that, and by the fact that it exists for another self-consciousness; that is to say, it *is* only by being acknowledged or "recognized." The conception of this, its unity in its duplication, of infinitude realizing itself in self-consciousness, has many sides to it and encloses within it elements of varied significance. Thus its moments must on the one hand be strictly kept *apart* in detailed distinctiveness, and, on the other, in this distinction must, at the same time, also be taken as *not distinguished*.[330]

Instead of a subject who exists in *itself* and for *itself*, Levinas' subject exists "for another," since "I am defined as a subjectivity, as a singular person, as an 'I,' precisely because I am exposed to the other. It is my inescapable and incontrovertible answerability to the other that makes me an individual 'I.' So that I become a responsible or ethical 'I' to the extent that I agree to depose or dethrone myself – to abdicate my position of centrality – in favour of the vulnerable other."[331] In his translator's introduction to *Otherwise Than Being*, which exposes in abridged form, Levinas' thought in the work, Alphonso Lingis is of the opinion that Levinas is attempting a phenomenological reduction and revival of responsibility "as a determinative structure of subjectivity," which "Husserl personal subjectivity" and "Heidegger's authenticity" initiated but did not conclude.[332]

The face of the other, that component of the other which is the base of social interaction, signifies nearness to the other person. Levinas calls this "proximity"

---

328 Ciaramelli, F., "Levinas's Ethical Discourse Between Individuation and Universaity" in *Re-Reading Levinas*, Bernasconi R. & Critchley S., (eds) Bloomington, IN: Indiana Unviersity Press, 1991, p. 88.

329 Pada, S. Roland Theuas, The Paradox of Ipseity and Difference: Derrida's Deconstruction and Logocentrism, in *Kritike*, 1*(1)*, 2007, (32–51), pp. 36–37. http://www.kritike.org/journal/issue_pada_june2007.pdf

330 G.W.F. Hegel, *Phenomenology of Mind*, trans. J.B. Baillie, New York: Dover Publications, Inc., 2003, pp. 104–105.

331 Levinas, E. & Keamey, R., "Dialogue with Emmanuel Levinas" in *Face to Face with Levinas*, pp. 26–27.

332 Lingis, A., Translator's introduction to *Otherwise than Being or Beyond Essence*, Pittsburgh: Duquesne University Press, 1997, pp. xvii–xxii.

which designates something excessive for Levinas. Proximity here "signifies a community that exists before any established fact, where the other's singularity is experienced by me as unparalleled, overwhelming, extreme, exclusive and pre-assigned."[333] Proximity is experienced as prompt connection or contact. Levinas explains:

> ...the proximity of the Other is not simply close to me in space, or close like a parent, but approaches me essentially insofar as I feel myself—insofar as I am—responsible for him. It is a structure that in no wise resembles the intentional relation which in knowledge attaches us to the object—to no matter what object, be it a human object. Proximity does not revert to this intentionality; in particular it does not revert to the fact that the other is known to me.[334]

Since proximity must be thought of as a burden upon me that comes from the outside, Levinas finds the possibility of ethics or the ground whereupon ethics first shows itself. Not only that, the self also, now, takes on a different feature.

> The self is a *sub-jectum*: it is under the weight of the universe, responsible for everything. The unity of the universe is not what my gaze embraces in its unity of apperception, but what is incumbent upon me from all sides, regards me in the two senses of the term, accuses me, is my affair.[335]

This is a drastic rethinking of identity by Levinas. All my tussles to ground and establish myself are ruined efforts to eliminate the other who appears as an obstacle between me and becoming myself. These efforts are ruined because the other has enmeshed him or her self so profoundly that I cannot get rid of the other without annihilating myself. In other words, the other turns into me, a state which Levinas calls substitution, that is, "the other is in me and in the midst of my very identification."[336] This is indicative of a sacrifice of self, because looking at it this way, the self or subject is under blockade, under siege. Fred Alford explains that "substitution is fusion without attachment, a binding, self-defining connection to the other in the absence of a relationship with the other, a relation without relation."[337] In contrast to Husserl, it is not necessary, according to Levinas, to conceive the Other under the form of an ego, rather, "in substitution my being that belongs to me and not to another is undone, and it is through substitution that I

---

333 Hand, S., *Emmanuel Levinas*, New York: Routledge, 2009, p. 55.

334 Levinas, E., *Ethics and Infinity: Conversation with Phillipe Nemo*, p. 97.

335 Levinas, E., *OB*, p. 116.

336 *Ibid*., p. 125.

337 Alford, F. *Levinas, the Frankfurt School and Psychoanalysis*, London: Continuum, 2002, p. 29.

am not 'another,' but me."[338] Equally Bernasconi explains that "Substitution offers a reexamination of the Western philosophical concept of identity, which Levinas associates with self-coincidence, self-possession, and sovereignty."[339] The other calls me, in my sovereignty, my singularity, taking me hostage and demanding my response and no other's. This call to responsibility for the Other "...commands me and ordains me to the other...and makes me approach him, makes me his neighbour. It thus diverges from nothingness as well as from being. It provokes this responsibility against my will, that is, by substituting me for the other as a hostage."[340] This is substitution, as initially formulated: "me for the other."

> In 'Substitution' the identity of the 'I' is under challenge, specifically in the context of responsibility. Unlike the conventional conception of responsibility, whereby I am primarily responsible only for what I have chosen or for what I have done, Levinas conceives of a responsibility to which one is elected and by which one finds oneself answerable for everything and everyone, even for one's persecution."[341]

In this challenge, it becomes very clear that "one is held to bear the burden of others; the substitution is a passive effect, which one does not succeed in converting into an active initiative or into one's own virtue."[342] In this passivity, prior to the other, I am also in a state of being prior to myself. I am freed from both. The self is freed from itself for it is not yet a self. In this passivity, the unique self is substitutable for every other. Without the distance required to identify itself, to distinguish itself from the other, the self is substitutable for this other. Spelt out in lucid terms, Levinas' argument is like this:

> To be an I with full subjectivity is to be aware of one's own ipseity, to be aware of oneself as a unique identity, or as a "unicity." The I, then, must be completely separated from what is not itself, and furthermore, must have an awareness of this separation. To say the same thing differently, the I must be aware of the limits of its own being.[343]

---

338 Levinas, E., *OB*, p. 127.

339 Levnias, E., "Substitution," in Peperzak, A., Critchley S., & Bernasconi, R., (eds.) *Basic Philosophical Writings*, Bloomington, Illinois: Indiana University Press, 1996, p. 79.

340 Levinas, E., *OB*, p. 11.

341 Levinas, E., *Basic Philosophical Writings*, p. 79.

342 Lingis, A., Introduction to *Otherwise than Being*, p. xxxvii.

343 Nuyen, A. T., "Altruism As The Condition of Subjectivity" in Kapur Basant, Kim Chong Chong, (eds.), *Altruistic Reveries: Perspectives from the Humanities and Social Sciences*, Dordrecht, The Netherlands: Kluwer Academic Publishers, 2002, p. 38.

To sum up, on the subject of this substitution, I am unique and no one can replace my responsibility. And this responsibility for the other originates from the alterity of the other. An ethical relation from the I to the other is asymmetrical or lopsided, and no one can take my place to be responsible for the other. The uniqueness of the I is the uniqueness of being irreplaceably matchless. In his dialogue with Mortley, Levinas says:

> When I talk about responsibility and obligation, and consequently about the person with whom one is in a relationship through the face, this person does not appear as belonging to an order which can be 'embraced', or 'grasped'. The other, in this relationship of responsibility, is, as it were, unique: 'unique' meaning without genre. In this sense he is absolutely other, not only in relation to me; he is alone as if he were the only one of significance at that moment. The essence of responsibility lies in the uniqueness of the person for whom you are responsible.[344]

The irreplaceability of the 'I' as substitution for the other as an absolute other is Levinas' fundamental teaching on ethical responsibility. Levinas tied the notion of substitution to the question of responsibility. In an interview in 1988 during which Levinas sets forth the central features of his ethical philosophy, he states that "to substitute oneself does not amount to putting oneself in the place of the other man in order to feel what he feels;" rather, "substitution entails bringing comfort by associating ourselves with the essential weakness and finitude of the other; it is to bear his weight while sacrificing one's interestedness and complacency-in-being, which then turn into responsibility for the other."[345]

2.5.5. *The Ontological Priority of Ethical Responsibility*

The examination of ethics and responsibility is made within the background of Levinas' critical appraisal of Husserl and Heidegger, driven under the impact and consternation of the Holocaust as well as out of a movement away from ontology. The question of ethics is the question of responsibility. The access to ethics—which for him should be raised to first philosophy—and to responsibility takes place in such a break with ontology; obviously, in a split with Heidegger. Levinas' ethics, therefore, starts with answering the call of responsibility from the face of the Other, paying special attention to the call of the Other that shapes the identity of the "I" as a side-effect. Ethics for Levinas is not a relation to the universal but to the unique, that is, the other as unique. "When I speak of uniqueness, I am

---

344 Mortley, R., *French Philosophers in Conversation*, New York: Routledge, 1991, p. 16.
345 Robbins, J., (ed.), *Is it Righteous to Be? Interviews with Emmanuel Levinas*, Stanford, Calif: Stanford University Press, 2001, p. 228.

also expressing the otherness of the other."[346] Ethics is a phenomenological reality—one attends to the Other, not out of hypothetical maxim, but from the call of a phenomenological reality witnessed as a trace in the face of the Other.

In *Otherwise Than Being*, Levinas presents responsibility as the pivotal component of subjectivity. A person becomes a subject when he or she is ethically responsible to and for the Other. The unique experience is the intersubjective relation. Simon Critchley explains this intersubjectivity as follows:

> When I totalize, I conceive of the relation to the Other from some imagined point that would be outside of it and I turn myself into a theoretical spectator on the social world of which I am really part, and in which I am an agent. Viewed from outside, intersubjectivity might appear to be a relation between equals, but from the inside that relation, as it takes place at this very moment, you place an obligation on me that makes you higher than me, more than my equal.[347]

Accordingly, subjectivity begins in reaction to the proximity of the Other. It does not arise out of the inventiveness or ingenuity of the subject, but crops up in reaction to the subjugation of the subject by the Other. In a word, Levinas makes two key claims: "first, that rather than being the remit of an already existent subject, ethics precedes ontology; and second that our responsibility is always to the other, without limit and prior to judgment."[348] The self is not a subject anymore, but a hostage to the other. This being-hostage, namely, a non-chosen, unsolicited responsibility for the other attests, as Derrida illustrates in his *Adieu a Emmanuel Levinas*[349] to the drastic deprivation or vulnerability of the subject which Levinas names the face of the Other. In a word, I have to face up even to that which does not interest me, because the Other comes to me as a face. As Levinas notes, "I understand responsibility as responsibility for the Other, thus as a responsibility for what is not my deed, or for what does not even matter to me; which precisely does matter to me, is met by me as face."[350] I am responsible for acting for the Other, beyond my own actions, beyond my autonomy. This obligation to, and

---

346 Levinas, E., *Entre Nous: On Thinking-of-the-Other*, trans. Michael B. Smith and Barbara Harshav, New York: Columbia University Press, 1998, p. 205.

347 Critchley, S., "Introduction", *Cambridge Campanion to Levinas*, Bernasconi, R., & Critchley S., eds. Cambridge: Cambridge University Press, 2004, pp. 13–14.

348 Shildrick, M., "Ethics at the Limits: Levinas, Politics and Responsibility," in Robin, M. Schott & Kirsten, Klercke, (eds.), *Philosophy on the Border* Copenhagen: Museum Tusculanum Press, 2007, p. 64.

349 Derrida, J., *Adieu a Emmanuel Levinas*, Paris: Galilée, 1997.

350 Levinas, E., *EI*, p. 95.

responsibility for, the Other is infinite. Very often Levinas insisted that "in ethics, the Other's right to exist has primacy over my own."[351]

Furthermore, Levinas insists that the relation of responsibility for the Other is essential and unavoidable. As an essential structure of human subjectivity, responsibility is prior to any rational calculation and it must be taken for granted even when we lay emphasis on the realities of conflict and violence among human beings. He unequivocally addresses this question in an essay titled *Ideology and Idealism* where he says:

> I have been speaking about that which stands behind practical morality; about the extraordinary relation between a man and his neighbour, a relation that continues to exist even when it is severely damaged. Of course we have the power to relate ourselves to them as to an object, to oppress and exploit him; nevertheless the relation to the other; as a relation of responsibility, cannot be totally suppressed, even when it takes the form of politics, or warfare. Here it is impossible to free myself by saying, "It's not my concern." There is no choice, for it is always and inescapably my concern.[352]

This inescapable concern for the other is suggestive that responsibility fears for the other: I fear for the death of the other. Somewhat, this fear for another is the starting point of responsibility (and recalls Hans Jonas's notion of an "heuristics of fear" in his thinking of responsibility as we shall see later in Part II). I fear for the other's affliction, but then again for my personal possible cruelty as a being-in-the-world that can create a paradise rebuffing and eliminating all others in distant land. Levinas goes so far as to raise this suffering of the other to the level of the "supreme principle of ethics."[353] Ethics is not grounded on the universal moral law but on the anguish of the other person. It is therefore an affliction, suffering, poverty and a vulnerability that command and call me. And why am I called? "I am he who finds the resources to respond to the call."[354] How, for any reason, could I find myself held hostage by another's vulnerability? This is of course the very essence of responsibility.

This idea of responsibility offered by Levinas makes unique sense of the fundamental insights of the duty of care: that we must put others first, and that this responsibility is not a regrettable imposition on our obviously individual and autonomous subjectivity, but entrenched in the idea of responsibility, and the *source*

---

351 Cohen, A. R., *Face to Face with Levinas*, New York: State University of New York Press, 1986, p. 24.

352 Levinas, E. "Ideology and Idealism" in *Modern Jewish Ethics*, ed. Marvin Fox, Columbus: Ohio State University Press, 1975, p. 137.

353 Levinas, E., *Entre Nous: On Thinking-of-the-Other*, p. 94.

354 *Ibid.*

of our individuality. This is because one cannot be responsible, even self-responsible, without the other.

> Positively, we will say that since the Other looks at me, I am responsible for him, without even having *taken* responsibilities in his regard; his responsibility is *incumbent on me*. It is a responsibility that goes beyond what I do. Usually, one is responsible for what one does oneself. I say, in *Otherwise than Being*, that responsibility is initially a *for the Other*. This means that I am responsible for his very responsibility.[355]

This excessive claim can only be understood in terms of Levinas' doctrine on subjectivity. Having defined subjectivity as responsibility, being responsible for the other is being responsible for his responsibility. Levinas puts this as follows: "Responsibility in fact is not a simple attribute of subjectivity, as if the latter already existed in itself, before the ethical relationship. Subjectivity is not for itself; it is, once again, initially for another."[356] The other gives me the inner distance that allows me to be for myself. Subjectivity, in other words, is posterior to responsibility. My responsibility precedes me. I have no choice.

Concluding Remarks

What we have done so far is to examine and highlight some tenets of Heidegger's fundamental ontology which forms the initial formation to his thinking on technology. He rejects all attempts to give a precise definition of Being. For Heidegger, any endeavour to present a definition of Being would lead to associating Being to a specific concept. He often speaks of what Being is not, rather than of what it is. In *Being and Time*, Heidegger takes up the question of the meaning of being. In doing so, he contends that Dasein's understanding of being is a prerequisite for the possibility of existence. Such an understanding is, according to Heidegger, fundamental to experience and even to being itself in the Heideggerian scheme.

Being manifests itself in the revealing-concealing process. Consequently, Being cannot be understood by way of metaphysical or representational thinking. It can rather be understood simply in its affinity with the realm of *Ereignis* or event of appropriation, i.e., with respect to *zusammengehörigkeit*, the essential belonging togetherness of Dasein and Being, a realm of deeper experience rather than mere intellectual knowing. Such a deeper experience is arrived at through essential think of being, poetic dwelling in the fourfold, aletheia the unconcealing

355 LEVINAS, E., *EI*, p. 96.
356 *Ibid.*

process of the truth of being and language, all through which man becomes the shepherd of Being.

Because thinking transcendental is no longer a conscious option for Heidegger, his critique of metaphysics—which was there right from the start of *Being and Time*—needs to go historical if he is to succeed on his contention that traditional philosophy mistakes the ontological difference between Being and beings. He does this by sketching the sliding itinerary of metaphysics from the Greeks through the moderns to the 20th century, when metaphysics was "overturned" by Nietzsche and superseded by "technological thinking", or what Heidegger calls *Gestells* or "enframing". This will be part of our main focus in the next part of our study.

Meanwhile, Levinas dedicated a quarter of a century studying Husserl and Heidegger in addition to setting the ground work for a critique of phenomenology. In doing so he detached the ethical gap required in the repression of the Other. Levinas advised us to think along the existential plane and to keep in mind that we have an ethical responsibility that pre-exists us. In criticizing Heidegger's fundamental ontology, Levinas presents the subjectness of the subject as residing on face to face with the Other. He clearly presents his understanding of subjectivity as a reversal of the conventional subject. For instance: "subjectivity as hostage. This notion reverses (*renverse*) the position from which the presence of the ego to itself appears as the beginning or the accomplishment of philosophy."[357] Amongst several illustrations of this reversal of the modern convention in philosophy are: (i), the subject is not a for-itself, but a for-the-other; (ii), the subject is not a freedom, but a passivity; (iii), the subject does not hypothesize or establish the meaning of the other, but is "affected" by the other. "The 'I' is not a nominative, but an accusative, a 'me'; the subject does not instigate, but can only respond. The subject is not a freedom, but a receptivity, the subject does not thematize, but is exposed to the transcendence of the infinite…the subject is precisely not an active subject, a spontaneity, but is subject*ed* to the other as a hostage."[358] Responsibility does not designate the action of the subject anymore, but is overturned into a proportioned passivity or inaction. Responsibility for the other is "the defeat of the I think", the defeat of "the originary activity of all acting, source of the spontaneity of the subject, or of the subject as spontaneity."[359] This is what Levinas calls substitution. For Heidegger, no one can substitute himself for me in my care for my being. Such

---

357 Derrida, J., *Adieu a Emmanuel Levinas*, p. 202.

358 Raffoul, F., *The Origins of Responsibility*, Blommington, Indiana: Indiana University Press, 2010, p. 217.

359 Derrida, J., *Adieu a Emmanuel Levinas*, p. 220.

care creates my being. For Levinas, no one can substitute for me in my responsibility for the other. Without this I have no unique subjectivity or ipseity. Authentic existence is here reconstrued in terms of responsibility.

However, as one can see, Levinas' critique of Heidegger, oddly enough, arises out of a position that Heidegger had previously deconstructed; that of Descartes' subjectivity and humanism. As Françoise Dastur notes, "just as in the case of Sartre…in the last analysis, it was also a Cartesian motif that Levinas opposed to Heidegger's thought."[360] But while Heidegger cheers the end of Western metaphysics with its subjectness of the subject, Levinas conceives himself as a (competent) advocate of Western metaphysics.

Regardless of Levinas' critique of the Heidegger, his effort is scrupulously hypothetical and as Critchley observes, "Levinas' big idea does not suffice for the solution of all our pressing and often conflicting ethical problems, and surely it would be nothing short of miraculous if it did."[361] So his call for ethical responsibility can merit an accolade for the idea of subjectivity with its transcendental undertone, but it becomes problematic to find in Levinas how we can actualize this responsibility. Hans Jonas's critique of technology in *The Imperative of Responsibility* will help us answer this question later in Chapter five of section three.

But having examined Heidegger's fundamental ontology, the next section of our study will open up Heidegger's critique of technology which springs from his fundamental ontology. This he did, through his call for the unmaking of metaphysics, with its technological mindset, which has led modern philosophy to the forgetfulness of being.

360 Dastur, F., "The Reception and Non-Reception of Heidegger in France," in *French Interpretation of Heidegger*, Albany, N.Y: Sunny Press, 2008, p. 100.

361 Critchley, S. *Leaving the Climate of Heidegger's Thinking*, p. 55.

Section Two
Towards Unmaking of Metaphysics and Questioning Technology

Chapter III: The Age of Technological Being and Emergence of Subjectivity

Introduction

In the last two chapters of our previous section, we have shown how Heidegger made a transition from situating his project within the metaphysical tradition to positioning it in opposition to that tradition. Although his claims that his focus upon the question of Being remains uninterrupted across his work, the apparent shift is noticeable. We equally showed that the forgetting of Being manifest itself traditionally, according to Heidegger, in the representational thinking of identity and metaphysics of subjectivity, which will culminate in modern technology as the complete oblivion of the ontological difference. Thus, it is obvious that there is a discontinuity between the early (Heidegger I) and later (Heidegger II) work, each hiding a subconscious continuity.

In this section of our study, we will see how such continuity is manifested in Heidegger's philosophy of technology which stems directly from his fundamental ontology and on the question of being. Particularly, his notion of technology as the enframing destining of the modern society is a disclosure of being itself as technology. The history of being gives Heidegger a standpoint from which to understand how in our modern world things have been turned into *objects*. The relationship of Heidegger's ontology to his philosophy of technology is one of the notions at stake in understanding the stimulations for his philosophical inquiry. Thus he proposes a history of being which climaxed in the technological conception of being, with the aim of helping us recognize and surmount our modern system of dealing with things as things and resources. Heidegger's philosophy of technology reveals not only the intrinsic danger in the technical, but also how this danger can be surmounted.

Despite the fact Heidegger had ample allusions to technology; a completely structured exposition of his idea of technology cannot be situated in any one specific work. But our analysis in this part of our study would turn to *The Question Concerning Technology* and *The Turning*, (without losing sight of "*The Age of the World Picture*" and "*Science and Reflection*") interpreted in conjunction with one another, as the two major works on Heidegger's philosophy of technology. He designed both works to serve as phenomenological description of the way Dasein comports itself towards technology. They supply the grounds for inaugurating

Heidegger's philosophy of technology. His work on this subject seems to be aggravated by a quest of the understanding of being.

In the opening passage of his 1950 lecture *The Thing*, Heidegger brings to attention the far-reaching transformation of the way in which the world makes itself present to humans through technology. Not only do humans have the aptitude to manipulate their environment through exceptional technological capabilities, but the world itself becomes understood in and through modern science. Now more than ever, media and mass communication make it possible to know the events and conditions of the world in such a manner that determines the way in which we understand being.

> All distances in time and space are shrinking. Man now reaches overnight, by plane, places which formerly took weeks and months of travel. He now receives instant information, by radio, or events which he formerly learned about only years later, if at all. The germination and growth of plants, which remained hidden throughout the seasons, is now exhibited publicly in a minute, on films. Distance sites of the most ancient cultures are shown on film as if they stood this very moment amidst today's street traffic [...]. The peak of this abolition of every possibility of remoteness is reached by television, which will soon pervade and dominate the whole machinery of communication.[1]

Heidegger fathoms our current interpretation of being by looking at one of its highest accomplishments, *scientific research*. He calls this prevailing conception of being, technological.

The Question Concerning Technology, as we shall see, contains his critique of modern technology in addition to his views concerning the possibility for a genuine philosophy of technology. Even though Heidegger specifically addresses the true nature of technology within this essay, his ideas pertaining to a philosophy of technology are anything but mundane; he is primarily interested in how we, as Dasein, can have a free relationship with technology.

In his final analysis of technology, we shall see that Heidegger is critical of those who, still caught in the subject/object picture, think that technology is dangerous because it embodies instrumental reason. Modern technology, he insists, is "something completely different and therefore new."[2]

This part of our study, therefore, strives to understand science and technology such that they have come to define existence in the modern era. To acquaint ourselves with the direction of Heidegger's overarching analysis, we first examine the nature of modernity and the question of science and technology before

---

1 Heidegger, M., The Thing in *PLT*, p. 63.

2 *QCT*, p. 5.

Heidegger's concerns. Revolving around the question of being, Heidegger's project strives to show how any questioning into the essence of technology must step back from the scope of technology. In so doing, he shows how science and technology did not emerge from the heavens, nor have they always been present as epistemological possibilities. Rather, he argues that the advent of science and technology happened in accordance with a particular way in which human being conceived itself as subject which came into prominence with Descartes' Cogito. The dominance of science and technology in modernity is based on the condition that human being is understood as a subject. His critique of traditional metaphysics will be the first step to lunch his grievances against modern technical mentality. To do this effectively, we shall first look at the historical milieu from which modernity evolved.

3. The Question and Meaning of Modernity

3.1. Medieval Background

Karl Rahner described philosophy as "the basic formula of man's self understanding in so far as the later is not simply an effect of revelation"[3]. It is certainly true that philosophy does yield a self understanding, but knowledge of an object other than the subject is epistemologically prior to self knowledge. Consequently, it cannot be accepted that philosophy can be defined in terms of self understanding. It is by knowing something other than 'himself' that man can come to perceive himself as a knower. Rahner's definition is too limited, too anthropological and too rooted in modernity to be able to give an adequate definition of philosophy.

St. Thomas spoke of the philosopher as the wise man who aims principally at the truth of things and this serves to orientate metaphysics away from the consciousness of modern philosophy towards its object—being. Thomas gave a historical summary of the progress of man towards the consideration of being as such: There was the first stage of the early Greek philosophers who identified being with corporeal nature. There was the second stage when Plato and Aristotle explained the reality of things by universal causes. There was the third stage when others advanced further and raised themselves to the consideration of being as being.[4] St. Thomas did not indicate who these others were.

3 Rahner, K., *Theological Investigation*, vol. 6, Baltimore: Helicon Press, 1969, p. 74.

4 Aquinas, T., *Summa Theologiae*, 1a, q. 44, a. 2, c. English translation, Introduction, Notes, Appendices & Glossary by Thomas Gilby, Cambridge: Blackfriars, 1967.

At the beginning of the second millennium, the five centuries of turmoil and chaos generated by the invasion of the tribes and finally ended through the evangelizing work of the church, the civilization of the Middle Ages was in a very special way a religious creation. Bereft of law and devoid of peace, the people of the empire entered upon centuries of darkness and anarchy. Yet, in the midst of this darkness, St. Augustine survived as the great teacher for more than a thousand years. More than any political leader, more than any military commander, more than any barbarian warlord, St. Augustine constituted the intellectual opening into a new epoch.

The church also survived and eventually evangelized the tribes. From the intellectual contribution of St. Augustine, from the evangelizing contribution of the church, the Christian reality of the middle ages emerged. Of this period of history, Romano Guardini said: "the medieval achievement was so magnificent that it stands with the loftiest moments of human history"[5] The vision behind that civilization was a Christian vision. The Middle Ages was based upon an ecclesial not a political unity. In the words of Dawson:

> The church was the universal society and the state was weak and barbarous and divided. The only true citizenship that remained to the common man was his membership of the church [...] the church was a world in itself with its own culture, its own organization and its own law. In so far as civilization survived, it was directly dependent on the church.[6]

The Christian mystery is based on a sacred history. The religious idea of the Christian is not the Greek-like contemplation of the self-thinking-thought, for whom creation and activity were unbecoming. The Christian mystery is about the God who has stepped outside himself in the free act of creating and further stepped outside himself by sending the Second Person of the Trinity into the ultimate extremity of death and abandonment. The Christian vision rooted in the incarnation and sacred history is always pushing towards the real as the proper location of the act of redemption. Although *Stoicism* reached into the Middle Ages and beyond into the metaphysics of Spinoza, the only philosophy which was available to the Christian scholars of the Middle Ages was the Neo-Platonism coming from St. Augustine. Neo-Platonism is always an intellectual movement away from the real, concrete world of existence. Nevertheless, the orientation towards the real continued. Rediscovery of the real world of the natural order was emphasized. The

5 Guardini, R., *The End of the Modern World*, Chicago: Henry Regnery Co. 1956, p. 23.

6 Dawson, C., *Progress and Religion*, New Jersey: Doubleday Image, 1960, p. 166.

Devotio Moderna[7] which was beginning to take shape in the Middle Ages indicates that the anthropological concerns were ushering themselves into Christian spirituality. Christmas for instance begins to replace Easter as the central liturgical event. The crucifix itself ceases to carry the image of the triumphant risen Christ but substituted with the suffering Christ. Benediction and the worship of Christ began to emerge, etc. The prevailing view was one of being sheltered and sustained in the God in-dwelt reality of the world. The world of nature was perceived as a spontaneous manifestation of God. "Even in the sub-lunar region, where individuals creatures are born, grow, decay, and die, some resemblance to God's immutability is preserved in the fixity of species".[8] This was the primordial insight of medieval man. It was pre-reflective intuition over which every subsequent philosophical and theological conclusion would be formed. Even the strictly astronomical sciences were taken into that pre-understanding. The universe consisted of the earth and its nine concentric spheres to which were attached the planets, the stars and galaxies. But beyond the ninth sphere there was the *Empyrium* which was the place of God.[9] In the medieval understanding then, the universe always had a place for God. Cosmology always opens up into theology. It was the era of St. Albert, St. Thomas Aquinas, St. Bonaventure, St. Dominic, St. Francis of Asisi and St. Bernard of Clairvaux. Every least thing was seen in its metaphysical and theological relationship to God. It was in meditation, philosophical, theological and contemplative reflection that the medieval man sought the truth of being. The medieval emphasis on the transcendence, on the metaphysical and theological

---

7 *Devotio Moderna* is a movement founded by Geert Grote of Deventer (b. 1340), in which groups of women and men formed households organized as communes and a lifestyle centered on devotion, but refusing to profess vows as religious or to acquire spouses and property as lay citizens. As devout men and women, over against indifferent parish activities, and those who vowed religion but lived in affluence, they embody "piety" in the "present-day."
"At its core the word meant 'of this era' as distinct from a 'earlier era,' thus *via moderna* to designate the newer way of philosophizing in the fourteenth century over against an older thirteenth-century way (*via antiqua*, the Thomist school)." For more detailed historical and complicated analysis and origins of *Devotio Moderna* from the 1380s to the late medieval times, in the Low Countries, the present day Belgium and Netherland, see, John, H. Van Engen, *Sisters and Brothers of the Common Life: The Devotio Moderna and the World of the Later Middle Ages*, Philadelphia, Pennsylvania: University of Pennsylvania Press, 2008.

8 Dicker, G., *Descartes: An Analytical and Historical Introduction*, New York, Oxford: Oxford University Press, 1993, p. 5.

9 Guardini, R., *The End of the Modern World*, p. 11.

did not exclude for him other forms of philosophical knowledge. According to Romano Guardini,

> Medieval anthropology for example, in both principle and application, is superior to its modern counterpart. In morality [...], medieval life had a firmer yet richer hold on reality than is possible for modern man; [...] In social philosophy and jurisprudence, medieval thought encompassed and ordered its concrete cultural situation to its own time, yet it offers insights which have basic validity for man at any time[10]

The truth of being emerges in philosophical meditation and not by means of empirical investigation. Medieval man did not seek to describe how things worked or to uncover the predictability of antecedents and consequents. He sought to know the meaning of things.

But the Middle Ages contained the germ of its own dissolution. The basic Medieval insight of man as a part of a God in-dwelt world was open to alternative interpretation. In the first instance, it could be understood that man was part of the universe; in the second instance, it could be understood that the world was an extension of man. Varied indications throughout the 14th and 15th centuries indicated that the second interpretation was beginning to take hold and that the medieval world-view was being substituted by an emphasis on man. Cosmological and the metaphysical-theological perspective were being replaced by an anthropological emphasis.

In the sphere of metaphysical and theological knowledge, Duns Scotus (1266–1308) had begun to be pre-occupied with the object of philosophy in the face of theology. The medieval synthesis of philosophy and theology was being prised open in a self-conscious attempt to attain the meaning of philosophy. To achieve this and to make of it an original science independent of theology, Scotus hit upon an idea of being which was pure identity and is prior to every concrete determination whether of the finite or of the infinite. For a being to be produced, it is absolutely necessary that it has a cause. In metaphysical demonstration, the major must express necessity in the order of *quidditative* being, the minor, a necessary relation between essence and existence, and the conclusion, the necessity of actual existence.[11] According to Scotus, "by positing the less noble extreme of some being, we can conclude that the more noble extreme is realised in some other being."[12] For example, "if some being is finite therefore some being is infinite, and

10 *Ibid.*, p. 16.

11 Duns Scotus, J., *Ordinatio*, 1, d. 2, q. 2, n. 1. In *Philosophical Writings*, edited and translated by A. B. Wolter, Edinburgh: Nelson's Philosophical Texts, 1962.

12 *Ordinatio*, 1, d. 39, q. 1, n. 13.

if some being is contingent, therefore some being is necessary."[13] Being whose act it is to be which St. Thomas persistently maintained has now been set aside. Henceforth, being is indetermination, insubstantiality.

The metaphysical empty formalism of Scotus as insubstantiality invited philosophers to eliminate it from the lexicon of the philosophical world. With that elimination accomplished, it was only a short intellectual step to the empiricism of William Ockhalm (1300–1350). For William, there was the science of the general and the universal only. But the world of reality is the world of a singular thing. A world of things separated from one another and of these separated things, there can be no scientific knowledge. Probability alone could be achieved.

In the schools of the 14th century, Scotus was regarded as a representative of '*via antiqua*' while William of Ockhalm was conceived as representing the '*via moderna*'. Because being was reduced to insubstantiality, Ockhalm ignored and directed his attention to the *Haecceity*, the singularity of things. With being of metaphysics filleted, nothing remained but the empiricism of Ockhalm.

The '*via moderna*' was taken along by Nicholas Krebs of Cusa (1401–1464). With Nicholas of Cusa, the concept was of paramount importance. Human knowledge certainly requires concepts and henceforth, epistemology will tend to replace ontology as the central metaphysical problem. The interest of Nicholas of Cusa in the concept centers on the role of mathematical concepts.[14] Given the nature and certainty of mathematical concepts, the propositions of mathematics serve as guide to the investigations proper to philosophy. Here, the fundamental significance is that the philosophic score was being written which Descartes will re-arrange and orchestrate.

The Copernican Revolution together with the theories and experiments of Galileo (1564–1642) overthrew the astronomy of Ptolemy. As Georges Dicker opines:

> Galileo's experiments refuted Aristotle's view that all objects in the sub lunar region move in a straight line toward their "natural place" [...]. Finally, as Copernicus theory (that the earth is not static and at the centre of the universe but, instead, revolves around the sun) gained ever greater acceptance, it ruined the entire medieval cosmos.[15]

The trial of Galileo involved many questions and among those questions was the nature of experimental science. Many philosophers and theologians did not accept

---

13 *Ordinatio*, 1, d. 39, q. 3, n. 13.

14 Copleston, F. C., *A History of Medieval Philosophy*, Notre Dame & London : University of Notre Dame Press, 1990 edition, p. 319.

15 Dicker, G., *Descartes: An Analytical and Historical Introduction*, p. 6.

the new physics precisely because it is not the physics of Aristotle. But the physics of Aristotle is a philosophical science and the newly emergent science of physics was empirical, descriptive and non-philosophical. Consequently, the exponents of philosophy and the exponents of the new physics were incapable of communicating with one another.

It was by means of Ptolemaic astronomy that the averagely or even poorly educated Christian perceived the world as having a place for God. The sudden overthrow of this representation of the metaphysical and theological truths of profound moment would leave the Christian people with the perception of a suddenly 'un-godded' world. This valid pastoral concern did not arise out of any spiritual obscurantism. Whatever legal injustice that was involved in the trial of Galileo cannot find any justification in the genuine catechetical problem which the scientific theory presented. It is not even permissible to perform an evil act so that a good end might come from it.

From the aspect of the transition from the middle ages to the modern era, the significance of Galileo is easy to see. In the post-Galileo world, man no longer stands under the protective gaze of God. God was no longer at home with the world and the modern man progressively perceived himself as an autonomous subject and answerable only to himself. In this transition, Being had been emptied of actuality, dismissed in favour of experience and further down the road, Suarez neutralized being into the realm of the merely possible.

This philosophical turnaround can be distinguished in the separation of mind and body, an increasing skepticism concerning religion, the flowering of scientific enquiry; in a nutshell the intellectual revolutions of the Renaissance. All of these developments contributed to human-centered, secular, scientific, and progressive philosophical picture of the world. This is the dawn of European modernity, and along with its many benefits came an increasing sense of alienation, cynicism, disenchantment and emptiness.

Modernity is a series of attempts to come to terms with the terrifying uncertainty and insecurity engendered by the 'flight of the gods'. Both humanism and the Reformation are intimately tied up with efforts to find a solution to this problem. We want to state that Descartes' thought (whom many see as the father of modernity) is a third attempt to resolve this problem. But as we shall see in the subsequent section, Heidegger's critique of Descartes' project shows that he falls short of achieving this. At the moment, let us examine the nature of modernity, the intellectual climate in which the critique of technology is based.

3.2. The Nature of Modernity

When we speak of modernity we seem to be giving an encompassing name to a whole epoch in world history, the modern age, as distinct from, say, the medieval or classical antiquity. The adjective 'modern', from which the term 'modernity' was derived in the 19th century, refers to that which belongs to a recent time. It may mean 'current' or 'contemporary', and is opposed to 'early' or 'ancient' and medieval. Since the 17th century wrangle pitting the 'Ancient' against the 'Moderns', the term has had positive connotations. And since the term "Modern" is used to describe a wide range of periods, modernity must be taken in context. The modernists based their arguments on the assumption that humanity is in progress. Such a terminology makes it legitimate to discuss questions as to when exactly the modern age may be said to have come into existence; what its origins may have been, or, indeed, if it has now come to an end. Suffice it then to say that modern can mean all of the post-medieval European history, in the context of dividing history into three large epochs: Antiquity or Ancient history, the Middle Ages, and Modern.

Hundreds of books have been written on the blend of factors that lead to and constitute the close of the middle ages and the emergence of the Renaissance and Modernity. A lot of controversies and opinions have been given by various authors as to the exact date of the commencement of modernity.[16] According to Stephen Toulmin, for much of the 20th century, people in Western Europe and North America generally accepted two statements about the beginning of Modernity and the modern period: namely, that the modern age began in the 17th century, with Galileo Galilei in physics, René Descartes in epistemology as well as Thomas Hobbes in politics. As Toulmin states:

> The intellectual revolution was launched by Galileo Galilei, and by René Descartes. It had two aspects: it was scientific revolution, because it led to striking innovations in physics and astronomy, and it was the birth of a new method in philosophy, since it

16 "Some people date the origin of modernity to the year 1436, with Gutenberg's adoption of moveable types; some to A.D. 1520, and Luther's rebellion against the church authority; others to 1648, and the end of the Thirty Year s' War; others to the American or French Revolution of 1776 or 1789; while modern times start for few only in 1895, with Freud's Interpretation of Dreams and the rise of "modernism" in the fine arts and literature." Toulmin, Stephen, *Cosmopolis: The Hidden Agenda of Modernity*, Chicago: The University of Chicago Press, 1990, p. 5. See also, Iliffe, Rob, "The Masculine Birth of Time: Temporal Framework of Early Modern Natural Philosophy" in *British Journal of the History of Science*, 33, 2000, pp. 527–453.

established a research tradition in theory of knowledge and philosophy of mind that has lasted right up to our own times.[17]

However Toulmin rejects the above conventional account which slates the foundation of modernity in the 17th century with Descartes and Galileo. His deep-seated convinction is that modernity begins with the Renaissance. He finds in the Renaissance a political balance and a tranquil acceptance of diversity, indeterminacy and difference of opinion.

On equilibrium, the single and most significant standard for modernity theorists is the Enlightenment, with its affinity for rationality and social progress. Miles Ogborn, in *Spaces of Modernity*, writes: "Against the backdrop of the Enlightenment, modernity is associated with the release of the individual from the bonds of tradition, with the progressive differentiation of society, with the emergence of civil society, with political equality, with innovation and change. All of these accomplishments are associated with capitalism, industrialization, secularization, urbanization and rationalization."[18] According to Feenberg, the historicizing trend in the emerging biological and social sciences of the late 18th and 19th centuries was firmly rooted in the idea of progress, which found surest guarantee in the promise of technology. By the end of the 19th century, under the influence of Marx and Darwin, progressivism had become technological determinism.[19] Subsequent upon the widespread reading of these materialistic masters, scientific progress was believed to ground humanity's advance towards autonomy and contentment. Therefore, modernity associated with post-Enlightenment projects of emancipation, rationalism, 'progress', social reform, secularism and the like, brought about a major and inevitably fraught transformation in the culture and politics of Europe. Accordingly, modernity could be described as essentially a European phenomenon together with the values they projected. As Michael Gillespie states:

> Modernity is a secular realm in which man replaces God as the center of existence and seeks to become the master and possessor of nature by the application of new science and its attendant technology. The modern world is conceived as the realm of individualism, of representation and subjectivity, of exploration and discovery, of freedom, rights, equality, toleration, liberalism, and the nation state.[20]

---

17 Toulmin, S., *Cosmopolis,* p. 14.

18 Cited in Porter, Roy, *The Creation of the Modern World: The Untold Story of the British Enlightenment*, New York: W.W. Norton, 2000, p. 488.

19 Feenberg, A., *Questioning Technology*, London & New York: Routledge, 1999, pp. 1–2

20 Gillespie, M. A. *The Theological Origins of Modernity*, Ill, Chicago: The University of Chicago Press, 2008, pp. x–xi. Michael Kimaid observed that "much of the formative discourse that sprung up around the concept of modernity was conceived in Europe

Gillespie, like most others, place the origin of modernity as the product of seventeenth- century, a movement against the medieval religious scholasticism in favour of science and reason. Thus, modernity "was rooted in the philosophy of Descartes and Hobbes and the science of Copernicus and Galileo."[21]

For our purpose, philosophically speaking, Descartes is unmistakably a place to begin. His inspiration and insight justifies his designation as the father of modem philosophy. The clarity with which he separates the subject of experience from its object, and hence the realm of subjectivity from objectivity, expresses once and for all the perceptual and conceptual tendencies that crystallized into the modern point of view. That he further identifies the subject's experience as the field of research further consolidates his claim to modernity, it is not a convenience to begin with Descartes but an obligation because he so explicitly expresses the beginning of what needs to be understood.

Nevertheless, we can only deal with history as it unfolded. The modern project, it is generally recognized, can be traced to the attempt to ground the attainment of indubitable knowledge on irrefutable grounds.[22] The process of reasoning needs to be self-validating without having to appeal to authority or depend on faith. It has to be able to generate from itself a set of necessary, self-evident principles which, literally, no-one could possibly doubt.

> In Descartes' system, reason first clears away all preconceptions and then elaborates its own first principles, accepting only clear and distinct conceptions which can survive the most rigorous examination. For the system to work, the universe must be modeled on a deductive system, so that what happens in it must be deducible from the laws of its operation and its initial state.[23]

This aspiration to build from absolutely secure foundations an indubitable body of knowledge about the world which humans dwell in had an emancipatory

---

from the fifteenth through the eighteenth century, and as such is subject to the beliefs and biases of the culture in which it was imagined." Cf. KIMAID, M., *Modernity, Metatheory, and the Temporal-Spatial Divide: From Mythos to Techne,* New York: Routledge, 2015, p. 15.

21 *Ibid.*

22 West, D., *An Introduction to Continental Philosophy*, Cambridge: Polity Press, 1996, pp. 10–13.

23 Ward, K., "The Decline and Fall of Reason", in U. King, (ed.) *Faith and Practice in a Postmodern Age,* London: Cassell, 1998, pp. 22–23.

objective. Humanity would come to self-realization in the struggle "to separate truth from falsehood, reason from unreason, fact from fiction."[24]

This approach to knowledge—from a solid foundation building upwards—and the collective discoveries aggravated by the scientific method suggested an evolutionary, progressive or dialectical dimension to human history. Progress seemed to be the unavoidable accompaniment of a rational analysis of human problems in which the causes of the defects of human life could be dispassionately examined and put right in an ascending progression towards human perfection.[25] It is not unexpected that dreams of the future conquered the imagination, the delusion of a society of social harmony, free from oppression, ignorance and prejudice. Such was the hint of the utopias announced by the Marquise de Condorcet's "Sketch for an Historical Picture of the Progress of the Human Mind."[26]

Whatever the public mean by 'modernity', according to Ellen Wood, and whether they see it as good or bad or both, they frequently believe "it has something to do with what sociologist Max Weber called the process of *rationalization:* the rationalization of the state in bureaucratic organization, the rationalization of the economy in industrial capitalism, the rationalization of culture in the spread of education, the decline of superstition, and the progress of science and technology."[27] Also modernity, when seen from the cultural and institutional conception, "is a mode of social life or organization rather than a cultural or epistemological condition." Here, "it is characterized by institutional structures and

---

24 NORRIS, C., *Reclaiming Truth: Contribution to a Critique of Cultural Relativism*, London: Lawrence & Wishart, 1996, p. 141. The scientific technique, based on scrupulous observation and well-tested hypotheses, seemed to guarantee the fulfilment of the desire for indubitable knowledge, of a different order from mere opinion or belief. It had the double merits of being rationally accessible to anyone who grasped its methods of operating and universal in character, i.e., not reliant on factors (such as culture, situation, upbringing) which could relativise perspectives. Science, it is claimed, more than any other force within history, has the capacity to makes all equal, since it obeys a logic and set of rules that no-one can control but only submit to.

25 HUNTER, J. D., "What is Modernity? Historical Roots and Contemporary Features" in P. SAMPSON, V. SAMUEL & C. SUGDEN, Faith *and Modernity*, Oxford: Regnum Books, 1994, p. 21.

26 CAHOON, L., (ed.), *From Modernism to Postmodernism: An Anthology*, Oxford: Blackwell, 1996, p. 72–82.

27 WOOD, E. M., *The Origin of Capitalism: a longer view*, London & New York: Verso Publication, 2002, p. 182.

processes, such as industrialism, capitalism, rationalization, and reflexivity."[28] This is suggestive that there is an intrinsic historical link between modernity and capitalism; as early forms of capitalism emerge in association with techno-development from the 18th century on, so the discourse of the modern also emerges as a set of representations of capitalism's present and future. Fredric Jameson comments that

> if I recommend the experimental procedure of substituting capitalism for modernity in all the contexts in which the later appears, this is a therapeutic rather than a dogmatic recommendation, designed to exclude old problems (and to produce new and more interesting ones).[29]

But, man's rise to culture of modernity is not without attendant depravities. Among the first casualties of mans modern worldview is the general belief in God and religion. Religion and faith are confined to this phase of human society, superstitious, gullible and less-critical. Indeed, "religion, faith and rationality present themselves as three successive layers in a historical process, as human instruments that gradually unfold and become distinct."[30] And as such, according to Gillespie, "modernity came to be as a result of the displacement of religious belief from its position of prominence at the center of public life into a private realm where it could be freely practiced as long as it did not challenge secular authority, science, or reason. The authority of religion to shape private and public life thus was replaced by a notion of private belief and ultimately personal "values."[31]

Therefore the "death of God" advocates present themselves as champions for the emancipation of man, and the world; they claim to be man affirming and world asserting. Their project is simply—Secularism! Materialistic Worldliness! and Indiscretions! Man is called upon to emancipate himself from forces and powers which is exemplified in Nietzsche's *Übermann* who is a symbol of man 'come of age'. Thus armed with science and the technology to harness the potentials of nature and his discovery, modern man launched the project of unlimited earthly ideology of progressivism; a programme of creating paradise on earth through man's proper effort. Science and technology was seen to cure all

---

28 Brey, P., Theorizing Modernity and Technology, in *Modernity and Technology*, Op. Cit., p. 35.

29 Jameson, F., *A Singular Modernity: Essays on the Ontology of the Present,* London: Verso, 2002, p. 143. Cited from, D. Punter, Modernity, New York: Palgrave Macmillan, 2007, p. 50.

30 Hoedemaker, B., *Secularisation and Mission*, Harrisburg: Trinity Press International, 1997, p. 18.

31 Gillespie, M. A. *The Theological Origins of Modernity*, p. x.

man's ill, which will usher in a worldly paradise. The is the spirit which animated economism in which all that matters for man is capitalistic economic well-being. The price to be paid is the depersonalization and de-humanization of man. In some cases, he becomes a mere object rather than a subject, a thing rather than a person. It is almost equal to saying that he is emptied of his humanity and his personality.

> Functionalization of man depersonalized and dehumanizes him. It stifles and oppresses him. He is no longer a person but a commodity to be used and exploited by some mass movements; whether social, industrial and political.[32]

Through functionalization, modernity has 'thingified', 'massified' and 'commodified' man. The situation becomes messier when he is taken up by the culture of economism. He becomes measured in terms of cash value, market price, by a "Greedy civilization" that puts premium on the economy rather than the human person. With Descartes' distinction between *res cogitans* and *res extensa,* and when human relationship in the world makes no more appeals to any transcendental values and becomes subsumed in materialistic categories, the relationship moves from "I and thou" to "it and it". This is the discovery of Gabriel Marcel and other theistic existentialists that "when man ceases to regard his fellow man as a genuine thou, when he calculatingly views him in a highly personal manner as an "it", an object to be used or manipulated then we are faced with a technomanic and technolatry."[33]

Beyond this spiritual loss of place, another unforeseen consequence of modernity became apparent. The destructive power or what (following Weber and Marx) Theodore Adorno and Marx Horkheimer of earlier Frankfurt School called "instrumental reason" in *Dialectic of Enlightenment*[34] showcases this consequence. Instrumental reason is the objectification of modernity, i.e., the transformation of the philosophical perception of the world from that which defines and controlled us into something else. The world becomes that which we can and must control, and indeed we define it in relation to ourselves, very much in opposition to the pre-modern ideas of defining ourselves in our relationship to the divine or the world. Global imperialism, modern armaments and warfare, and environmental degradation have been seen as upshot of this re-definition of

32 Lescoe, J. F. *Existentialism: With or Without God*, New York: Alba House, 1972, p. 17.

33 *Ibid.*, p. 18.

34 Adorno, T. W., Horkheimer, M., & Gunzelin S., N., *Dialectic of Enlightenement: Philosophical Fragments*, Trans. Edmund Jephcott, Stanford, Calif.: Stanford University Press, 2002.

humanity and its relation to the world. Reason became the sole player in the field! It became an autonomous power which carried other overbearing clouts within its ambience—capitalism, imperialism, technology and state bureaucracy—but without sufficient power of discernment to see the inherent dangers: "The idea that the free market is self-stabilizing is an archaic, curious relic of Enlightenment rationalism."[35] By pretending to be the measure of the knowable, reason became reductionist. Having accepted that the main end of human existence was reachable to reason and having realized its limitations in practice, "there developed an increasing divorce between the objective world created by technological rationality and the subjective world of meaning and purpose."[36] Here, as in Heidegger, the administered world is none other than the worldly actualization of the metaphysics of subjectivity that culminates in Hegel's *Logic*. Unlike Heidegger, however, at least in the early stages of Critical Theory, criticism is pursued for the sake of a *future* thought of as objective reason, and for the sake of some past.[37]

This is symptomatic of Weber's notion of rationalization of society which is the prevalent acceptance of rules, effective productivity, and pragmatic results as the correct way to encounter human activities and the instituting of a social organization around this notion. This notion has been very influential today in modernity and social theory as modernity displays an extreme dynamism and globalizing latitude. Jürgen Habermas, under the influence of Weber and Marx, analyzed modernity as an "unfinished project."[38] Habermas made distinction between the early phase of modernity which witnessed the rise of the "bourgeoisie public sphere," which mediate between state and the public sphere, and late phase of modernity in which the state and private corporations took over vital functions of the public sphere, as a result of which the public sphere became a sphere of

35 Gray, J., *False Dawn: The Delusion of Global Capitalism*, London: Granta, 1998, p. 198.

36 Touraine, A., *Critique of Modernity*, Oxford: Blackwell, 1995, p. 5. Here, the person is simply reduced to choosing between objects in the outside world put there by the harnessing of instrumental reason to the domination of the book of nature. But pure choice, when there is no ultimate reason for choosing, because the meaning of existence is unknowable through reason alone, is degrading; it shrinks the complexity of the full potential of humanness.

37 Ferry, L. & Renaut, A., *Heidegger and Modernity*, trans. Franklin Philip, Chicago and London: University of Chicago press, 1990, p. 83. Originally, published as *Heidegger et les Modern*, Paris: Grasset & Fasquelle, 1988.

38 Habermas, J., "Modernity: An Incomplete Project," in Hal Foster, *The Anti-Aesthetic: Essays on Postmodern Culture*, Port Townsend, Wash.: Bay Press, 1983, pp. 3–15.

domination.[39] The consequence of this has been a colonization of the life-world by a merged system of economy and state, technology and science that carried out its functional laws in all spheres of life.

In his work "*The Consequences of Modernity*," among others, Anthony Giddens analysed modernity as resting on four major institutions, namely: industrialism, capitalism, surveillance, and military power.[40] Also, Giddens, in *Modernity and Self-Identity*, uses his typical identification of the contemporary period as one of 'late modernity' to put what may seem at first glance to be a disarmingly simple series of questions:

> What to do? How to act? Who to be? These are focal questions for everyone living in circumstances of modernity—and ones which, on some level or another, all of us answer, either discursively or through day-to-day social behaviour.[41]

From this perspective, our sociological definition of modernity above comes into play. The disembedding of social life, the elimination of social relations and institutions from local contexts requires a reflexive appropriation of knowledge, which is the production of methodical knowledge about social life that is then reflexively applied to social activity. Together, these developments create a social dynamism of displacement, impersonality, and risk. Bauman puts it this way:

> ...modern states did not emerge from inter-communal conferences, or as federations of parishes and township. They were born and grew up in dog fights with 'local particular isms' and at the expense of expropriating the locally based powers of a greater part of their pre- modern authority. One wonders whether the same operation won't have to be repeated, two centuries later—but this time on a global scale.[42]

---

39 Habermas, J., *The Structural Transformation of the Public Sphere*, Cambridge, Mass: MIT Press, 1989. Reprinted in 1991. See also, Brey, Philip, "Theorizing Modernity and Technology, in Misa, J. Thomas, Brey, P. & Feenberg, A., *Modernity and Technology*, Cambridge, Mass.: The MIT Press, see esp. PP. 34–45.

40 Giddens, A., *The Consequences of Modernity*, pp. 55ff. Here, Giddens takes on the question of the institutional dimensions of modernity as well as the globalizing of modernity. He described industrialism as the transformation of nature, the development of "created environment"; Capitalism is the capital accumulation in the context of competitive labour and product markets; While Surveillance has to do with the control of information and social supervision, Military power is the control of the means of violence in the context of the industrialization of war.

41 Giddens, A., *Modernity and Self-Identity: Self and Society in the Late Modern Age*, Cambridge: Polity Press, 1991, p. 70.

42 Leighton, D., 'Searching for Politics in an Uncertain World: Interview with Zygmunt Bauman', *Archive*, 10, 1, 2001, p. 11.

Ulrich Beck had long taken up this issue of risk when he projected a theory of "reflexive modernity" wherein the role of technology is unambiguously acknowledged and deliberated in terms of revolutions in the nature of rationality. Beck made distinction between two stages of modernization which he dubbed "simple modernization" and "reflexive modernization." Simple modernity creates a technology that is both exceptionally dominant and totally fragmented. Beck argues that today a "risk society" is emerging and is particularly manifest in the environmental realm. According to Beck, "risk society are produced because the certitudes of industrial society dominate the thought and action of people and institutions in industrial society....It arises in the continuity of autonomized modernization processes which are blind and deaf to their own effects and threats. Cumulatively and latently, the latter produce threats which call into question and eventually destroy the foundations of industrial society."[43] This will be our point of emphasis in the third part of our study when we move towards the technological rationality in risk society.

The idea of modernity has in recent times been superseded with the idea of post-modernity. The postmodern epoch has been variously described, but always, of course, in relation to modernity. According to Ellen Wood, "post-modernity generally represents a phase of capitalism marked by certain distinctive economic and technological characteristics (the 'information age', 'lean production', 'flexible accumulation', 'disorganized capitalism', consumerism, and so on. But more particularly, it is marked by certain cultural formations summed up in the formular 'postmodern*ism*', whose single most outstanding feature is its challenge to the 'Enlightenment project."[44] Ellen explains further that postmodernism perceives the world as deeply fragmented and uncertain, rejects and 'totalizing' discourse, any complete and "universalistic theories about the world" and equally "rejects any universalistic political projects, even universalistic emancipator projects—in order words, projects for a general *human* emancipation rather than very particular struggle against very diverse and particular oppressions."[45]

Having looked at the question of modernity from different perspectives as enunciated by different authors at different eras, let us now examine the place of Descartes at the origin of modernity and see how Heidegger used his criticism of Descartes' subjectivity to lunch his critique of modern technology.

---

43 Beck, U., "The Reinvention of Politics: Towards a Theory of Reflexive Modernization" in U. Beck, A. Giddens, S. Lash, *Reflexive Modernization: Politics, Tradition and Aesthetics in the Modern Social Order,* Cambridge: Polity Press, 1994, pp. 5–6.

44 Wood, E. M., *The Origin of Capitalism: a longer view,* p. 190.

45 *Ibid.*, p. 191.

3.2.1. *Cartesian Doubt and Reformulation of Metaphysics*

The forces which Luther has unleashed in the life of the church were given their first philosophical formulation by Descartes. Luther has set the individual over and against the church and has concentrated theology upon the requirements of the individual: the private interpretation of the scriptures by the individuals denied the teaching role of the church; Christian faith was concentrated upon the sinfulness and the salvation of the individual. In the sphere of morality, the private conscience was the supreme standard. That which in Luther (theology of Luther) was the anguish of the self became in Descartes the certainty of the 'Cogito'.

Descartes was as ambitious in philosophy as Luther was in theology. In the wreckage of the medieval synthesis, Descartes intended to lay the foundation of human knowledge and to develop a method suitable for such knowledge. As the medieval synthesis disintegrated, the empirical sciences of observation and the mathematical sciences began to make rapid progress. The philosophical sciences by contrast had disintegrated into futile and ceaseless disputatiousness. "Of philosophy I will say nothing except [...] that yet there is not a single matter within its sphere which is not still in dispute, and nothing, therefore, which is above doubt..."[46]

Cartesian science, we want to say, is an attempt to construct a bastion, a fortress of reason that could provide not only individual certainty and security, and not only mitigate and eliminate the deficiencies of nature, but also bring an end to the religious and political strife as well as intellectual riot that were ripping Europe to pieces.

At the core of the Cartesian scientific revolution stands an epistemological revolution directed primarily against the scholastic-Aristotelian and their sense based epistemology. By repeatedly attacking the reliability of sense-experience as a means of knowing the nature of bodies and their properties, Descartes sought to undermine scholastic accounts of nature and the substantial forms and real qualities invoked in those accounts. In the words of Daniel Garber, "Descartes begins with an account of the nature of body, which, as he says, 'does not consist in weight, hardness, colour, or the like, but in extension alone', and an account of the modes of body, shape, size, and motion...[47]

46 Descartes, R. *Discourse on Method*, 1.

47 Daniel, G., "Semel in Vita: The Scientific Background to Descartes' Meditations" in Amélie Oksenberg Rorty (ed.) *Essays on Descartes' Maditations*Los Angeles: University of California Press, 1986, p. 83–84.

The fundamental principle and foundation of Descartes' science is *Cogito ergo sum*. This is the thought that he believes everyone or nearly everyone could experience if only they follow the path he laid out. "The search for truth, which is also the search for knowledge turned into search for certainty. It did so because the pursuit of certainty is the only possible road for the *pure* search for truth, the project of improving the truth-ratio which is not constrained by any other limitations at all."[48] According to Martial Gueroult:

> Descartes wishes to examine the whole sphere of certainty. He does not wish merely to have an illusion of certainty, to have blind faith in a certainty that is not itself controlled —in brief, in a certainty that is not certain of itself—not knowing whether this certainty is established and in what way it is established.[49]

Thus, if we wish to end up in complete certainty, we must not admit in ourselves anything that is not absolutely certain, that is to say, we must doubt everything that is not certain with absolute certainty, and remove from ourselves everything that is stricken by doubt.

Starting *a priori*, from within his own mind and before experience, Descartes tries to devise a series of logically indubitable assertions which would permit him to establish the existence of the world around him. A rapid transition to the Method of Doubt occurs in the *First Meditation*:

> Reason persuade me already, that I should without assent no less carefully from things which are not clearly certain and indubitable, as from things which are evidently false; so if I find some reason for doubt in each of them, this will be enough for me to reject them all.[50]

Descartes pointedly reconstructs the formula he followed. First, everything must be doubted, even the most apparently self-evident perceptions. We are regularly deceived by visual illusions and hallucinations, so how can we be certain that everything we see around us is not deceptive? Indeed, there is no assessment with which we can establish that all we experience is not happening in a dream. Even

---

48 Bernard. A. O. Williams, *Descartes: The Project of Pure Inquiry*, New York: Routledge, 2005, p. 34.

49 Gueroult, M., *Descartes' Philosophy Interpreted According to the Order of Reason, vol. 1: The Soul and God,* trans. Roger Ariew, Minneapolis: University of Minnesota Press, 1984, p. 15. Originally published as *Descartes selon l'ordre des raisons, vol. 1, l'âme et Dieu,* 1952, 1968.

50 Descartes, R. *Meditations on First Philosophy with Selections from Objections and Replies*, trans. John Cottingham, with Introduction by Bernard Williams, Cambridge: Cambridge University Press, 1986, Med. 1, p. 17.

the truths of mathematics and logic may be doubted if we consider the likelihood of an evil god who incessantly tries of deceive us. Once we have doubted the whole lot that can be doubted, the second step of Descartes' project is to look for something that can't be doubted. This he finds in the acknowledgment that every time he thinks, he must necessarily exist as a thinking thing. This is absolutely certain: *I think, therefore, I exist.*[51] The third step in Descartes' program is to establish the existence of the world on the foundation of his new discovery. A number of the ideas that he, as a thinking thing, possesses are too immense to have originated from him, for the source of an idea must include as much reality as the idea itself. Ideas or representations are mental entities, or contents of consciousness.[52] So, since he has the idea of God as a perfect being, God must exist and he should not be a deceiver since deception is a sign of imperfection. Eventually, then, it is God who guarantees that the world as it presents itself to the senses is pretty much the way that it appears to be. There is a world outside of the human mind, and so long as we don't allow our imagination to run faster than our judgment, we will not be misled.

From Descartes, we can deduce threefold necessity namely:

1) The necessity for initial or preliminary doubt.
2) The necessity to exclude nothing from doubt on the assumption that doubt is not radically impossible.
3) The necessity to treat provisionally as false the things touched by doubt—which carries the necessity to reject them entirely.

There correspond to this threefold necessity three characteristics of Cartesian doubt: it is *methodological*, it is *universal*, and it is *radical*.[53] From the time when Descartes recognized that he may perhaps not scrutinize, in whichever genuine space of time, each of his thoughts of which he had great doubts, he exclaimed; "Nor therefore need I survey each opinion individually, a task that would be endless. Rather, because *undermining the foundations will cause whatever has been built upon them to crumble of its own accord*, I will attack those principles that supported everything I once believed."[54] This is the basis of Cartesian wisdom, and the great power point on which he believe humanity can stand to move the world.

As he was composing the Replies to the Objections submitted to his *Meditations*, Descartes confided to his close friend Marin Mersenne in 1641:

---

51 Descartes, R., *Meditations*, 2.

52 Gorner, P., *Heidegger's Being and Time, An Introduction*, p. 50.

53 *Ibid.*

54 Descartes, R. *Meditations*, 1 (2).

> I may tell you, between ourselves, that these six Meditations contain the entire foundation of my physics. But is not necessary to say so, if you please, since that might make it harder for those who favour Aristotle to approve them. I hope that those who read them will gradually accustom themselves to my principles and recognize the truth in them before they notice that they destroy those of Aristotle.[55]

The above statement is indicative that "Descartes is absolutely clear that the program of the Meditations is not an autonomous philosophical project, but prelude to a larger scientific program;"[56] his letter to Mersenne is suggestive that the motivation for the Meditations cannot be merely the refutation of skepticism, it is, as it were, a Greek gift that Descartes is attempting to send behind the lines of Aristotelian science.

The next stage in the process of 'purgation' involves the search for some metaphysically certain beliefs. This, Descartes did, through mathematical rigor as we shall in the next sub-section.

3.2.2. Mathematical Demonstration and Deduction

Descartes titled his book *Meditation on the First Philosophy*. Now first philosophy was a term used by Aristotle to designate the science of being as being. Even at the beginning of modern philosophy, Descartes was still dealing with metaphysics. But it was evident that Descartes was a metaphysician who was about to betray metaphysics. Descartes states: "...All philosophy is like a tree of which metaphysics is the root, physics the trunk and all other sciences the branches that grow out of this trunk which are reduced to three principles, namely, medicine, mechanics and ethics."[57] According to Jean-Luc Marion, "it does not appear that a metaphysics is any longer possible, nor really required, in as much as the univocal rationality of divine mathematics has taken its place."[58]

The betrayal is evident already. No longer is metaphysics the supreme and the ultimate science, it is the beginning of knowledge rather than the end of knowledge. Metaphysics merely fastens the tree of knowledge to the soil but the fruits of knowledge resides in sciences—medicine, mechanics and ethics. The degrading of metaphysics is further noticed in the contrast which Descartes made between

---

55 ADAM, C., TANNERY P. (eds.) *Œuvres de Descartes*, vol. 111, Paris: Vrin, 1996, pp. 297–298.

56 Garber, D., *Descartes Embodied: Reading Cartesian Philosophy Through Cartesian Science*, Cambride: Cambridge University Press, 2001, p. 223.

57 DESCARTES, R. *Principles of Philosophy*, Preface.

58 MARION, JEAN-LUC, *On Descartes Metaphysical Prism*, trans. JEFFERY L. KOSKY, Chicago & London: University of Chicago Press, 1999, p. 21. Originally published as *Sur le prisme métaphysique de Descartes*, Presses Universitaires de France, 1986.

the speculative philosophy of the schools and the new practical sciences which enabled man to master nature. "...In room of the speculative philosophy usually taught in the schools, to discover a practical [...] we might also apply them in the same way to all the uses to which they are adapted and thus render ourselves the Lords and possessors of the knowledge."[59]

Philosophy could only give probable arguments and conclusion. Because philosophy could only furnish probability, Descartes turned to mathematics in search for certainty.

> I was especially delighted with the mathematicians on the account of their certitude and evidence of their reasoning [...]. I was astonished that foundations so strong and solid should have no loftier superstructure reared on them.[60]

Like Nicholas of Cusa, Descartes himself was a celebrated mathematician. Of all the sciences, it was mathematics alone which was able to reach certain and demonstrable conclusion. If philosophy was to reach any certain truth, it must therefore become mathematical. There will be as much certainty in philosophy as there is in mathematics in it. The role of mathematics for the Cartesian reformed philosophy is not far to seek:

> ...Arithmetic, Geometry and other subjects of this kind, which deal with the simplest and most general things, regardless of whether they really exist in nature or not contains something certain and indubitable. For whether I am awake or asleep, two and three added together are five, and a square has no more than four sides.[61]

Descartes understood that mathematics deals with an object so pure and uncomplicated and it does not makes assumptions of any kind which experience might render ambiguous. Mathematics consists of the coherent deduction of consequences for mathematical principles and axioms. Descartes did not intend that man should limit his studies to mathematics, but he did intend that the method of mathematics would be the method of scientific knowledge in order that philosophy might obtain certitude equal to that of mathematics. In the *Rules for the Direction of the Mind,* he wrote: "We should busy ourselves with no object about which we cannot attain a certitude equal to that of the demonstrations of Arithmetic and Geometry."[62]

---

59 DESCARTES, R. *Discourse on Method*, 6.

60 DESCARTES, R. *Discourse on Method*, 1.

61 DESCARTES, R . *Meditations*, 1 (8).

62 MARION, J.-L. (ed. & trans.), *Règles utiles et claires pour la direction de l'esprit*, The Hague : Nijhoff, 1977, p. 2.

In the universal science which Descartes aimed to achieve, reality was relegated and the object of this new universal science was the idea of concept. It is a return to the conceptual in the manner of mathematics that Descartes hope to attain certainty and truth. He was not concerned with sciences such as physics, astronomy or medicine. Such sciences lack the intellectual purity of mathematics because they had for object physical material things and consequently, were "of doubtful character".[63] But mathematics which does not consider the real at all is certain beyond any doubt.

When Descartes made the mathematical knowledge the none-philosophical knowledge, certain things followed. The first principles of the mind were henceforth no longer first principles of being. The principle of identity, for example, lost its position as the principle of being and became instead the principle of logic and of knowledge. Being nothing but mathematical entities, the obviousness, the intuitive character derived from the concepts themselves, the mathematical sciences do not derive from reality with all its obscuring burden of matter or materiality.

Now mathematics is largely concerned with entities of reason, the relations between them and the consequences which necessarily follow. What mathematics knows are these entities. The unknown and the excluded is the factor of existence. A philosophy then which is to be re-organized along the line of the mathematical will inevitably be a philosophy which is primarily concerned with the conceptual. According to Aquinas, "the science of mathematics treats its object as though it was something abstracted mentally whereas it is not abstract in reality"[64]

Over the last hundred years, there has been a considerable debate about Descartes' originality, leading up with a debate about the origin and nature of modernity. Modernity in this sense is understood as the consequence of the Reformation and the development of human inwardness. Starting from the end of the nineteenth century, a number of scholars, often neo-Thomist in inclination and antagonistic to modernity and Descartes, tried to show that Descartes' thought was not as original as he had contended by showing the many ways in which he drew upon scholastic thought. Descartes in their view concealed the medieval foundation of his new house, building a mechanic's workshop on a cathedral's foundation.[65]

---

63 Descartes, R. *Meditations* 1 (8).

64 Aquinas, T. *Summa Theologiae*, 1a, q. 44, a 1, ad 3.

65 Gillespie, M. A., *The Theological Origin of Modernity*, p. 317.

> Aristotelian doctrines are presented in Descartes' work, but not always or mostly by adoption. Here as in so many other instances the robe is scholastic but the flesh under it is not.[66]

However, Heidegger and some later postmodernist thinkers argue, on the contrary, that although there were undoubtedly medieval echoes in Descartes' thought, it was profoundly modern because it had the doctrine of subjectivity as its basis. Nevertheless, contrary to the past tradition that climaxed in Hegel, Heidegger maintained that the subjectness of the subject emanating from Descartes was not the moment of humankind's coming of age, his arrival or as his homecoming, but the moment of its most forceful estrangement, its "world-midnight".[67]

Nonetheless, the goal of Descartes' scientific project is to make man master and possessor of nature and in this way to prolong human life (perhaps infinitely), to eliminate want, and to provide security. This is probably why Jean-Luc Marion opines that "among the numerous difficulties raised by the novelty of Cartesian thought, one of the greatest and yet one of the most secret, is whether Descartes at all deserves the name 'metaphysician', or whether he even intended to be called one"[68] According to him, when we speak of a Cartesian metaphysics, we in fact mean an epistemology, a theory of consciousness, some proofs of the existence of God, or a foundation of physics, and so forth.[69]

3.2.3. The Primacy of Consciousness (Subjectivity)

Descartes fundamental principle is often understood as self-consciousness or subjectivity. For him, cogito signifies consciousness. "By the word thought, I understand all that which so take place in us, that we of ourselves are immediately conscious of it…"[70] What Descartes' means by self-thinking, however, is quite different from the ordinary notion of self-consciousness. Thinking for Descartes is clearly reflexive. He writes to Mersenne in July 1641:

---

66 SECADA, J., *Cartesian Metaphysics: The Late Scholastic Origins of Modern Philosophy*, Oxford: Oxford University Press, 2000, p. 35.

67 HEIDEGGER, M., *The End of Philosoph: Overcoming Metaphysics,* p. 60.

68 MARION, J-L., "On Descartes' Constitution of Metaphysics", in Tom Sorell, ed. *Descartes,* Aldershot, England: Dartmouth Pub. Comp. 1999, p. 21.

69 Ibid.

70 DESCARTES, R. *Principles of Philosophy*, pt. 1, 9.

> I have demonstrated that the soul is nothing other than a thing that thinks; it is therefore impossible that we can ever think of anything without having at the same time the idea of our soul, as of a thing capable of thinking all that we think about.[71]

Everything then, of which we are conscious, be it the operation of the will, the imagination or the sensibility, are thoughts because we are immediately conscious of them. It is within the certainty of consciousness that Descartes discovered the certainty of his own existence. The certainty of his existence did not depend on things because all things are yet included in his doubt. "This proposition, I am, I exist, is necessarily true each time it is expressed by or conceived in my mind"[72]

The certainty of his existence lies in consciousness. This is not merely a certainty consisting of the assurance that "in order to think it is necessary to exist"; it is a certainty based upon the assurance that every thought contains the certainty of the self. At the heart of every operation of the cogito, there is the consciousness of the existence of the self. The existence which is assured in consciousness, which is given in every thought, opens up to be "a substance whose essence or nature consists in thinking".[73] The self is a substance whose essence is consciousness and in that consciousness, the certainty of the self is intuitively given.

Having secured the certainty of the self, having determined the condition of truth, Descartes set out to discover the extra-mental things which he had included in his doubt. Post-Cartesian philosophy is critical of him for this. It is said that he saw the Promised Land from afar but failed to enter in and take possession of it. It is also argued that a residual scholastic philosophy remained in him directing him towards the extra-mental things.[74]

The significance of Descartes for modern philosophy then is difficult to exaggerate. At the head of modernity, Descartes articulated the permanent themes of post Cartesian philosophy. There is the primacy of consciousness at the core of which is the self. Consciousness in modern philosophy is henceforth self-consciousness. Philosophy subsequent to Descartes has taken possession of the vision glimpsed by Descartes. Along the pathways to modernity, little of substance has been added to the Cartesian insight. What is new in post-Cartesian philosophy is that the consequences which flowed from the cogito of Descartes were implacably drawn out. In the words of Ernst Behler: "A self-reflective consciousness of time combined with a need for self-assurance accompany the modern age through all

71 *CSM*, I : 195.

72 Descartes, R. *Meditations* II.

73 Descartes, R. *Discourse on Method*, 4.

74 Depré, O. & Lories, D., *Lire Descartes Aujourd'hui*, Louvain-Paris : Édition Peeters, 1997, p. 4.

its phases and this quite naturally. For being modern means essentially a departure from exemplary models of the past, a decentering of habitual ways of viewing the world, and the necessity for producing normative standards out of oneself."[75] Also as Maesschalck observed, Descartes *cogito* is at the base of the problem of the connection between theology and modernity. God from now on, may not be met by someone on the road to autonomy.[76] Emancipating ourselves from our obligations as God's creatures, we look to ourselves as the ground for our beliefs. The ideal of certainty about our salvation, which was based on knowledge of a being other than ourselves, becomes the ideal of certainty about us and our own unequivocal knowledge. In effect, the medieval certainty of salvation was not based on explicit knowledge at all but rather on the protection of the religious rituals and routines of daily life in the Middle Ages, many of which was centered around the cathedrals.

Heidegger comments that the claim to an absolutely unwavering foundation for thinking originates in that emancipation of man in which he frees himself from obligations to Christian revelation and church doctrine to a legislating for himself that takes its stand upon itself. "We, not God, are responsible for our knowledge, and we "take a stand upon" it by evaluating and criticizing it. Dasein's transcendence as "standing towards" Being undergoes a radical change. Now "self-liberating man himself posits what is obligatory," and, because of this, what is obligatory "can henceforth be variously defined."[77]

At the end of the enlightenment, Kant took the decisive step in developing the Cartesian attempt to determine the condition of knowledge and being. Kant did not deny the object but his entire concern was to set out the cognitional structure of the subject to which the object must conform if it is to be known at all.

---

75 Behler, E., *Irony and the Discourse of Modernity*, Seattle & London: University of Washington Press, 1990, p. 3.

76 Maesschalck, M., *Le Principe d'Autonomie: Introduction aux auteurs modernes*, Louvain-la-Neuve: Éditions de l'insitut supérieur de philosophie, 1992, p. 61. "Le cogito cartésien n'est pas le pretext d'une refonte de la démarche ontothéologique, mais un premier pole d'une tension qui, posée en relation avec l'infini comme second pole, définit la position modern du problem de l'exitence humaine.... En arriere-fond, une telle relecture conduit au probleme du rapport entre théologie et modernité et, plus précisément, au problem de la reception théologique des conditions imposes au langage par la modernité. Dieu désormais ne peut se rencontrer que sur le chemin de l'autonomie.

77 Heidegger, M., *QCT*, p. 148.

> Hitherto, it has been assumed that all our knowledge must conform to object [...]. We must therefore make trial whether we may not have more success in the task of metaphysics if we suppose that object must conform to our knowledge.[78]

The knowing subject is henceforth inserting itself into the world of things in order to say of those things that they are knowable only to the extent that they can be placed within the cognitional horizon and coordinates of the self. It is the subject who knows. It is within the immanence of that subject that the criteria of truth, that the condition for the possibility of all knowing are to be formed. The question at hand was to determine the transcendental horizon of consciousness to which the object must conform and to which the object can be known. "I entitle transcendental all knowledge which is occupied not so much with the object as with the mode of our knowledge of our object."[79]

The *a priori* forms of sensibility and understanding constitute the subjective conditions of being known within which the object must appear if it is to be known. According to Kant, "there are only two stems of human knowledge, namely, sensibility and understanding, which perhaps springs from common root."[80] Through sensibility objects are given to us, through understanding, they are thought. The conclusion which Kant eventually reached from this was a dramatic curtailment of knowledge.

Knowledge is limited to that which falls within the *a priori* forms of sensibility. Anything which falls outside that which cannot be brought within the network of *a priori* conditions cannot be object for human knowledge. But the subject matter of metaphysics is not given in the sensible intuition. For instance, God does not belong within the space-time coordinates of the *a priori* forms of sensibility. Consequently, God is unknown and unknowable and human knowledge cannot demonstrate his existence or non-existence.

Thus the consequences deriving from the certainty of the Cartesian cogito emerged thematically in Kantian philosophy. The betrayal of metaphysics in Descartes became in Kant the rejection of metaphysics as a mode of knowledge. The certainty of consciousness and the existence of the self as the abiding core of every cogitation was with Kant expanded with great vigour. What Descartes had hinted, Kant was able to accomplish. In Kantian philosophy, the self inserted itself into the sphere of things, secured its dominant position among things by determining the conditions in which those things would be knowable and accessible

78 Kant, I, *Critique of Pure Reason*, preface.

79 Kant, I., *Ibid., Introduction*, p. 25.

80 *Ibid*, A 15, B 29.

to knowledge. This was something which Descartes attempted, but only Kant achieved. The Kantian investigation of pure reason ended as an interrogation of man: "The whole interest of my reason, whether speculative of practical is concentrated in the three following questions: what can I know; what can I do; what can I hope."[81]

In Cartesian philosophy, and indeed modern philosophy, consciousness is always self-consciousness. In medieval philosophy, consciousness was always consciousness of being and it was always within the knowledge of that being that consciousness of self emerged. While medieval and ancient philosophies dealt with the consciousness of being, it would be untrue to conclude that they lacked any understanding of self-consciousness. St Thomas possessed an elaborate theory of what he called reflection. There is an incomplete reflection wherein the intellect is aware of itself and its existence in its operation. "The philosopher says… we sense that we sense, we understand that we understand; and because we sense this, we understand that we exist."[82]

The cogito of Descartes is situated at the level of incomplete reflection wherein the intellect reflecting upon its operation knows that it exists. Beyond this there is complete reflection which takes place in the judgment. Here, the intellect knows itself and knows that it is its nature to conform to reality. Associated with this presence of intellect to itself, there is indirect reflection. In this, the intellect knows its act of apprehension and sees all the way down to the phantasm of imagination; "To perceive the universal nature existing in the individual".[83]

The entire theory of consciousness is permanently situated within the intellectual knowledge of an object distinct from the intellect itself. Thus:

> No one perceives himself as understanding except from this: that he understands something; for to understand something is prior to understanding oneself as understanding; and therefore, the soul arrives at actually perceiving that it exists through the fact that it understands or senses.[84]

---

81 Kant, I., *Critique of Pure Reason*, A 15, B 32. (But this Kantian humanism or anthropology is already blurring the aspiration of man to God.)

82 Aquinas, T., *Truth (Quaestiones Disputatae De Verite)*, q. 10, art. 8, trans. By Robert W. Mulligan, J.V. McGlynn, & R.W. Schmidt, Chicago, IL: Henry Regenery Company, 1952–54.

83 *Summa Theologiea*, 1a, q. 84, a. 7, ad. C.

84 *De Veritate*, q. 10, a. 8, C.

Originally, the intellect as human is a capacity to know. Now, that which is in potency can be brought to actuality only through the intervention of something which is already in act.

> ...The cognitive soul is in potentiality both to the images which are in the principles of sensing, and to those which are the principles of understanding. For this reason, Aristotle held that the intellect by which the soul understands has no innate species but is at first in potentiality to all such species.[85]

Precisely because the intellect is originally in potentiality to know, consciousness cannot be the primitive datum of metaphysics. The intellect can only be known in so far as it is in act and it is in act in so far as it knows something other than itself. The intellect as human is a potentiality in the order of intelligible being. The intellect as human does not understand itself by its essence but by its act. It is not present to itself by virtue of its ontological constitution because it is only a possible intellect. It is only in so far as a thing is in act that it can be known. Consequently, the human intellect can come to know itself only in so far as it is actually knowing and such knowing for the human intellect is knowledge of the thing other than itself.

This characteristics of the intellect as human can be stated in another way. A power is specified by its act but the act of the intellect is the act of knowing. But the act itself is specified by its object. It is in the act of knowing the object that man can come to know himself, i.e., to attain to consciousness.

> A power as such is directed to an act. Wherefore we seek to know the nature of a power from the act to which it is directed and consequently the nature of a power is diversified as the nature of the act is diversified. Now the nature of the act is diversified according to the various natures of the object.[86]

It is only in the divine and angelic orders that intellect and thing understood are constitutionally identical. "The essence of an angel is as an act in the genius of intelligible things and therefore it is both intellect and the thing understood."[87]

The Cartesian method is based upon the primacy and certainty of consciousness. As such, no matter how modern philosophy may diverge, the primacy and certainty of consciousness is the basic principle in which they all agree. This modern decision for consciousness is not a minor technical change of perspective. The origin of modern philosophy in the certainty of consciousness was a major

85 *Summa Theologiae*, 1a, q. 84, a. 3, ad. C.
86 *Summa Theologiae*, 1a, q. 77, a. 3, ad. C.
87 *Summa Theologiae*, 1a, q. 77, a. 3, ad. C.

dislocation of intellect as the faculty of being and it has interiorly determined philosophy of modernity to this day.

Every cogitation of the cogito postulates the subject. The subjectness of the subject resides in the self certainty of a consciousness that is already oriented to autonomy. This subject which is at the heart of cogito took upon himself the un-goddedness of consciousness and became for himself a further factor in the development of modern atheism. In the words of Fabro;

> Modern philosophy has constituted the most daring attempt of the human spirit to attain total and radical autonomy and to mould thought within the confines of its own self. Neither ancient philosophy, with its Olympian calm and serenity, nor even Christian thought, bulwarked by revelation, can sustain comparison with the restlessness, vitality and inexhaustibility of the modern systems, insofar as speculative boldness and variety of fields of investigation in the whole sphere of the spirit are concerned.[88]

Finally, from the cogito of Descartes to the transcendental ego of Husserl, the subject who derives the certainty of existence in the certainty of consciousness has sought to constitute itself as the centre before which everything must present itself. It is only in relation to the subject that the objectness of everything can be legitimated. The very knowability of the object is dependent upon submitting to the *a priori* forms which reason imposes upon it before that consciousness. From being an affair of the certainty of knowledge, consciousness in Cartesian philosophy has been transformed from question of cognitional certainty into being a source of power and it is this source of power which has become a dominant theme in the philosophies of Marx, Nietzsche and Freud.

3.3. Heidegger's Approach / Critique of Western Metaphysics

The philosophical enterprise is one that thrives and grows on criticism. Though philosophers have the same object of study-everything that is; *reality*, each philosopher perceives this differently. Thus, there are as many philosophies as there are philosophers. Each up-coming philosopher, filled with a noble ideal of how he conceives reality, often writes off the efforts of his predecessors as having no substance.

Thus an Aristotle would get up and sum up the effort of his predecessors in explaining the nature and meaning of reality (being), as akin to that of ...untrained man in fight, "...for they go round their opponents and often strikes fine

---

88 Fabro, Cornelio, *God in Exile: Modern Atheism From its Roots in the Cartesian Cogito to the Present Day*, trans. Arthur Gibson, Westminster, Maryland: Newman Press, 1968, p. 68.

blows but they do not fight on scientific principles, and so too these thinkers do not seem to know what they say…"[89] In the same vein, Heidegger came up later and accused his predecessors of the forgetfulness of Being. According to him, Western metaphysics inquires after entities as entities, entities in their being and hence, it fails to consider Being *qua* Being, the Beingness of Being.

In this discussion, one must think through the history of metaphysics to show how the ontological difference has become subject to certain "forgetfulness" (*Vergessenheit*) proper to "Western thinking in its entire nature."[90] With the forgetfulness of Being and the 'grounding' of beings, metaphysics for Heidegger became onto-theological and the problem is to determine how his happened: "Where does the essentially onto-theological constitution of metaphysics come from? To take-up the question thus posed means, at the same time, to carry out the step back. In this step we now contemplate the essential ancestry of the onto- theological structure of all metaphysics."[91]

Iain Thomson, in the paper on *Undestanding Heidegger's Destruktion of Metaphysics* restated what he described as the "deepest problem" as follows. "How did the metaphysical project of 'grounding' beings come to have onto-theological structure"?[92] Thomson goes on to explain and answer this question in detail that:

> Heidegger situates his account of metaphysics as onto-theology within the context of ancient Western philosophy. As the first Western metaphysicians investigated the 'primordial matter [*ursprüngliche Sache*] of thinking', what Heidegger calls 'the primal matter,' [*die Ur-sache*] they attempted to put this *prōtē archē* into language. Heidegger translates *prōtē archē* as "the first ground", and argues that it was as a result of this quest for such a first ground that the earliest Western metaphysicians postulated two different kinds of beings as the *prōtē archē*: an ontological universal and first being 'and a theological 'highest and ultimate being'. In other words, the first Western metaphysicians pursued the *prōtē archē* in terms of two different kinds of grounding beings, attempting both a bottom-up ontological 'ground-giving' based on a "universal and first being" and a top-down theological 'founding' from a 'supreme and ultimate being'. Here, then, Heidegger provides a historical analysis in support of his thesis that 'since the beginning of Western thought, Being has been interpreted as the ground in which every being as such is grounded'.[93]

---

89 ARISTOTLE, *Metaphysics*, BK,1, Ch. 4, 15.

90 HEIDEGGER, M., *ID.*, p. 51.

91 *Ibid.*, p. 56.

92 THOMSON, I., Ontotheology? Understanding Heidegger's Destruktion of Metaphysics, in *International Journal of Philosophical Studies*, 8(3), 297–327, 2000, p. 311. http://www.tandf.co.uk/journals.

93 *Ibid.* See also HEIDEGGER, M., *ID*, p. 61. "…beings as such is the universal and primal at one with beings as such in the highest and ultimate. The unity of this One is of such

Furthermore, as Heidegger understands the history of metaphysics, 'Western European thinking is guided by the question: "What are beings?" (or "What is that which is?"—"*Was ist das Seinde?"*) This is the form in which it ask about Being (*Sein*).[94] 'What is a being?' asks about the Being of beings by searching both for what makes a being a being (the *quiddity*, the essence or 'whatness' of beings) and for the way in which a being is a being (the existence or 'thatness' of beings). Given the conceptually 'two-fold' ambiguous form of the question, both are legitimate and historically pervasive ways of understanding 'the Being of beings'. Heidegger puts it elaborately clear:

> Metaphysics states what beings are as beings. It offers a logos (statement) about the οντα (beings). The later title 'ontology' characterizes its nature, provided, of course, that we understand it in accordance with its true significance and not through its narrow scholastic meaning. Metaphysics moves in the sphere of the ον η ον: it deals with beings as beings. In this manner, metaphysics always represent beings as such in their totality; it deals with the beingness of beings (the ουσια of the ον). But metaphysics represents the beingness of beings (*die Seiendheit des Seienden*) in a twofold manner: in the first place, the totality of beings as such with an eye to their most universal traits (ον καθολον, κοινον;) but at the same time also the totality of beings as such in the sense of the highest and therefore divine being (ον καθολον,ακροτατον, θειον). In the metaphysics of Aristotle, the unconsciousness of beings as such has specifically developed in this twofold manner.[95]

This beingness of being and the heighest and divine being interpretations are the two stems which Heidegger described as ontology and theology from which he derived his description of metaphysics as having become onto-theological. Thus;

> If we recollect the history of Western-European thinking once more, then we will encounter the following: The question of Being, as the question of the Being of beings is double in form. On the one hand, it asks: What is a being in general as a being? In the history of philosophy, reflections which fall within the domain of this question acquire the title ontology. The question 'What is a being?' (or 'What is that which is?) simultaneously ask: Which being is the highest (or supreme) being, and in what sense is it the highest being? This is the question of God and of the divine. We call the domain of this question

a kind that the ultimate in its own way accounts for the primal, and the primal in its own way accounts for the ultimate."

94 HEIDEGGER, M., Kant's Thesis about Being, KLEIN AND POHL., (trans.), in *Southwestern Journal of Philosophy*, 4(3), 1973, p. 10.

95 HEIDEGGER, M., "The Way Back into the Ground of Metaphysics" in *Existentialism from Dostoevsky to Sartre*, p. 275; *Wegmarken*, p. 373.

> theology. This duality in the question of the Being of beings can be united under the title onto-theology.[96]

Both ontology and theology are 'logies' (*Logien*) since they provide the 'reason' in 'statements', about entities. In the introduction to *Being and Time* Heidegger comes down hard on the ontological tradition for thematizing being into "substances" in the case of ancient and medieval philosophy, and into "objects" in the case of modern philosophy. The fault of the philosophers within these traditions is that they have missed the ontological import of this fact. Tradition declared Being indefinable and therefore raising the question of the meaning of Being is superfluous. Even when this question has been seemingly raised, it has been a matter of 'what is Being?' or 'what are Being as beings?'. It was in the raising of the question of Being in the mode of 'what' that tradition erred in the quest for Being. This is because; the question of Being is not an ordinary question. It is not as if we are asking what is man, what is a stick, or what is God. For, according to Heidegger:

> "Being"—that is not God and not a cosmic ground. Being is farther than all beings and is yet nearer to man than every being, be it a rock, a beast, a work of art, a machine, be it an angel or God. Being is the nearest. Yet the near remains farthest from man. Man at first clings always and only to beings.[97]

Any attempt to answer any of these questions above would normally, positively or negatively imply some differentiating qualities. This is because, by implication, we are asking for what is one among many, for something determinable, objectively identifiable, an entity among other entities. This was the fault of tradition. It laid emphasis on the "what" in formulating the Being question and for its answer, got particulars. This is so because 'in what is asked about, there lies also that which is to be found by the asking'.

To get to the real question of Being in its profundity, Heidegger opted, not to the modern philosophy as elaborated from Descartes, but for a reversal to the Greek origin. As we have already seen from the beginning, everything about Heidegger's thought is Greek. The virtues and defects of his works are those of the Greek philosophy. The word 'philosophy' is of Greek origin. To raise the fundamental question of philosophy is to belong to the origin of philosophy, to be claimed by the Greek thing. This return to the Greek source of philosophy was enhanced for Heidegger by the spiritual power of the Greek language. The return

---

96 Heidegger, M., *Kant's Thesis about Being*, pp. 10–11. Cf. also *Identity and Difference*, p. 59, "Thus they are more precisely called onto-logic and theo-logic. More rigorously and clearly thought out, metaphysics is: onto-theo-logic."

97 "Letter on Humanism" in *BW*, p. 234.

is not a return to some primitive condition of philosophy. It is a return to the greatest of the beginning. Heidegger discarded the notion that the beginning of philosophy was a primitive one.

> But whatever is great can only begin great. In fact, its inception is always what is greatest. Only the small begins small [...]. The great begins great, sustains itself only through the free recurrence of greatness, and if it is great, also comes to an end in greatness. So it is with the philosophy of the Greeks.[98]

This is a return to philosophy in its original purity undamaged by the subsequent history of metaphysics. Traditional ontology has preceded by way of categories which condition and objectifies their contents. In so doing, it has always been a theoretical investigation into entities. Heidegger calls this mode of investigation "ontical" which he distinguishes from "ontological" inquiry, which aims at the meaning of Being itself. This is the error that modern philosophy found itself and it was unable to emancipate itself from the shackle of this tradition. In fact, modernity with its exaltation of Cartesian subjectivity blocked any remaining possibility of inquiry into Being.

3.3.1. Heidegger's Critique of Descartes' Project

In *Being and Time*, Heidegger momentarily undertakes a destructuring of the philosophy of Descartes. In the treatise, *Modern Science, Metaphysics, and Mathematics*, Heidegger identifies modern philosophy as beginning with Descartes. Prior to Descartes, above all in the middle ages, "philosophy stood—if it stood independently at all—under the exclusive domination of theology, and gradually degenerated into a mere analysis of concepts and elucidation of traditional opinions and propositions."[99]

By challenging the Cartesian legacy in epistemology, Heidegger is attempting to demonstrate that there is no subject detached from the outside world of things, since Dasein is fundamentally Being-in-the-world. Nonetheless, Being-in-the-world does not mean that it is like beer in a bottle. Being-in as the most indispensable and existential characteristics of Dasein, signifies the expression of such terms as "dwelling," "being familiar with," and "being present to" as we saw earlier. By discarding the distinction between the subject and the object he takes Dasein as the starting point because only Man is the Being who is conscious of himself, of the world around him, and of Being. Dasein is the only Being which

---

98 *EM*, p. 12; *Introduction to Metaphysics*, pp. 16–17.

99 Heidegger, M., "Modern Science, Metaphysics, and Mathematics," in *Basic Writings*, p. 297.

can gain access to this problem. Dasein is always aware of itself as being in a world. Dasein, which raises the question of Being, must be disclosed in its Being because it is different from other beings. Therefore, unlike the idea that man can be understood in terms of the concepts "res extensa", i.e., in terms of its physical and mechanical facets in the Cartesian replica, Heidegger tries to swing the idea of subject from the Cartesian mechanistic interpretation of man to its more primordial sense of Being.

It is therefore not surprising that one of the most important aspects of Heidegger's philosophical thoughts is his condemnation of Descartes' philosophy of subjectivity. When Descartes regarded the *cogito sum* as the underlying proof of his own existence as well as the foundation of everything, he, Heidegger observes, diminishes every entity to ideas or symbols whose legitimacy is verified and substantiated by the regulations forced on them by the subject self-worth. This turn to subjectivity stands as a distinctive mark of the modern era; where prior to Descartes, man identified himself as "*zoon logon echon*, an *animal rationale*, a child of God, etc., into a subject.[100] According to Kockelmans' interpretation of Heidegger, the shift from animal rationale to a subject signifies a shift in the essence of man himself. "As Heidegger sees it, the decisive factor in the constitution of the modern era is not so much the fact that man has freed himself from his previous obligations and, thus, freed himself onto himself, but rather the fact that the essence of man himself has changed."[101] Thus, Descartes initiated a paradigm shift in the tradition with the *Cogito*, namely, moving man out of the world and making him the ground for the world.

Heidegger analysts acknowledge the significance of the question of subjectivity in his philosophy. However the difficulty for Heidegger is not merely the anthropocentric posture as such. The major complexity is that Descartes' idea, which envisions the ontological centrality of humanity, directs the entire modern philosophy on the path into subjectivism. In fact it has been argued that "Heidegger is concerned to show that Descartes' cogito ergo sum, the founding principle of modern metaphysics, seals human being's interpretation of itself as a subjectum, as the ground of being and truth, and, in the same stroke, of the world as a picture."[102]

---

100 Kockelmans, J. J., *Heidegger and Science*, Washington DC: University Press of America, 1985, p. 178. See also, Heidegger, M., *Introduction to Metaphysics*, p. 151.

101 *Ibid.*

102 Hans, L., "Collective Self-Legislation as an *Actus Impurus*: a response to Heidegger's critique of European nihilism," in *Continental Philosophy Review*, vol. 41, No. 3, Sept. 2008, p. 325.

Furthermore, the Descartes that Heidegger encounters enjoys the status of being phenomenologically motivated if not that of being a phenomenologist.[103] The Cartesian self is by no means able to surmount the ontological dichotomy which Heidegger identifies at the hub of emerging process of modernity. As a substitute for the *Cogito Sum* he posits Dasein's Being-in-the-world.

In reading Descartes as a metaphysician of modernity, Heidegger makes further case on the question of duality created by Descartes and modern philosophy as it emerges in subject versus object (the world). Heidegger, in its place, unifies the duality of modern philosophy by arguing that subject and object (world) belong together in the single entity, namely, Dasein. Subject and object are not two beings, because they are the central determination of Dasein in the unity of the uniqueness of Being-in-the-world. Heidegger turns around Descartes' *Cogito Sum*, and he holds that "Sum" must be stressed first. Thus he formulates "I-am-in-the-world" as an understanding of Being: In this sense, Dasein is not a *cogito*. Dasein and its world can never be separated. Consequently, "I-am-in-the-world" go before the "*cogito sum*." The truth of cogito is substituted in the disclosedness of Being which is essentially primordial truth. Unlike Descartes and others, he breaks the sequence of the tradition in terms of an understanding of world. His understanding of Being is Being-in-the-world, but the world of the Being of Dasein is not the physical world. It is the world of Dasein. As Heidegger assert early in SZ:

> In taking over Descartes' ontological position Kant made an essential omission: he failed to provide an ontology of Dasein. This omission was a decisive one in the spirit (*im Sinne*) of Descartes' ownmost tendencies. With the 'cogito sum' Descartes had claimed that he was putting philosophy on a new and firm footing. But what he left undetermined when he began in this 'radical' way, was the kind of Being which belongs to the res cogitans, or—more precisely—the meaning of the Being of the 'sum.' By working out the unexpressed ontological foundation of the 'cogito sum', we shall complete our sojourn at the second station along the path of our destructive retrospect of the history of ontology. Our Interpretation will not only prove that Descartes had to neglect the question of Being

103 MARION, J-L., "Heidegger and Descartes" in *Critical Heidegger*, ed. CHRISTOPHER MACANN, London & New York: Routledge, 1996, p. 70. In this essay, Jean-Luc explains that while Heidegger was expounding and critiquing Descartes at Marburg, Husserl was expounding and approving Descartes at Freiburg in a course of lectures given in the winter term 1923/24. Hence, at first, Descartes appears for Heidegger in a positively phenomenological light through the arbitration of Husserl. In fact the sameness between Descartes and (Husserlian) phenomenology can be deployed in two directions; either Descartes is a phenomenologist because he anticipates Husserl; or Husserlian phenomenology is not completely phenomenological, because it remains ensnared in Cartesian positions which have not been criticized, or even recognized.

altogether; it will also show why he came to suppose that the absolute 'Being-certain' ("*Gewisssein*") of the cogito exempted him from raising the question of the meaning of Being which this entity possesses.[104]

Taking this neglect into consideration, Heidegger reasons that Descartes takes and interchanges the philosophical interpretation of modern ontology into modern physics with its mathematical abstractions. This mathematization conceals the real meaning of Being.

3.3.2. On Mathematics

The ambition of Descartes' method is to arrive at certainty. The catchphrase "Cogito ergo sum" creates the indubitable point of departure of the entire mathematics that Descartes wants to construct. Heidegger says that "The radicalism of Cartesian doubt and the rigor of the new founding of philosophy and of knowledge in general is an illusion, and thus the source of fateful delusions that are hard to root out even today."[105]

It is no coincidence that modern science emerged at the same time as the metaphysical panorama of subjectivity. Heidegger argues that, once our reason sets itself up as the judge of what-is as Descartes suggests above, "the Being of beings must (inevitably) become thinkable in the pure thinking of mathematics. Being as calculable in this way, Being as set into calculation, makes beings into something that can be ruled in modern, mathematically structured technology, which is essentially something different from every previously known use of tools."[106] Heidegger maintains that technology is not just the set of novel techniques for achieving particular purposes but rather a way of revealing what-is. As such, it is the ground and drive force for modern, numerically-oriented science, not a side-effect of it.

Mathematics itself, or even the value accorded to it, is not a new development. Plato gave mathematics a prominent place in the curriculum of the Academy for the education of the free mind. In Heidegger's understanding, mathematics is not just the science of numbers but has played a crucial role as the realm of

---

104 SZ, p. 24; *BT.*, p. 46.

105 HEIDEGGER, M., *Being and Truth*, translated by G. FRIED & R. POLT, Bloomington, Indiana: Indiana University Press, 2010, p. 32. Published in German as Heidegger, M., *Sein und Wahrheit*, Frankfurt am Main: Vittorio Klostermann, 2001.

106 HEIDEGGER, M., *Introduction to Metaphysics*, p. 207.

what we can know "in advance"[107] from the beginning of metaphysics. But what is new in the modern era is that the "in advance" comes to refer to clear-cut conscious knowledge of objects in advance of actual experience. Heidegger states: "Only where thinking thinks itself is it absolutely mathematical, that is, a taking cognizance of that which we already have."[108] Self-conscious reflection about the principles that govern our universe, in particular the things present-at-hand that appear to such conscious reflection, is the hallmark of science and the key to prediction. As such, Heidegger concludes that the process of doubt is an illusion:

> The foundation of philosophy should be something purely simple, and purely perspicuous. This demand is justified only if one presupposes that the knowing and questioning of philosophy is subordinate to the "mathematical" method. But this is an arbitrary presupposition that Descartes does not ground in any way. Descartes does not even make an attempt to ground it. He lacks every motive for such an attempt. *The rigor of his process of doubt is a mere illusion*, and not, as it were, just because he comes to a secure standpoint afterwards, but because *behind* the process of doubt there stands the completely ungrounded opinion that the method of philosophical questioning and grounding is the 'mathematical' method. This presupposes a prior decision that the basis on which all knowledge of philosophy is to be grounded can only be what has the character of the indubitably *present at hand.*[109]

Heidegger remarks that modern natural science turns nature into an object "by way of mathematical projection." When ancient and medieval science used numbers, measurements, observation, and instruments, it did so in a very different way than modern science does.

> For in all this, that which is decisive about the experiment is completely missing. Experiment begins with the laying down of the law as a basis. To set up an experiment means to represent or conceive (vorstellen) the conditions under which a specific series of motions can be made susceptible of being followed in its necessary progression, i.e., of being controlled in advance by calculation.[110]

For man to be able to predict what will happen is for him to be able to control what will happen and take it over in his self-legislated obligation. Roger Bacon and other medieval scientists observed nature but they did not try to bind it to their own leading-strings, as Descartes and Kant would have had them do.

---

107 HEIDEGGER, M., *What is a Thing*, trans. W. B. BARTON, JR. & VERA DEUTSCH, Indiana: Gateway edition, 1969, p. 73.

108 *Ibid.*, p. 104.

109 HEIDEGGER, M., *Being and Truth*, p. 33.

110 *QCT*, p. 121.

The metaphysical undertaking for Descartes remained the emancipation of man designed for man's own autonomy as self-confidence and sovereignty with the aim of constructing a metaphysical base. As Heidegger explains: "the essence of the modern age can be seen in the fact that man frees himself from the bonds of the Middle Ages in freeing himself to himself."[111] Based on this understanding, Heidegger concludes that; "Descartes does not bring philosophy back to itself and to its ground and basis, but drives it still farther away from the asking of its fundamental question."[112] This was the problem which all modern philosophies tried to tackle in their various epistemologies. But to answer this fundamental question, Heidegger is of the opinion that we have to go back to the origin and raise the Being question in its ontological difference, and this require the re-interpretation of metaphysics.

3.4. Towards the Unmaking of Metaphysics

3.4.1. Metaphysics

In the beginning of its history "Being opens itself out as emerging (*physis*) and unconcealment (*aletheia*).[113] But with Plato this changed. It was with Plato that philosophy first entered on the way of science and became a subject in the school curriculum. Plato did not, however, discard everything about the original Greek ontology. He retained an insight into Beingness as presence. But Plato responded to this claim of Being by interpreting it as Idea, as *eidos*. "The Being of being rests in beingness but this—the *ousia*—Plato calls Idea."[114]

The Platonic understanding of beingness as Idea has clearly lost one aspect of the ontological difference. The individual being is never truly being. The actual being is incomprehensible in its beingness when thought in terms of Idea. The individual being has been consigned to unreality.

But while Plato had lost the ontological difference, there is already in him a preparation for the metaphysical difference. The ontological duality has turned about into 'lostness' and oblivion. The philosophy of Plato contains a rudiment of the distinction between essence and existence, between the standing-out of concealment and that which stands out of concealment. According to Heidegger,

---

111 Heidegger, M., "Age of the World Picture," in *QCT*, p. 127.

112 Heidegger, M., *Being and Truth*, p. 32.

113 Heidegger, M., *The End of Philosophy*, p. 4.

114 Heidegger, M., *What is Philosophy*, p. 55. "What Plato thought as the true, and for him sole, beingness (*ousia*) of beings, presence in the manner of idea (*eidos*), now moves to the secondary rank within Being." Cf. *The End of Philosophy*, p. 8.

"here is the concealed origin of the later distinction between *existential* and *essentia*."[115] But this metaphysical distinction between essence and existence was only a preparation in the philosophy of Plato. It remained for Aristotle to bring the metaphysical distinction to explicit statement and concept.

The object of first philosophy in Aristotle was existence as existence. "…he whose subject is the existing things qua existing things, this is the philosopher."[116] This statement of Aristotle indicates for Heidegger that Greek philosophy had definitively changed over from ontology to metaphysics. The first philosophy of Aristotle was no longer concerned with Being but rather was preoccupied with the existing thing. In this transition to metaphysics, the question of Being remained unthought because the ontological difference had been forgotten and recast.

In pre-Socratic philosophy, Being revealed itself in the ontological difference; in the philosophy of Aristotle however, Being has turned about into oblivion of this duality. This duality is replaced by a distinction which effectively conceals the ontological difference. The primary meaning of ousia in Aristotle is 'the man there' or 'the horse there'. "Presence in the eminent and primal sense is the persisting of something which lingers of itself, lies present, the persisting of the individual in each case […] The This, The singular."[117] In the primary sense, *ousia* signifies that-it-is. In the secondary sense, 'there was' signified what a thing is. In this way, the metaphysical distinction of what is and that it is has replaced the ontological difference and covered it all.

In the turning-about of Being into the metaphysics of Plato and Aristotle, Aristotle was more true to the original intuition of pre-Socratic thought. The *eidos* of Plato however could "never admit the individual being as what is truly in being." The ontological difference has then been eliminated in Plato. Despite the Aristotelian pre-occupation with the existing thing, his distinction between the primary and the secondary modes of *ousia* was more in keeping with the ontological difference of pre-Socratic philosophy. But this remained an echo. The distinction which flared for a moment in Aristotle's thinking became firmly waged into the existing thing as a distinction between what-a-thing-is and that-a-thing-is. In both Plato and Aristotle, the ontological duality is hidden. Aristotle saw that which "is," but the hidden itself—Being—remained concealed from him.

Ousia, that which is present, was *eidos* in Plato and *Energeia* in Aristotle. In the transition to the Latin world, *eidos* gradually became *concept* and *Energeia*

---

115 HEIDEGGER, M., *Introduction to Metaphysics*, p. 193. "The history of Being as metaphysics begins with this distinction and its preparation." Cf. *The End of Philosophy*, p. 2.

116 ARISTOTLE, *Metaphysics*, Bk. 4, Ch. 3, 1005B, 10.

117 HEIDEGGER, M., *The End of Philosophy*, p. 7.

became *actualitas.* As Heidegger states: "Ever since the transformation of *energeia* to *actualitas* (reality), the real is truly what is in being and thus decisive for everything possible and necessary."[118]

Given the wide spread influence of the politically empirical element of Rome, the Christian element of the Roman church, the transition of this Greek metaphysics into Latin enframed the oblivion of Being attained in Plato and Aristotle ever more securely and universally. This entrapping of metaphysics in Christian faith was "the baneful destiny of Being."[119] The Latin *actualitas* did not in itself add any further degree of hiddenness and concealment to Being. Such hiddenness was already contained in *Energeia*, but it certainly spread this hiddenness across the entire geographical and temporal sphere of inference belonging to the Latin language and Roman sway.

The real enframing, the real difficulty was the association between Greek metaphysics and Christian faith. This was the association that deepened the oblivion of Being and worsened its already wounded condition. By coming to language as *actualitas*, Being has become what-is and what-is is the real. By coming to language as *actualitas,* this real itself is no longer that which is present. It is now seen in terms of causal making. In such a metaphysics, Christian faith was able to "secure metaphysical justification for itself"[120] and its understanding of creation. Christian metaphysics in adopting the Latin *actualitas* reached the point where Being was self evident and taken for granted. That which is "the mystery…the one mystery"[121] is no longer able to withstand or resist human knowing. That which is primordially hidden is overcome by the academic business of knowing. With *energeia* understood in terms of the metaphysical distinction of essence and existence with its attendant demand for causes, metaphysics "developed into pile of distortion, no longer recognizing itself, covering up the primordial essence of Being. Herein lies the reason for the necessity of the "destruction" of this distortion, when a thinking of the truth of being has become necessary."[122]

---

118 *Ibid.*, p. 13.

119 Heidegger, M. "Anaximander's Fragment" in *Early Greek Thinking*, p. 40.

120 Heidegger, M., *The End of Philosophy*, p. 14.

121 Heidegger, M., On the Essence of Truth, in *Basic Writings*, p. 130. Cf. also *Existence and Being*, p. 341.

122 Heidegger, M., *The End of Philosophy*, p. 14–15. "Being as reality attains an assumed self-evidence which has remained decisive ever since…Being as reality, now immediately comprehensible to everyone, get firmly fixed so that soon…the primordial Greek essential character of Being is once and for all misunderstood and made inaccessible"

3.4.2. Overcoming of Metaphysics

Since metaphysics has constituted the forgetfulness of Being, Heidegger was, therefore, determined to 'destroy' the entire history of metaphysics. He did not want to fall into the same trap as Nietzsche who, after criticizing the metaphysical tradition, not only was unable to emancipate himself from it, but also solidified it as the utmost possibility. Early in *SZ*, Heidegger made his intention clear: "...we are to destroy the traditional content of ancient ontology until we arrive at those primordial experiences in which we achieved our first ways of determining the nature of Being."[123]

We have seen that Heidegger's criticism goes to the core of Cartesianism and of modernity, puncturing their ontological foundation, concept of truth, and central character. But we should not regard this criticism as aimed at destroying the tradition. Rather its intent is to broaden its base. Thus, '*Destruktion*'[124], the name the early Heidegger gave to his project of working through the history of philosophy, may in retrospect be a misleading choice of term.

In SZ Heidegger expanded the term in §6 under the title "*The Task of Destroying of History of Ontology*". Here, Heidegger speaks in great details of a need to 'destructure' *the tradition.* The *Destruktion* can be seen as one side of a recollection process, '...as the rupture from the forgetfulness of Being, as the manifold probing that thought must perform in gaining access to each epoch and its configuration—over and beyond their contemporaneous subsistence in our memory'.[125]

The term *Destruktion* which already appears early in Heidegger's thinking, undergoes significant metamorphoses throughout his writings. As Kisiel points out, the term appears in a course on a full-fledged 'destruction' of Natorp's concept of constitution, presented in the summer of 1920.[126] It also appears in a letter to Karl Löwith during 1920, even though the letter was only published in 1946.[127] Here the word is used in the context of the need for a Destruktion of 'culture' or ruin.[128]

---

123 *SZ*, p. 22; *BT.*, p. 44.

124 *SZ*, pp. 23–27; *BT.*, pp. 44–49.

125 Janicaud, D., 'Overcoming Metaphysics?' in Dominique Janicaud and Jean-Francois Mattéi, *Heidegger: From Metaphysics to Thought*, trans. Micheal Gendre, New York: Sunny, 1995, p. 4.

126 Kisiel, T., *Heidegger's Way of Thought*, London: Continuum Books, 2002, p. 30.

127 Moran, D., "The Destruction of the Destruction: Heidegger's Versions of the History of Philosophy" in *Martin Heidegger: Politics, Art and Technology*, eds. Karsten Harries and Christopher Jamme, New York: Homes and Meier, 1994, pp. 175–196.

128 Barash, J. A., *Martin Heidegger and the Problem of Historical Meaning*, New York: Fordham University Press, 2003, p. 98.

The term also appears in Heidegger's review of Jasper's *Psychology of World Views*, where Heidegger refers to a '*geistgeschichtliche Destruktion der Überlieferten*', underscored in an affirmative sense.[129] Furthermore, according to John van Buren, Heidegger first used the term in GA 58 in connection with Luther's attack on Aristotle and scholasticism.[130]

As Gadamer has pointed out, the German word does not have the negative connotation of its English or French cognates.[131] In Gadamer's finding also, Heidegger uses the term to refer above all to an activity that characterized all authentic thinking; the rejuvenation of conceptual language.[132] Heidegger's ambition is not to destroy the tradition but to widen it and to reinterpret its main principles in light of what, in his view, has been neglected as the tradition was shaped.

The destruction of the history of metaphysics certainly involves the technical task of 'overcoming' metaphysics in its very manifestation. But it also involves the wider undertaking of overcoming those things which has followed from metaphysics. To overcome these consequences which derive from metaphysics is the very first task which must be undertaken. Thus, to 'overcome' (*überwindung*) means to go beyond in degree or quality, to dominate, to pass over a limit or an obstacle etc.[133] By dismantling or destructuring the reified "technical" vocabulary

---

129 MORAN, D., "The Destruction of the Destruction," p. 183.

130 VAN BUREN J, "Martin Heidegger, Martin Luther" in KISIEL, T., & VAN BUREN, J., (eds),. *Reading Heidegger from the Start: Essays in is Earliest Thought*, New York: Sunny, 1994, (pp. 159–174), p. 167; See HEIDEGGER, M., Gesamtausgabe, Vol. 58: *Grundproblem der Phänomenologie*, Frankfurt a, M: Vittorio Klostermann, 1993, p. 139.

131 HANS-GEORG GADAMER, "Hermeneutics and Logocentism," in DIANE P. MICHELFELDER & RICHARD E. PALMER, eds. *Dialogue and Deconstruction: The Gadamer-Derrida Encounter*, Albany: State University of New York Press, 1989, p. 121.

132 HANS-GEORG GADAMER, "Letter to Dallmayr," in *Dialogue and Deconstruction*, p. 99.

133 In the words of Schlawin, „Etwas überwinden heißt: darüber hinausgehen, es unter sich bringen, es hinter sich bringen, es in ein Vergangenes verwandeln, das nun gegen das Neue zurückbleibt…" SCHLAWIN, H., „Heidegger Überwindung der Metaphysik" in *Zeitschrift für philosophische Forschung*, 8, 1954, p. 586. Also „Although Heidegger uses the familiar word *Überwindung* for "overcoming," he means it in the sense of the less familiar word *Verwindung*. When something is overcome in the sense of being *überwunden*, it is defeated and left behind. This is not the sense Heidegger intends here. When something is overcome in the sense of being *verwunden*, it is so to speak, incorporated. For example, when one "overcomes" a state of pain, one does not get rid of the pain. One has ceased to be preoccupied with it and has learned to live with it. Thus, to overcome metaphysics would mean to incorporate metaphysics, perhaps with the hope, but not with the certainty, of elevating it to a new reality." Cf. HEIDEGGER, M., "Overcoming Metaphysics" in *The End of Philosophy*, footnote 1, p. 84.

of philosophy as it has become "sedimented" in the tradition, it becomes possible for the original insight of living language to reappear. This reflects the hermeneutical turn in Heidegger's application of phenomenology to the history of philosophy.[134]

Heidegger's intention then was to salvage those pieces of the tradition which most modern philosophers overlooked or forgot. He seeks to expand the foundation of the tradition by salvaging its forgotten Greek origins. For him, Greek philosophy is much more than a basis to ascertain where the tradition deviated from what he understood was the true path. As he sees it: "Ancient ontology…is fundamentally not unimportant and can never be overcome, because it represents the first necessary step that any philosophy at all has to take, so that this step must always be repeated by every actual philosophy."[135] In this way, Heidegger's *Destruktion* does not abandon the ontological tradition. But as Moran explains;

> Heidegger's willing of the destructive question is not a phenomenological paying attention, a way of gaining sharper focus in the description of the phenomenon—rather it acts as a will to power over the phenomenon, wresting, rooting up the phenomenon from its historic- temporal bed, a breaking apart of the phenomenon's inner coherence in order to find what Heidegger regards as the essence of the phenomenon.[136]

In *The Basic Problem of Phenomenology*, Heidegger describes destruction as "a critical proccss in which the traditional concepts, which at first must necessarily be employed, are deconstructed down to the sources from which they were drawn. Only by means of this destruction can ontology fully assure itself in a phenomenological way of the genuine character of its concepts."[137]

According to Heidegger, western civilization in all its relations with beings is in every aspect sustained by metaphysics. Every era, every human epoch, however diverse they are—Greece after Plato, Rome, the middle ages, modernity, the age of technology—is established in some metaphysics and is positioned thereby in a specific relationship to what-is as a whole. But metaphysics is nihilism proper and the metaphysics of Plato is no less nihilistic than that of Nietzsche. Insofar as metaphysics thinks the being of beings, it reduces Being to a being. It does not think Being as Being. His attempt to overcome metaphysics consists in thinking back Being to the primordial beginning of the west—the early Greek experience

---

134 Walter, Lammi, "Hans Georg Gadamer's Platonic Destruktion of the Later Heidegger," in *Philosophy Today*, Vol. 4, No. 3/4, 1997, p. 394.

135 Heidegger, M., *The Basic Problems of Phenomenology*, p. 111.

136 Moran, D., "The Destruction of the Destruction," p. 190.

137 Heidegger, M., *The Basic Problems of Phenomenology*, p. 23.

of what is present in presencing—and in repeating this beginning, so that the western world can begin anew.

Hence, Heidegger urges that we go into the 'ground' of metaphysics that has hitherto remained unthought. The 'step back' (*Schritt zurück*) into the ground of metaphysics is a step out of metaphysics or an overcoming of metaphysics.[138] As Heidegger writes: ...Being, since the beginning of Western thought, has been interpreted as the ground in which every being as such is grounded."[139] In his 1967 lecture in the *Academy of Arts and Sciences* in Athens Heidegger writes:

> Step back means withdrawal of thought from the world civilization, at a distance from her, not in their denial, to engage in what would in the beginning of Western thought still remain unthought, but are nevertheless already known, and so our thinking is dictated.[140]

Given that the metaphysical tradition of the West has been caught up in a relentless misunderstanding of entities with Being, it has maintained itself in the forgetfulness of Being. Therefore the overcoming of metaphysics is the overcoming of the forgetfulness or withdrawal of Being in Western history. The thinking that gives precedence to beings and turns away from Being is what the metaphysics of Western tradition basically is.

Modern metaphysics is a transcendental philosophy which has become epistemology.[141] Deep at the heart of that epistemology however, there is anthropology. "Philosophy in the age of completed metaphysics is anthropology...philosophy has becomes anthropology and in this way a prey to the derivatives of metaphysics."[142] In modern metaphysics which has become an anthropology, the investigation of man has been handed over to physics in the widest sense—the physics of life and the physics of man. The philosophy of man has become a victim of biology and

138 ID, p. 118; *Identity and Difference*, p. 52 "Mit diesem Hinweis sei andere naheliegende Mißdeutung des Titels „Schritt zurück" ferngehalten, die Meinung nämlich, der Schritt zurück bestehe in einem historichen Rückgang zu den frühesten Denkern der abendländischen Philosophie. Das Wohin freilich, dahin der Schritt zurück uns lenkt, entfaltet und sich erst durch den Vollzug des Schrittes."

139 ID, p. 96; *Identity and Difference*, p. 32. Here we need to recall that "Being" means always and everywhere the Being of beings. p. 61.

140 Heidegger, M., *Denkerfahrungen*, Frankfurt a. Main: Vittorio Klostermann, 1983, p. 139 (translation is mine) „Schritt zurück heisßt: Zurücktreten des Denkens vor der Weltzivilisation, im Abstand von ihr, keineswegs in ihrer Verleugnung, sich auf das einlassen, was im Anfang des abendländischen Denkens noch ungedacht bleiben müßte, aber dort gleichwohl schon genannt und so unserem Denken vorgesagt ist."

141 Heidegger, M., "Overcoming Metaphysics" in *The End of Philosophy*, p. 88.

142 "Overcoming Metaphysics" in *The End of Philosophy*, p. 99.

psychology. Heidegger even mentioned briefly the theological impart of this anthropology. The anthropology of man has changed the religious relation to God "into mere religious experience". "When this occurs, then the gods have fled. The resultant void is compensated for by means of historiographical and psychological investigation of myth."[143]

The completed metaphysics is expressed in technology. "The name "technology" is understood here in such an essential was that its meaning coincides with the term 'completed metaphysics.'"[144] According to Bruce Foltz, "As the completion of metaphysics, and therefore as something that is essentially metaphysical, technology specifies and establishes the being of whatever may be counted as an entity; in instituting this ground, technology determines the ontological basis for an era—in this case, our own era."[145] In this completed metaphysics, technology makes everything a standing reserve for future use. Ordered into standing reserve, "beings loses the previous uniqueness of their authoritative claim."[146] In this loss, there requires the collapse of the world and the desolation of the earth. This loss is further aggravated by the consumption of Being. Being is enframed and abandoned. "The 'world wars' and their character of 'totality' are already a consequence of the abandonment of Being."[147] With the abandonment of Being, technology presses forward to 'a constant form of using things up'. In this context, war and peace are both modes of this consumption. Even the peace subsequent on a war is only a condition in which war-like consumption of beings is no longer experienced as hostile. "War is no longer anything which could terminate in peace. War has become a distortion of the consumption of beings which is continued in peace."[148]

This very factor which derives from completed metaphysics required to be overcome and unmade. This can be done only when man himself is unmade and then comes 'to the part of thinking, poetizing building'. With this in mind, we will enter into the next chapter with the view of examining Heidegger's thesis on technology, his critiques and propositions.

---

143 Heidegger, M., "The Age of the World Picture" in *QCT.*, p. 117.

144 "Overcoming Metaphysics" in *The End of Philosophy*, p. 93.

145 Foltz, V. B., *Inhabiting the Earth: Heidegger, Environmental Ethics, and the Metaphysics of Nature*, p. 6.

146 Heidegger, M., Overcoming Metaphysics" in *The End of Philosophy*, p. 86.

147 *Ibid.*, p. 103.

148 *Ibid.*, p. 104.

Concluding Remarks

From what he have seen in this chapter, Descartes sees in modern subjectivist metaphysics the predominance of man among all beings and hence the predominance of a being over Being. This is for Heidegger one of the most extreme forms of the Western forgetfulness of Being.

In Descartes' interpretation, tradition and faith should prostrate before a more compelling benchmark of knowledge: "certainty." Fortified with a method of attaining certainty, the subject will be capable of deciding for himself what amounts to knowledge and therefore what sincerely exists. Having banished God from the centre of the universe, before long, the human subject arrogated God's creative role in all aspects of life. Man became his own God. According to Heidegger, "Nietzsche is the final and most radical spokesperson for this one-sided elevation of human subjectivity. For Nietzsche, man must shape every perspective in advance: he must will it. Nietzsche will not let Being be. He regards human creativity as divine and is completely blind to Being in its *self*-presencing. By bringing modern subjectivism to its completion he brings to a close Western metaphysics."[149] Thus, The Will to Power became the overriding principle only in modern times which was manifested in modern scientific-technological thinking. Such thinking has transformed not only the world, but mankind itself, into a governable object. "The overman is left to impose his subjective values on the beings he masters. Nietzsche thereby leaves contemporary man lost in the midst of beings abandoned by Being."[150] Scientism and its positivistic spirit were proclaimed as the only valid method to truth. Through the scientific discovery of the forces of nature and through his technological control over this power, man no longer feels so stuck by the mystery of nature' powers, which formerly forced him to appeal to the awesome might of the divinity. He now desacralizes nature and harnesses its power to build a kingdom for himself. And the consequence of this is what Heidegger pointed out long ago in his critique of technology. This will be our focus in the next chapter.

149 SMITH, G. B., *Nietzsche, Heidegger and the Transition to Postmodernity*, p. 229.
150 *Ibid.*

Chapter IV: Heidegger, Technology, and Human Destiny

Introduction

As the title of this chapter suggests, we shall be examining the question of technology in the light of Heidegger's interpretation and analysis. In our previous chapter, Descartes famously asserts "I think, therefore I am." Upon this allegedly indubitable proposition he attempted to construct a philosophical system which would serve as the foundation of all sciences, ushering in the spirit of modernity. Heidegger criticized modernity with its primacy of consciousness over Being. His interrogation of Being involves not only the investigation of the being of beings but also the questioning of the functionality of the knowledge of being, and man's use of that knowledge. In other words investigative ontology, appropriative epistemology and praxique utility are three sides of the triangular human venture vis-à-vis material existence.[1] The main focus of this chapter is mostly on the praxique: the operation that man makes in beings in view of self-satisfaction in being, growth and development. This is in other words called technology.

Heidegger claim that, in the West, technology does not merely designate a set of instruments or a mode of production: it designates a cast of mind. However, the Western technological mind-set is not something new. It can be traced back to the very beginning of Western philosophy, back to Plato. Heidegger's insinuation suggests that Platonic metaphysics only hints at a technological worldview which comes to full blossom two thousand years later in the thoughts of Descartes who rang the bell that gathered the wits together, the result of which is the modern technological Age.

We shall see in this section that, according to Heidegger, technology is "fundamentally" indistinguishable from modern metaphysics, providing for us a kind of metaphysical constitution of the world that is taken for granted as the essential core of any human possibility or opportunity of action or potentiality of world-making. In technology, humans deal with nature, they bend it at will and they do so according to a fundamental necessity of anticipatory confrontation, where instrumentation offers the conditions or requirements for human action, even survival. Thus, technology, with its ever-increasing growth and velocity, seems not

1 Iroegbu, P. O., *Enwisdomization and African Philosophy*, Owerri, Nig: International Universities Press, 1994, p. 49.

to remain a mere universe of instruments. Instead, it seems to impose itself as an essential attribute or feature of modern world-making at least if we are thinking about changing the world or restructuring its character.

It is therefore not surprising that while many championed and continue to advocate humanity's new role as arbiter of the cosmos, Heidegger bewails it. For him, Cartesian "certainty" and Nietzsche's "Will to Power" has transformed not only the world, but mankind itself, into a governable object. Today, the human mind and body can be operated and manipulated with computerized and military precision. Our genetic inheritance can be "reprogrammed," our social condition reengineered. As such, our environment is subject to profound alteration. According to Heidegger, as this chapter will show us, we regard our environment as merely "standing reserve" of energy which can be tapped at will. The resultant effect is the worldwide environmental degradation in the form of oil leakage, desertification, deforestation, global warming, and most frightening of all, the dread of nuclear extermination.

In lieu of discussing particular environmental concerns and possible solutions, this chapter would take a step back for a moment with an eye towards attaining a better understanding of Heidegger's philosophy of technology and his critique of modern technological mind-set. In doing so, we will first examine the situation in which we currently find ourselves by way of Martin Heidegger's 1953 lecture "*The Question Concerning Technology.*" Herein we shall find that technology, as Heidegger sees it, has become an essential way of our dealing and coping with the world. That said, we will uncover Heidegger's claim that a certain *destining* is at work at the core of modern technology and in so doing investigate back into the inauguration of the modern era in an attempt to detect initial signs of this destining. But before we examine this Heidegger's perspective, we shall first have a cursory look at what technology means in its modern parlance, with particular emphasis on its counter-productivity in modern society. Such negative effects calls for a closer look at technology from various schools of thought. It becomes therefore imperative to also delve into different trajectories on the question of technology, both existential and critical views, which will give us the picture of the intellectual environment in which Heidegger posits his theory of technology. While some existential and critical theories of technology are in accord with Heidegger, others are not. For instance, Don Ihde's earliest book on philosophy of technology; *Technics and Praxis: A Philosophy of Technology* (1979) was dedicated to Heidegger, but more recent works have increasingly turned critical, largely from dissatisfaction with Heidegger's romanticism, but even more with the inapplicability of Heideggerian analysis to the wider variety of different technologies. This is also our problem with Heidegger's outlook on modern technology.

4.1. Towards a Definition of Technology

Before elaborating further, the term "technology" needs to be clearly defined. To do this however, will depend on the perspective from which one is looking at the question. A renowned historian of technology Thomas Hughes pointed out that "defining technology in its complexity is as difficult as grasping the essence of politics. Few experienced politicians and political scientists attempts to define politics. Few experienced practitioners, historians, and social scientists try to exclusively define technology."[2] Most historians writing on technology have defined the term mostly by presenting and discussing pertinent examples. Here, we shall learn something about technology by considering few definitions that scholars in the field have given and their connection to different philosophical school of thought.

Generally speaking, technology is the application of scientific and theoretical knowledge to the concrete and practical goals of human life with the aim of altering, manipulating and controlling the environment for human benefit. Technology consists of the use of materials, implements, technical skills to make life better or more enjoyable and enhance work productivity. While science concentrates on why and how things come to pass; technology centers on making things to happen.

Bernard Stiegler in *Technics and Time* defines technology in two ways: as "the pursuit of life by means other than life", and as "organized inorganic matter."[3] John Kenneth Gilbraith, however, incorporated social organizations and value systems when he defined technology as "the systematic application of scientific or other knowledge to practical tasks.[4] Taking a gleam from this, Pacey in *The Culture of Technology*, characterizes technology as "the application of scientific or other knowledge to practical tasks by ordered systems that involve people and organizations, living things and machines."[5] Taken cognizance of diversified definitions of technology, Val Dusek categorized the definition into three schools, (a) technology

2 Hughes, P. T., *Human-built World: How to Think about Technology and Culture*, Chicago: University of Chicago Press, 2004, p. 2.

3 Stiegler, B., *Technics and Time, 1: The Fault of Epimetheus*, trans. Richard Beardsworth, & George Collins, Stanford, California: Stanford University Press, 1998, pp. 17, 82.

4 Gilbraith, J. K., *The New Industrial State*, New York: New American Library, 1967, chapter 2.

5 Pacey, A., *The Culture of Technology*, Cambridge, MA: MIT Press, 1983, p. 6.

as hardware; (b) technology as rules; and (c) technology as system.[6] From the foregoing definitions, one can say that technology is closely related to science and engineering to the extent that they interlock each other.

> Science became so identified with practical benefits that the dependence of technology on science is commonly assumed to be a timeless relationship and a single enterprise. Science and technology, research and development—these are assumed to be almost inseparable twins [...]. The belief in the coupling of science and technology is now petrified in the dictionary definition of technology as applied science, and journalistic report under the rubric of "science news" are, in fact, often accounts of engineering rather than scientific achievement.[7]

From this account, another way of describing the scope of the word is to recognize that its current significance is linked to the capacity of the hard sciences to create systematic knowledge about the physical world we live in; this knowledge, when put to practical use and applied to material nature, through various techniques and processes, leads to the transformation of material reality through the creation of new structures, materials and entities. As such, technology might be also described as the technique of transforming material reality through the application of scientific knowledge and method to the objects and entities of the world.

Whereas 19th Century philosophers such as Karl Marx were philosophically interested in tools and techniques, among the most prominent 20th century philosophers to directly address modern technology was Martin Heidegger as we shall see in this chapter.

4.2. Counter-Productivity of Modern Technology

By counter-productivity, we imply the contradictions, absurdities, paradoxes and destructive effects of modern technology on the human person, society and the universe at large. While man was in earlier times mainly threatened by nature, he now also has to face the threat of technology. Modern technology is at the moment capable of eradicating human life on earth through a number of separate paths. The technological development moreover experiences threats from within. Large-scale technological developments over and over again attest to be at risk as

---

6 Dusek, V., *Philosophy of Technology: An Introduction*, Malden, MA: Blackwell Publishing Ltd., 2006, pp. 31–33.

7 McClellan III, J. E., & Harold, D., *Science and Technology in World History: An Introduction*, 2nd Edition, Baltimore, Maryland: The John Hopkins University Press, 2006, p. 1.

a result of human blunder or due to bad technical functioning, and confront us with extensive consequences.

Perhaps in ten centuries to come when archaeologists will be excavating our cities and examining our bones for categorization, they will see our modern achievements in science and technology and what we did to reach such giant strides. Then, maybe, they will create a new categorization or nomenclature of hominid for this period in man's evolution and call us *Homo Destructibus.* Homo Destructibus will be described as having dominance during the Modern era and met his zenith during 1900–2000 AD and beyond. Distinct rather *via negativa,* this most modern of our ancestors, they would say, had a large cerebral aptitude for thought and reason, but evidence shows little ability for commonsense that has become uncommon or even consciousness outside self.

Take for instance the events of the two World Wars. World War II was a time of invention and a time of hostility. On August 6, 1945, the United States used a massive, ten-foot atomic bomb, baptized "Little Boy" by American soldiers, against Hiroshima, Japan. This bomb flattened the city, and while intransigent Japan was still trying to comprehend this devastation, United States thought that "Little Boy" was not man enough and struck again, this time, with a plutonium bomb, "Fat Man" on Nagasaki which exploded 1,650 feet above the city. Whether the atomic bombings of Hiroshima and Nagasaki, three days later, were warranted or not is still very much debated today, but one thing for sure, it led to the surrender of the once combative nation of Japan, formally ending World War II.

After the bomb was dropped, two-third of Hiroshima was destroyed. Within three miles of the explosion, 60,000 of the 90,000 buildings were demolished. Captain Robert Lewis, the co-pilot of The Enola Gay, a B-29 Superfortress aircraft that carried the bomb, announced: "Where we had seen a clear city two minutes before, we could no longer see the city. We could see smoke and fires creeping up the sides of the mountains."[8]

---

8 Robert, L., in Ronald T. Takaki, *Hiroshima: Why America Dropped the Atomic Bomb*, New York, Boston: Little, Brown and Company, 1995, p. 43. The bombing of Hiroshima was one of the pivotal events of the twenthieth century, yet this controversial question remains unresolved. At the time, General Dweight Eisenhower, General Douglas MacArthur, and the chief of staff Admiral William Leahy all agreed that an atomic attack on Japanese cities was unnecessary. All of them believed that Japan had already been beaten and that the war would soon end. Was the bomb dropped to end the war more quickly? Or did it herald the start of the Cold war? The historian, Takaki in this book, explores these factors and more. Relying on top secret military reports, diaries, and personal letters, Takaki relates international policies to the individuals involved. For

It was estimated that around 140,000 and 70,000 died from radiation poisoning and burns in Hiroshima and Nagasaki respectively in 1945 alone. Who knows how many thousands more have died from injuries and sickness attributable to radiation exposure released by the bombs in the succeeding decades. A survivor described the damage to people:

> The appearance of people was [...] well, they all had skin blacked by burns [...]. They had no hair because their hair was burned, and at a glance you couldn't tell whether you were looking at them from front or in back [...].They held their (forward) like this [...] and their skin- not only on their hands, but on their faces and bodies too- hung down [...]. If there had been only one or two such people [...] perhaps I would not have had such a strong impression. But wherever I walked I met these people [...]. Many of them died along the road—I can still picture them in my mind—like walking ghosts.[9]

Whether directly or indirectly, deliberate or accidental, through negligence or even with due diligence, humans are particularly dexterous at creating disasters, a truth that is quite evident throughout history. "Critics of technology see such

---

more on Hiroshima and Nagasaki atomic bomb, see, also, Poolos J., *The Atomic Bombing of Hiroshima and Nagasaki,* Chelsea House Pub., 2008; Estrin, B.L., *The American love lyric after Auschwitz and Hiroshima*, New York: Palgrave, 2001; Kort, Michael, *The Columbia guide to Hiroshima and the bomb,* New York: Columbia University Press, 2007. In this volume, Kort describes the wartime circumstances and thinking that form the context for the decision to use these weapons, surveys the major debates related to the decision, and provides a comprehensive collection of key primary source documents that illuminate the behaviour of United States and Japan during the closing days of World War II.

9 In Lifton, J. R., Death *in Life: Survivors of Hiroshima*, New York: Random House, 1967, p. 27. See also, Ogura, T., *Letters from the end of the world: a firsthand account of the bombing of Hiroshima*, New York: Kodansha International, 1997; Barker, R., *The Hiroshima Maidens: a Story of Courage, Compassionm and Survival,* New York: Viking, 1985. Another eyewitness, Yuka, a young Japanese speaking through the voice of Edita Morris states: "...partout autour de moi, il y a des gens qui courent, qui courent...Ils me poursuivent, avec leurs visages carbonisés, avec des lambeaux de chair arrachés de leurs épaules...Cette fille au visage rongé par les flammes, cet home qui porte sa femme morte sur son dos...Ici, c'est un groupe d'écoliers écroulés les uns sur les autres, tout morts. Lá, c'est un chien, les pattes prises dans l' asphalte fondue. C'est ce qui nous attend tous si nous ne courons pas assez vite...Loin devant moi, j'aperçois la ligne noire du fleuve et des ombres qui plangent dans ses eaux. Comme des torches vivantes, les cheveux en flammes, les femmes s'élancent du rivage en grappes serrées..." Cf. Edita Morris, *Les Fleurs D'Hiroshima*, trans. Suzanne Lipinska, Paris: René Julliard, 1961, pp. 13–14.

disasters as an inevitable result of measuring progress in terms of our ability to manipulate the environment through technology."[10]

Thus we will be relegated to chronological record. Not remembered for our intellectual statures, our imaginative, inventive or resourceful immensity or our impressive scientific and technological achievements as *Homo Technologicus*. Instead, we might be remembered as those over-ambitious humans who, due to intellectual over-maturity and spiritual backwardness; capitalistic mentality with active money-making faculty devoid of responsibility, moved from barbarism to degradation without the usual interval of civilization. One salient critics of technology of that war era was Romano Guardini.

While there are scattered criticisms of what we now call technology in ancient and medieval authors, sustained criticism can be traced only from the late 18th Century until the present. Writings like Goethe's *Faust* (1808 and 1832), Shelley's *Frankenstein* (1818), and Huxley's *Brave New World* (1932 and 1958) have been part of the popular awareness of the ways through which humanity has benefited the most from the rise of industrialization and the technological mind-set. These works of literature however calls attention to the gloomy side of technology, an aspect that must be controlled in order to avert the worst of disasters.

4.3. Trajectories in the Philosophy of Technology

Before examining the question of technology in Heidegger, we shall mention different trajectories in the philosophy of technology. This, we will do from two angles; the ***existential perspective,*** which is more anthropological, and the ***critical perspective***, which is more sociological. We shall do this in the form of literature review. This will enable us to see the existing philosophical environment in which Heidegger examined the question of technology, the early, later and current efforts to bring to modern man's consciousness, the counter-productivity of technology.

4.3.1. Existential Perspective

a) Romano Guardini (1885–1968)

Reviewing the wreckage of that encounter after the war, the Catholic theologian Father Romano Guardini saw instead the end and the collapse of the modern

10 Gruen, L., Technology, in C., Hanks, (ed.) *Technology and Values*, p. 423. See also Jamieson, D., (ed.), *A Companion to Environmental Philosophy*, Malden, MA: Blackwell Publishing, 2003, p. 339.

project. The aspirations that had stimulated the founders of modern thought—the subjugation of nature through science, the conquest of human nature through science and technology and the emancipation of power from moral restraint and achievement of personal desires—have been achieved beyond anyone's wildest imaginations, and they had turned to ruins before that success could be enjoyed. As he states: "From modern science, technology has grown, and technology is a concentration of processes allowing man to posit ends in conformity with his own desires."[11]

Romano Guardini, sometimes referred to as a "Philosopher of the Christian World," wrote a book, *The Letters from Lake Como* in order to confront the dominance of technology. The book comprises of a series of letters which he wrote while reflecting upon the impact of technology on the Como landscape in Italy, a countryside which symbolizes the old way of life before technological advancement. His letters were a reflection on what he observed as Southern Europe industrialized; thoughts on the changing nature of tools and the idea of a craft, and what the changes were doing to the landscape. In this letter, according to Pattison, "Guardini contrasts the kind of human invention manifested in the architecture and agriculture of the Lake Como region with the products of modern technology. The former are not, of course, 'natural'. They are 'man-made', yet they cohere with the landscape, the rhythm of the seasons and the encompassing world of nature."[12] His premise is that technology "has become a destiny that subjugates its human creators as much as their creation."[13] He offers a way forward by suggesting that "we must recover a sense of the sacred before the sacred name can be heard again."[14] He focused on the potential for salvation, if technology could only be focused on the morally correct objectives. He stressed that machines give us freedom, but freedom 'for what' and toward what 'meaning.'

Guardini begins with the question of how machines are influencing, shaping and dominating culture, nature, and humanity. In the decades prior to World

11 Guardini, R., *The End of the Modern World*, p. 42.

12 Pattison, G., *Thinking About God in an Age of Technology*, Oxford: Oxford University Press, 2005, pp. 41–42.

13 Guardini, R., *The Letters from Lake Como: Explorations in technology and Human Race*, trans. Geoffrey W. Bromiley, Grand Rapids, Mich.: Eerdmans Pub. Co., 1994, p. xiv; Originally published in German as Romano Guardini, *Die Technik und der Mensch: Briefe vom Comer See*, mit einem Vorwort von Walter Dirks, Mainz: Matthias-Grünewald-Verlag, (1927) 1981.

14 Ibid., p. xv.

War II, and in the years of that war, the individual motivated by technology broke into the field of history and took possession.

> The technological mind sees nature as an insensate order, as cold body of facts, as a mere given, as an object of utility, as raw material to be hammered into useful shape; it views the cosmos similarly as a mere space into which objects can be thrown with complete indifference."[15]

Therefore, Guardini made his laments about modern technology along six different lines:

Firstly, Technology gives rise to a kind of *artificiality of existence*: Our human planet, according to Guardini, "seems from the outset to have about it something alien to nature, something unreal and artificial."[16] Specifically, in connection with an understanding of the present era as an era of technology, Guardini says that "Our age is not just an external path we tread; it is ourselves."[17] That is to say, technology is not just something that mediates between our mental intentions and the physical world about us, technology gets inside our heads and affects the very way in which we conceive our reality.[18] Undoubtedly there is a problem of culture since we have never had a relationship with nature in its unscathed form. Guardini uses the illustration of a normal sailing ship against a modern ocean-liner. The ship, although it still has affinity with the technological, at any rate, respects the natural laws, given that it maneuvers in consonant with the currents and the waves. But the ocean liner is symptomatic of how our technology infringes on the natural order.

> We have now broken free from the living order of times: morning and evening, day and night, weekday and Sunday, changes of the moon and seasons. We live in an order of time that is our own making, fixed by clocks, work, and pastimes. The sphere in which we live is becoming more and more artificial, less and less human, more and more [...] barbarian. The profound sadness of this whole process lies over Italy."[19]

This painted scenario is the artificiality of existence that technology brings with it.

Secondly, the price for this mastery over nature is the *abstraction* of our planet where everything becomes concepts and formulas and as such removes us from

---

15 GUARDINI, R., *The Essential Guardini: An Anthology of the Writings of Romano Guardini*, selected and with an Introduction by Heinz R. Kuehn, Chicago: Liturgy Training Publications, 1997, p. 17: See ROMANO GUARDINI, *The End of the Modern World*, p. 73.

16 GUARDINI, R., *The Letters from Lake Como*, p. 11.

17 *Ibid.*, p. 81.

18 PATTISON, G., *Thinking About God in an Age of Technology*, p. 3.

19 GUARDINI, R., *The Letters from Lake Como*, p. 17.

the proximity to nature and eradicates the particular in support of the universal. Thus, Guardini observed that:

> In this new sphere things are no longer directly detected, seen, grasped, formed, or enjoyed; rather, they are mediated by signs and substitutes.... We want to move away from the particulars that occupy us to what is structured, so that we may...master the whole of the reality around us. In this process we move away from direct encounter with things, from direct grasping and being grasped, and arrive outside a connection between the ego and things..."[20]

The overbearing hunt for knowledge in this innovative technological culture has become conceptual, mechanical, formulaic and uncompromising. Thus, "on the basis of a known formula, materials and forces are put into the required condition: machines. Machines are an iron formula that directs the material to the desired end."[21] Guardini explains further that "what the concept is for knowledge of things, mechanisms, instruments, and machines are for practical action. Machines are steel concepts. They lay hold of many things in such a way as to disregard their individual features and to treat them as though they were all the same."[22]

Whereas culture in the past was natural and creative, there is now a mechanical desire. It's starting point according to Guardini is the inaccessible, rationally understood power of nature, which works through the machine. "If the object, that is, the factually contingent historical world, is absolutized, then it will be replaced by the modernistic concept of "nature", which becomes the all-supportive and infinite ground from which subjects, history, and culture now proceed."[23]

Hans Urs von Balthasar explains the end result this way: "In sum, the artificial world is no longer centered in man, and it fragments into individual spheres whose very lack of connectedness tears man apart and leads to 'deep-seated anarchy'."[24]

Thirdly, the ensuing feeling becomes one of *consciousness and awareness* of everything. What we observe emerging is "a global consciousness in its first outlines." Guardini made particular reference to print media here. "Newspapers are a technique of developing awareness. By them we today become aware of what is going on around us and to us and in us. Reporters are present at events to

---

20 Guardini, R., *The Letters from Lake Como*, p. 20.

21 *Ibid.*, p. 46.

22 *Ibid.*, pp. 23–24.

23 Hans, Urs von Balthasar, *Romano Guardini: Reform from the Source*, New Edition, San Francisco: A Communion Book, Ignatius Press, 2010, p. 39. (Title of the German Origin: Romano Guardini: Reform aus dem Ursprung, Einsiedeln, Freiburg: Johannes Verlag, 1995.

24 *Ibid.*, p. 13.

describe and integrate them. Cameras take pictures of them. Nothing happens anymore without being noticed."[25] He writes of this in the context of how people learn through awareness and consciousness in daily life, beyond the realms of schools and formal education. He adds films and literature and other artistic expressions to this "total life of learning."[26] This is apparently a good thing; however Guardini suggests that it is not. This is because:

> Life needs the protection of non awareness.... We cannot perform an intellectual act and at the same time be aware of it. We can only look back on it when it is completed. If we try to achieve awareness of it when we are doing it, we can do so only by always interrupting it and thus hovering between the action and knowledge of it."[27]

Corresponding to the way that plants grow only when their roots are sheltered in obscurity, "all life must be grounded in what is not conscious and from that root emerge into the brightness of consciousness."[28]

Fourthly, what technology aims at is *mastery*, which pays no attention to the traditional route of life and nature and eliminates or destroys true culture. According to Guardini,

> The person engaged today in the labour of 'technics' knows full well that technology moves forward in final analysis neither for profit nor for the well-being of the race. That person knows in the most radical sense of the term that power is its motive—a lordship of all; that person seizes the naked elements of both nature and human nature. His or her action bespeaks immense possibilities not only for creation but also for destruction, especially for the destruction of humanity itself."[29]

---

25 Guardini R., *The Letters from Lake Como,* p. 29. Guardini goes on to lament the emerging consciousness of people's lives through increased knowledge of geography, statistics, science, and psychology, via increased media. Although curious to us today, Guardini observes that the major discoveries have already been made, that only their "inner relations" remain to be sorted out: "I sense in many areas of a ntural science a feeling that the most decisive things, laws, forces, and forms are already there to see" (pp. 38–39). Guardini feels that excessive knowledge is destroying the magic and essence of our lives: "The root of life itself, what is innermost to it, is lit up. Can life sustain this? Can it become consciousness and at the same time remain alive?" (p. 32).

26 *Ibid.*, p. 31.

27 *Ibid.*, p. 31.

28 Guardini, R., *The Letters from Lake Como,* p. 32.

29 Guardini, R., *The Essential Guardini: An Anthology of the Writings of Romano Guardini*, p. 17; "These two phenomena, the not-human man and the not-natural nature, promise to be the foundation upon which the world of the future will be erected. Man will then face an existence in which he will be free to further his lordship of creation, carrying it even to its last consequences. This mastery will be open to him because he

The kind of knowledge used by land cultivators is a temperate mastery that follows the natural contours of the land. "The other form of knowledge and its mastery is very different. This knowledge does not inspect; it analyzes. It does not construct a picture of the world, but a formula. It's desire is to achieve power so as to bring force to bear on things....the basis and character of its dominion: compulsion arbitrary compulsion devoid of all respect."[30] This technological advancement has according to Guardini gone beyond human control and the eventual outcome is still yet to be determined.

> It has become an instrument of human power for the mastery of the world and as such has already outgrown man's restraining control, thus enslaving him; the person has become a thing. The individual who is aware of his powerlessness abdicates in the face of this alliance between technology and power, on the one hand, and between power and the omnipotent State, on the other."[31]

Guardini here has already anticipated the "challenging-forth" of nature into "standing-reserve" for future use that Heidegger will define as the dangers of technology as we will later see in this chapter. In its overpowering power, nothing on the way of technology is spared; even time and space are sometimes victims of technological undoing. "Time and space...are mastered by means of communication."[32]

Fifthly, the end result of this mastery of nature is the *autonomy* which results in the *dissolution of the organic*. A particular kind of desire has united with a particular understanding of nature to create a situation in which:

> Something new comes into being in this way, a whole world of works, goals, institutions, and orders that are no longer determined by our living constitution but by unleashed natural force, by the rational autonomy of this force which goes its own way and no longer worries about any organic standard. This new force is governed by a human attitude that no longer feels itself tied by living human unity and its organic compass and

---

has permitted himself utter freedom: the freedom to determine his own goals, to dissolve the immediate reality of things, to employ its elements for the execution of his own ends. These things he will do without any consideration for what had been thought inviolate or untouchable in nature. He will ignore that strong sense of the sacredness of nature which had endured within mankind's earlier vision of the world. *The End of the Modern World*, p. 73–74.

30 Guardini, R., *The Letters from Lake Como*, p. 44.

31 Hans Urs von Balthasar, *Romano Guardini: Reform from the Source*, p. 16.

32 *Ibid.*

> that regards as petty and narrow the limitation in which the earlier time found supreme fulfillment, wisdom, beauty, a well-rounded fullness of life.[33]

Formerly, man used tools, but according to its natural fabrications. At present, a border line has been traversed and "that has made it possible to break into the closed nexus of nature." "Individual forces such as steam, electricity, and chemical energy have been taken out of their natural context."[34] Observing the difference between our modern age and Renaissance and Middle Ages, Guardini states: "What has come since seems to be governed by a different basic attitude, by the desire to set goals independently of organic connections and on the basis of rationally emerging forces that are mechanically put in the service of this desire and its goals."[35] Hans Balthasar sees this autonomous desire as the emergence of modern subjectivity over against the objective world:

> The "modern age", which came to an end internally in our century, attempted to transcend the "polar" structure, and therefore the contingency, of the finite and to absolutize the constitutive factors of existence [...]. The created person posits himself in absolute terms as a *subject tout court*, which lays claim to the objective world and culture (the humanized world) as its product. This "conditional autonomy" is 'extended into one that is unconditional..."[36]

Lastly, technology facilitates population growth and discards art in the interest of enhanced productivity to counter the consumerism of the masses. Standardization and mass production emasculates artistic creativity and quality. "Numerical growth does not proportionately increase creative depth. Less numerous peoples in antiquity and the Middle Ages had more concentrated creative force than larger peoples today."[37] Guardini seems to be already pointing out to the dangers inherent in the process of economic globalization with its attendant consequences.

Guardini asks the following urgent question: "in all that is taking place, is a life supported by human nature and fully human work possible"?[38] Unfortunately the answer to this question must be no. As he elaborates;

> We originally confronted the task of having to assert ourselves vis-à-vis nature, which then threatened on all sides because it had not been mastered by us and was thus a chaos for us [...] taking possession of the world [...] we have released new forces that had

33 *Ibid.*, p. 72.
34 *Ibid.*, p. 71.
35 *Ibid.*, p. 75.
36 Hans Urs von Balthasar, *Romano Guardini: Reform from the Source*, p. 38.
37 Guardini, R., *Letters from Lake Como*, p. 53.
38 *Ibid*, p. 78.

> not yet been released by our own attitude and the form of the world. These forces have increased, and now they have unleashed a new chaos."[39]

The dread of what technology may possibly do to us has become for the most part heightened in the modern world, in which we dread weapons of mass destruction, or wonder about the threatening effects on society of altering the human chromosome. Guardini is here, in effect, echoing what would later be referred to as manufactured risks in a second modernity or risk society by Ulrich Beck as we shall see later in the next chapter.

The answer, declares Guardini, is not to shrink back into idealism and reverert to nature, but rather, to subject technology and the anarchy it generates under human control. Thus, there is a way forward in which "technical events and unleashed forces can be mastered only by a new human attitude that is a match for them."[40] In this way

> We must not oppose what is new and try to preserve a beautiful world that is inevitably perishing. Nor should we try to build a new world of the creative imagination that will show none of the damage of what is actually evolving. Rather, we must transform what is coming to be. But we can do this only if we honestly say yes to it and yet with incorruptible hearts remain aware of all that is destructive and nonhuman in it."[41]

Although it is produced by human beings, what technology itself produces is neither human nor natural. "What takes place here is not human, at least if we measure the human by the human beings who lived before us. It is not natural if we measure the natural as it once was."[42] Thus, we do not need a lesser amount of technology but extra. We need more mature; more intellectually planned, more responsible technology. This can only be possible "if living people first make their influence felt in the sphere of objective nature. If they relate this nature to themselves and in this way create a "world" again."[43]

Role of Christianity:

Guardini's systematic questions for Christians are insightful. He sees Christianity as contributing to the problem but also as expressing hope for a solution.

> It is Christianity that has made possible science and technology and all that results from them. Only those who had been influenced by the immediacy of the redeemed soul to

39 *Ibid.*, pp. 83–84.
40 *Ibid.*, p. 80.
41 *Ibid.*, p. 80–81.
42 *Ibid.*, p. 73.
43 *Ibid.*, p. 83.

> God and the dignity of the regenerate, so that they were aware of being different from the world around them, could have broken free from the tie to nature in the way that this has been done in the age of technology.[44]

Since the technological worldview is where the machine is the symbol of fulfilled culture, there is a loss in religious enthusiasm. Hence, we should think through the technological process mostly through the question of religion. Probably we may see that "the center of a faith approach must be found profoundly than before in what is truly personal, in the venture and fidelity of decision."[45]

Notwithstanding his concerns regarding technology, Guardini expresses an optimistic and positive hope for the future and recommends the modern man to think about the pressing tasks of technology. This rigorous task of merging the technologizing of culture with rejuvenation in Christian religious existence is especially needed and required for such a time as this. The forces of technology created by human hands have broken free from the hands that created them. Therefore we need to subject technology under our control, as against subjugation of nature.

The modern age was prone to come to grips with necessary innovations by means of rational intellect and organization. The predicaments we face today are so enormous that we must reach for a deeper hold. Guardini proposes that a "new humanity must emerge of more profound intelligence, new freedom, new inwardness, new form" to "create a 'world' again" – a new order, a new universe, as in the days when we were confronted with the duty to "Fill the earth and subdue it" and "shape chaos into a human world."[46] Guardini concludes that "true education… rooted in being, not in knowledge"[47] is the path to this new humanity, though we might achieve it only in eternity. In brief, now that science has begun to disintegrate the natural elements, something parallel must take place on the human stage: man must scrutinize the fundamental facts of his existence. If he does not, events will overtake him, leaving him an ever greater alien on earth.

Generally speaking, men have the same opinion that technology, economics and politics must be directed "realistically," but what they mean is in a mode which completely disregards ultimate values: man's personal destiny and all that is God's due. Guardini as a result emphasizes the need to see further than the realistic, utilitarian, rational modes of life to awareness and realization of the

44 *Ibid.*, pp. 81–82.
45 *Ibid.*, p. 111.
46 *Ibid.*, pp. 83–84.
47 *Ibid.*, p. 88.

interplay of the sacred and the human in a technological/scientific age. Guardini is not a pessimist, but he is critically concerned about the prospective loss of human dignity and individual responsibility in a world culture dominated by technological utilitarianism.

Although Guardini bemoans the effect of technology on people, he does not consider how objectionable and unwarranted or misguided technology might be to God. Further, he does not think our responsibility to creation by taking into consideration our responsibility to God, the creator. Instead, he talks about our responsibility toward the use of technology in purely an abstract sense: "We have seen that machines give us constantly increasing power.... If we have power, we have to use it, and that involves conditions. We have to use it with responsibility, and that involves an ethical problem."[48]

When Guardini states that we can only achieve new humanity in eternity, we find him echoing an idea that Heidegger would later on articulate and orchestrate, that the source of liberation from the destructive force of technology lie not in rejecting technology but within the very realm of technology itself. His "call for resolute decision and responsibility vis-à-vis technology anticipates a key element in Heidegger's thought."[49] Heidegger's problem was not how to move beyond technology, but our ability to have a different understanding and attitude towards technology. Akin to Heidegger, Guardini thinks this must involve art. According to him, there are evidences everywhere that we are capable, even on the very brink, of achieving an artistic form for shaping technology such that it is truly expressive of our humanity.[50] Mastering chaos will necessarily involve placing ourselves at God's disposal, and operating with his freedom and power. "I believe that God is at work. History is going forward in the depths, and we must be ready to play our part, trusting in what God is doing and in the forces that he has made to stir within us."[51]

b) Jacques Ellul (1912–1994)

Jacques Ellul, in his book, *The Technological Society* (1964), originally titled: *La Technique: L'enjeu de siècle* (1954), is more interested in technological phenomenon from philosophical perspective and discusses its metaphysical implication under the name of technique. Ellul went further than a sheer consideration of

48 *Ibid*, p. 106.
49 Pattison, G., *Thinking About God in an Age of Technology*, p. 68.
50 Guardini, R., *Letter from Lake Como*, pp. 92–93.
51 *Ibid.*, p. 96.

technology to give a breakdown of the elementary method of technique, a mode he defined as "the totality of methods rationally arrived at and having absolute efficiency."[52] Jacques Ellul set forth seven characteristics of modern technology:

> The characteristics of technique which serves to make efficiency a necessity are rationality, artificiality, automatism of technical choice, self-augmentation, monism, universalism, and autonomy. The rationality of technique enforces logical and mechanical organization through division of labour, the setting of production standard, etc. and it creates an artificial system which "eliminates or subordinates the natural world.[53]

Incapable of instituting a reasonable existence separate from the simulated setting of a technological environment, human beings hang their final trust in it. "Whereas earlier social orders operated with a triad—humans/ tools/ nature—in technological societies, nature recedes and humans perceive themselves as living in a technical artifice, existing not in nature but in culture."[54] And as Ellul explains;

> Man does not any longer live in a natural environment but rather in a milieu composed of the products of his technology [...]. He can no longer take any significant action without technological intermediation. Technology constitutes an engulfing universe for man, who finds himself in it as a cocoon.[55]

What Ellul is saying in essence is that no technique is possible when men are free. "When technique enters into the realm of social life, it collides ceaselessly with the human being to the degree that the combination of man and technique is unavoidable, and that technical action necessarily results in a determined result [...]. Technique must reduce man to a technical animal, the king of the slave of technique.[56] The revolution today, therefore, is that the world has entered a technological civilization. Technology in essence, is not merely one more arena for philosophers and sociologists to investigate, but a new foundation for understanding the self, human institutions, and ultimate reality. And what is more, technological society wants to emancipate itself from its environment and the consequences, Ellul states, are

---

52 Ellul, J., *The Technological Society*, trans. John Wilkinson, New York: Knopf, 1964, p. xxv.

53 Fasching, D. J., *The Thought of Jacques Ellul: A Systematic Exposition*, Toronto: Edwin Mellen, 1981, p. 17.

54 Clifford, G. Christians, John, P. ferre & Fackler, P. Mark, *Good News: Social Ethics and the Press*, New York: Oxford University Press, 1993, p. 170.

55 Ellul, J., "Symbolic Function, Technology and Society," in *Journal of Social and Biological Structures*, 1(3), 1978 (207–218), p. 216. (Cited in Clifford, G. et al.)

56 Craig, Hanks, ed. *Technology and Values (Essential Readings),* West Sussex, UK: Wiley-Blackwell Edition, 2010, 70.

that "technique worships nothing, respects nothing. It has a single role: to strip off externals, to bring everything to light, and by rational use to transform everything into means."[57]

c) *Gilbert Simondon (1924–1989)*

Gilbert Simondon, although without mentioning Jacques Ellul or Heidegger, (who, within this same era, emphasized the negative impacts of technology on society and culture) wrote a book, *Du Mode d' existence des Objets Techniques,*[58] the *Introduction*, of which can be seen as a response to their ideas. His work can be interpreted as a humanist appeal for an end to alienation between humans and machine, and the inauguration of a technological culture wherein humans are no longer either masters or slaves of machines, but equals whose responsibility is to regulate technological collections. Even though we live in world inundated by technology, we nonetheless, have complications grasping the ontological position of technological entities.

Simondon's analysis is more or less divided into three parts: (i) the formation and advancement of technological entities or objects; (ii) the associations of homo sapiens to technological objects, taking into consideration historical along with contemporary interactions; (iii) the essence of what he calls "technicity," or the technical mode of being-in-the-world as compared to both religious as well as metaphysical world-relations which always entail ethical demands.

Contrary to more orthodox assumptions of technology, Simondon's thoughts on the formation and advancement of technological objects do not begin with mere tools, "for no fixed structure corresponds to a particular use."[59] They have a certainty which is independent of the user's posture and which can be ascertained by learning their history and evolution. Simondon understands technical objects as evolving compounds of relations rather than in terms of function, efficacy, material or form. Dumouchel, interpreting Simondon, observes that "the subsumption of technical objects under the category of use ignores their technical characteristics, their technicality. If its internal structure is what makes a technical object what it is, the concept of use bypasses the essential."[60] Specifically, in

---

57 Ellul, J., *The Technological Society*, p. 142.

58 Simondon, G., *Du Mode d' existence des objets techniques*, Méot, 1958; second edition, Paris: Aubier, 1989.

59 *Ibid.*, p. 19.

60 Dumouchel, P., "Gilbert Simondon's Plea for a Philosophy of Technology", in Feenberg, A. & Alastair H., (eds.), *Technology and the Politics of Knowledge*, Bloomington & Indianapolis: Indiana University Press, 1995, p. 257 (pp. 255–271).

connection with others, a realm in which technology has regularly been regarded as bitterly alienating; Simondon recommends attention to technological objects and actions as a means of questioning notions of individuality, self and subject.

> Through the intermediary of the technical object, an interhuman relation that is the model of transindividuality creates itself. This can be understood as a relation which does not put individuals in relation by means of their constituted individuality, which separates the one from the other, nor by means of that which is identical in each human subject, for example, the *a priori* forms of sensibility, but by means of this charge of pre-individual reality, this charge of nature which is conserved with individual being, and which contains potentials and virtuality.[61]

Simondon's remark opposes some forms of technological reductionism. Technology is not, according to him, as State and enterprise cultures of innovation and modernization often envisaged; some substance or vital force, which must be cultivated, controlled or assembled for political or commercial ends. Every now and then, it becomes that, but only when entangled in relations that fix, limit or take advantage of its potentials. Occasionally and to some extent, with an always wavering step, particular technical objects and ensembles overflow existing modalities of perception and progress on many different scales. As Pascal Chabot states, "the effects of the invention exceeds the initial formulation of the problem."[62] In such scenarios, the technical object forms an 'interhuman relation' of a particular kind that Simondon terms 'transindividual.' As Bruno Latour puts it, "far from

---

61 Simondon, G., *Du Mode d' existence des objets techniques,* p. 248: "Par l' intermédiare de l'objet technique se crée alors une relation interhumaine qui est le modèle de la transindividualité. On peut entendre par là une relation qui ne met pas les individus en rapport au moyen de leur individualité constituée les séparant les uns des autres, ni au moyen de ce qu'il y a d'identique en tout sujet huamin, par exemple les forms *a priori* de la sensibilité, mais au moyen de cette charge de réalité pré-individuelle, de cette charge de nature qui est conserve avec l'être individual, et qui contient potentiels et virtualité. The term individuation in Simondon's parlance demarks a process of genesis which leads to the constitution of an entity or a being. Is never conceived as a final stage, but "the individual is that what has been individualized and that continues individualizing itself." (Cf. Simondon G., *L'individu et sa Genèse Physico-Biologique,* Paris: Presses Universitaires de France, 1964, p. 197) The term preindividual demarks an oversaturated state, a state "more than unity and more than identity" out of which a dynamic process of dephasing which can be understood as a sort of stretching of this initial complex agglomerate might arise and thus unfold an operation of individuation. (Cf. Simondon, G., *L'individuation Psychique et Collective*, Paris: Aubier, (1989) 2007, p. 215.

62 Chabot, P., *La Philosophie de Simondon*, Paris: Librairie Philosophique J. Vrin, 2003, p. 20: "Les effets de l'invention excèdent la formulation initiale du problème."

primarily fulfilling a purpose, they [technologies] start by exploring heterogeneous universes that nothing, up to that point, could have foreseen and behind which trail new functions."[63]

Directly focusing on the makeup and development of such technical objects, the nature of their internal parts and the corresponding processes of interaction and exchange, Simondon identify the process of "concretization" as a major trait of technological development. "It is the historical process through which an abstract schema enters into the concrete world of material things and physical processes."[64] In a technological object that is still *conceptual*, i.e. which only starts to develop, the parts are functionally interconnected in such a way that, "like workers, they cooperate without knowing exactly what the others are actually doing." They work end-to-end, sometimes even against each other. According to Simondon, the concretized technological object is "no longer fighting with itself, no secondary effect infers with the function of its totality or remains outside of it."[65] Dumouchel explains concretization in these words:

> Concretization implies a form of technical judgment and intuition which cannot be equated with theoretical knowledge. This is precisely what we should expect of a good notion of technical progress. It should take into account not only the conceptual design of an object but also its technicality: the interaction of the various physical processes it involves, the evolutionary possibilities they contain, and the limitations they impose.[66]

Thus, Simondon's concept of "technological object" does not refer to single beings, but to a series or row of such beings. In order words, he is interested in the "individuality" of technological objects, not in their "singularity" because technical individuals reveal a central aspect of technology, also present, but somewhat obscured, in tools and technical systems.[67] As he explains, this individuality is linked to a "pure functional scheme", a diagram representing the invention of an object and at the same time giving guideline for its construction. Thus, the individuality of technological objects lies for Simondon in their respective functional diagrams and the sequence of matching or corresponding material concretizations. That is to say, technological objects never stand unaccompanied. They are all the time part of a line constituted by earlier and later examples of related objects.

---

63 Latour, B., "Morality and Technology. The End of the Means," In *Theory, Culture & Society*, vol. 19, 5/6, 2002, p. 250 (247–260).

64 Dumouchel, P., "Gilbert Simondon's Plea for a Philosophy of Technology", p. 259.

65 Simondon, G., *Du Mode d' existence des objets techniques*, pp. 21, 34.

66 Dumouchel, P., "Gilbert Simondon's Plea for a Philosophy of Technology", p. 263.

67 *Ibid*., p. 264.

d) Albert Borgmann (1937)

Albert Borgmann's work *Technology and the Character of Contemporary Life* can be perceived as a foremost rejuvenation of orthodox philosophy of technology, especially of Heidegger's argument. A neo-Heideggerian, Borgmann advanced an attitude to technology that begins from tangible, solid technological artifacts, rather than taking technology per se as a functional object or entity. He fathoms technology like a paradigm, a structure that has integrated into people's culture, permeated into everyday activities that its effects are hardly noticeable. This is what he calls 'device paradigm,' that is, a universal pattern followed by modern technology.

Thus, the notion of *device paradigm* means viewing technology entirely as a device (or set of devices) that convey a series of commodities, not a specific or unique thing, which is provided in a "commodious way"[68] and evaluating or even changing the technical features and powers of such devices, while keeping the device functionally the same. Technology, according to him, is "the new orthodoxy, the dominant character of reality."[69]

Borgmann's reaction to the device paradigm is to yearn for a re-establishment of what he calls "focal things" and practices. Focal things are "things that require our bodily engagement and provide a variety of benefits, rather than simply one commodity. Through that bodily engagement we develop skills which contribute to our character, and through sharing with other the tasks required by focal things we forge relationship."[70] In order words, "focal things require a practice to prosper within."[71]

Borgmann's philosophy of technology thus centers on the concern for engagement with reality not just things or devices. "If we are to challenge the rule of technology, we can only do so through the practice of engagement."[72] This is because, "devices ask for as little involvement as possible; they create the *availability* of commodities by keeping their machinery in the background as much as they can and putting their commodities in the foreground. Against this, 'things' do not

68 Borgmann, A. *Technology and the Character of Contemporary Life*, Chicago: University of Chicago Press, 1984, p. 42.

69 *Ibid.*, p. 189.

70 Chapman. A., *Democratizing Technology: Risk, Responsibility and the Regulation of Chemicals*, London: Earthscan, 2007, p. 17.

71 Borgmann, A. *Technology and the Character of Contemporary Life*, p. 196.

72 Borgmann, A. *Technology and the Character of Contemporary Life*, p. 207.

separate machinery from commodity. Rather, they *engage* people."[73] Technology threatens this engagement by inviting a consumptive way of taking up with reality and by replacing reality with technological information.

> Countering technology through a practice is to take account of our susceptibility to technological distraction, and it is also to engage the peculiarly human strength of comprehension, i.e., the power to take in the world in its extent and significance and to respond through an enduring commitment.[74]

Borgmann does not vilify technology per se. He merely summons us to set appropriate limits to technology. Using the beauty of the wilderness as an example of the "focal things," that should teach us to set such limits, he explains:

> Killing the wilderness or keeping it at bay. Technology kills the wilderness when it develops it through roads, lifts, motels, and camping areas. It keeps the wilderness at bay when, without affecting untouched areas permanently, it insulates us from the engagement with the many dimensions and features of the land, as it does through rides in jet boats or helicopters. Here we can see that technology with its seemingly infinite resourcefulness in procuring anything and everything does have a clear limit. It can procure something that engages us fully and in its own right only at the price of gutting or removing it. Thus the wilderness teaches us not only to accept technology but also to limit it.[75]

He counsels us to make room in a technological age for matters of ultimate concern—things and practices that engage us in their own proper right. According to Mumford "what is needed is a technology so varied, so many-sided, so flexible, so responsive to human need, that it can serve every valid human purpose. The only true multi-medium remains the human organism itself."[76]

In a later text, *Power Failure: Christianity in the Culture of Technology*, Borgmann attempts to sketch the relationships between Christianity and technology by exploring some of the "invisible" dangers of a technology-driven mode of life. Precisely, he demonstrates how devices and consumption have substituted physical things and practices in daily life and calls for a dynamic Christian practice in

---

73 Verbeek, P-P, *Moralizing Technology: Understanding and Desgining the Morality of Things*, Chicago: University of Chicago Press, 2011, p. 47.

74 Borgmann, A. *Technology and the Character of Contemporary Life*, p. 210. A device is an artifact or instrument or tool or gadget or mechanism, which may be physical or conceptual. It includes hardware and software.

75 *Ibid.*, p. 195.

76 Mumford, L., *The Myth of the Machine: Technics and Human Development*, New York: Harcourt Brace Jovanovich, 1967, p. 298.

a technological age as the most promising response to technology.[77] Borgmann concludes that technology is neither an unequivocal evil nor a downright good, and that we do not need to annihilate it or escape from it.

e) Don Ihde (1934)

Don Ihde, a contemporary American philosopher of technology is well known for his application of phenomenology to technology, the theory of human-technology relations. Beginning with *Technics and Praxis*,[78] and *Existential Technics*,[79] Ihde was downplaying some Heideggerian influences in favour of a Husserlian phenomenology. The aim of his theory, as he puts it, is to "show what is invariable in the way humans experience their technologies."[80]

In *Technics and Praxis: A Philosophy of Technology*, Ihde examines the capacity of technology artifacts to transform awareness, especially the moment it is embodied in scientific instruments and technologies, for instance, the telescopes, hearing aids, eye glasses or a blind man's cane. Don Ihde has termed the extraordinary connections and relationships between human beings and such artifacts *embodiment relations;* a situation in which artifacts are experienced as means through which one's immediate environment or surrounding is encountered or met. "When we humans use technologies, both what the technology 'is' or may be, and we, as users undergo an embodying process—we invent our technologies, but, in use, they 're-invent' us as well."[81]

Also in *Existential Technics*, Ihde emphasis more on human-technology relations where he shows that our connection with technologies radically influences our existential condition, mostly the manner in which we comprehend our world and our human race. Ihde states that "all self-interpretation takes its shape in a certain way with respect to some basic form of existential praxis which is projected upon the world and reflected back in ways which become dominant ways of understanding ourselves and our world."[82]

---

77 BORGMANN, A., *Power Failure: Christianity in the Culture of Technology*, Michigan: Brazo Press, 2003, p. 8.

78 IHDE, D., *Technics and Praxis*, Dordrecht: Reidel, 1979.

79 IHDE, D., *Existential Technics*, Albany: SUNY Press, 1983.

80 IHDE, D., *Philosophy of Technology: An Introduction*, New York: Paragon Press, 1993, p. 111.

81 IHDE, D., *Listening and Voice: Phenomenologies of Sound*, 2nd Edition, Albany, New York: State University of New York Press, 2007, p. 243.

82 IHDE, D., *Existential Technics*, p. 22.

Possibly Ihde's most extensively read book in the philosophy of technology, still, is *Technology and the Lifeworld: From Garden to Earth*, a work that integrates multicultural dimensions into an analysis of the lifeworld role of *technics*. Drawing from the traditions of phenomenology and hermeneutical philosophy, Ihde examines the diversity of human relations to technology, and then, makes the fascinating argument that despite the fact the none-neutrality of human-technology relations reveals in diverse ways in the framework of different traditions, diverse geographies, and time periods poles apart, it however remains repeatedly the case that "*human activity from immemorial time and across the diversity of cultures has always been technologically embedded*."[83] Ihde therefore calls for the transformation of modern technological science designed at preserving and protecting the inherited earth. "What is needed is a much radically demythologized story of the structures and limits of human-technology and of the non-technological possibilities of relation to an environment, or 'world'"[84] Thus in his later works, especially *Technology and the Lifeworld*,[85] Ihde has advocated an environmental activism that could only be implemented politically.

This existential perspective of technology, as seen above, is nonetheless much more critical towards technological advances as it affects humanity. Lewis Mumford, Ortega y Gasset, and Heidegger, are all exponents of the existential and phenomenological or humanities philosophy of technology.

4.3.2. Critical Perspective

Since World War II, the focus of technological activity has undergone a major change. While the old activities are still pursued, they have been largely superseded by applications of technology at the microscopic level, especially in Information Technology and genetic engineering. The propensity can be expected to continue into the foreseeable future. Thus, "technologies interact deeply with society and culture, but the interactions involve mutual influence, substantial uncertainty, and historical ambiguity, eliciting resistance, accommodation, acceptance, and even enthusiasm."[86] However, some are skeptical and have warned against the inherent dangers associated with such technological interactions

---

83 Ihde, D., *Technology and Lifeworld: From Garden to Earth*, p. 20.

84 *Ibid*., p. 17.

85 Ihde, D., *Technology and the Lifeworld: From Garden to Earth*, Bloomington: Indiana University Press, 1990.

86 Misa, J. Thomas, "The Compelling Tangle of Modernity and Technology" in *Modernity and Technology*, Misa, J. T., Philp B., & Feenberg, A., (eds.), Cambridge: The MIT Press, 2003, p. 3.

a) *Theodor Adorno (1903–1969) and Max Horkheimer (1895–1973)*

As the World War II was coming to an end, Theodor Adorno and Max Horkheimer of the Institute for Social Research, otherwise known as the 'Frankfurt School'[87] of philosophy wrote the *Dialectic of Enlightenment* (1944). This classic proclaimed the unique and more uncertain path of critical theory in the years following the war. They perceived the remarkable accomplishment of Enlightenment in throwing out the enchanted world in the advancement of modern physics, science and its technological offshoot, which climaxed absurdly by the 20th century in the chilling reappearance of disenchantment by way of despotism and social disequilibrium. Adorno and Horkheimer saw inherent dangers in technologies focused on the "means as ends" without a strong sense of connection with nature.

> At the moment when human beings cut themselves off from the consciousness of themselves as nature, all the purposes for which they keep themselves alive – social progress, the heightening of material and intellectual forces, indeed, consciousness itself – become void, and the enthronement of the means as the end, which in late capitalism is taking on the character of overt madness, is already detectable in the earliest history of subjectivity."[88]

This book, *Dialectic of Enlightenment* became an articulate attack on the modern conquest of untainted pragmatic wisdom and its attendant technology. The ambitious nature of human reason and skill, they claimed, had been filleted of any reference to human ends and reduced to an absolute tool of the powerful, thus perpetuating its self-destruction. "What men want to learn from nature is how to

---

87 The school originated in the Institute for Social Research (Institute) as part of the Frankfurt Institute in 1923. The Frankfurt school is best known for its program of developing a "Critical theory of society." "Critical theory is primarily a way of doing philosophy, intergrating the normative aspects of philosophical reflection with the explanatory achievements of social sciences. The ultimate goal of its program is to link theory and practice, to provide insight, and to empower subjects to change their oppressive circumstances and achieve human emancipation, a rational society that satisfies human needs and power...." Cf. *Cambridge Dictionary of Philosophy*, pp. 278–79. Prominent among the members included Herbert Marcuse, Max Horkheimer, Theodore Adorno, Friedrich Pollock, Erich Fromm, Leo Lowenthal and Jürgen Herbamas. Cf. Kellner Douglas, *Critical Theory, Marxism and Modernity*, Baltimore, Md.: John Hopkins University Press, 1989, p. 12. See also Fédéric Vandenberghe, *A Philosophical History of German Sociology*, trans. Carolyn Shread, USA and Canada: Routledge, 2009, p. 157ff.

88 Adornor T. & Horkheimer, M., *Dialectics of Enlightenment*, trans. E. Jephcott, Stanford, Calif: Stanford University Press, 2002, pp. 42–43.

use it, in order wholly to dominate it and other men."[89] Thus, they asked; "… why mankind, instead of entering into a truly human condition, is sinking into a new kind of barbarism".[90] The Enlightenment hope for a flourishing and prosperous society has become a paradox as the prospects of its fulfillment were increasingly an illusion. They conclude; "In the most general sense of progressive thought, the Enlightenment has always aimed at liberating men from fear and establishing their sovereignty. Yet, the fully enlightened earth radiates disaster triumphant."[91] This technophobic climate also attracted the attention of Herbert Marcuse and others, a later generation of the Critical Theorists.

b) Herbert Marcuse (1898–1979)

In the essay *Some Social Implications of Modern Technology*, (1941) written in collaboration with the Frankfurt School, Herbert Marcuse made distinction between *technology* and *technics*. According to him, technology is a "social process," or a "mode of organizing and perpetuating (or changing) social relationships, a manifestation of prevalent thought and behaviour patterns, an instrument of control and domination." Technics refers to the techniques of production which can "promote authoritarianism as well as liberty, scarcity as well as abundance, the extension as well as the abolition of toil"[92]

---

89 *Ibid.*, p. 3.

90 *Ibid.*, p. xi.

91 *Ibid.*, It should be recalled that it was at this time too, in 1949, that Heidegger wrote "*The Question Concerning Technology*." These two texts were the deepest theoretical expressions of the type of culture criticsm that came to prominence in the 1960s.

92 Marcuse, H., *Technology, War and Fascism, Collected Papers of Herbert Marcuse*, Vol. 1, ed. Douglas Keller, London and New York: Routledge, 1998 p. 41. In Marcuse's view, the Third Reich was a technocracy, meaning that technical rationality and the demands of efficiency, along with the primacy of a "matter-of-factness" mentality generated nonhuman values that superseded those pertaining to the welfare of the people. His description and analysis of "matter-of-fact" thinking, which he develops further in a later essay in this volume called "The New German Mentality," argues that the technological attitude mediates human thought and relationships such that human individuality is weakened. According to Marcuse, the "machine process" generates a "mechanics of conformity," whereby "individuals are stripped of their individualty, not by external compulsion, but by the very rationality under which they live." The "mechanics of conformity" governs "performance not only in the factories and shops, but also in the offices, schools, assemblies and, finally, in the realm of relaxation and entertainment" (p. 48).

Marcuse, in *One-Dimensional Man* (1964) also argues that "contemporary society seems to be capable of containing social change—qualitative change which would establish essentially different institutions, a new direction of the productive process, new modes of human existence."[93] There is no space for dissent, he argues, because technology has obliterated the boundary between public and private, where individual becomes the society and the society, the individual. The result is "a thoroughly static system of life,"[94] a patently unecological society in which forward progress is the ultimate casualty.

According to Marcuse, it is not knowledge or technological devices that are primary but the technological relation to reality that makes progress in science and technique possible in the first place. This is because, "advanced industrial societies create false needs that integrate individuals into the existing system of production and consumption via mass media, advertising, industrial management, and scientific-technological modes of thought."[95] Such false assumptions have led to technological domination and control of the individual and the society in which both cultural, economic and political opinions are shaped by a technical mind-set:

> Technology serves to institute new, more effective, and more pleasant forms of social control and social cohesion [...]. In the face of the totalitarian features of this society, the traditional notion of the "neutrality" of technology can no longer be maintained. Technology as such cannot be isolated from the use to which it is put; the technological

---

93 Marcuse, H., *One-Dimensional Man,*Boston: Beacon Press, 1964, p. xii.

94 *Ibid.*, p. 17. Marcuse states that such a static way of life is manifested when technological rationality leads to dullness of life since machines carry out what was normally done by men. As he states: "A man who travels by automobile to a distant place chooses his route from the highway maps. Towns, lakes and mountains appear as obstacle to be bypassed. The countryside is shaped and organized by the highway: what one finds en route is the by-product or annex of the highway. Numerous signs and posters tell the traveler what to do and think; they even request his attention to the beauties of nature or the hallmarks of history. Others have done the thinking for him, and perhaps for the better. Convenient parking spaces have been constructed where the broadest and most surprising view is open. Giant advertisements tell him when to stop and find the pause that refreshes. And all of this is indeed for the benefits, safety and comfort; he receives what he wants. Business, technics, human needs and nature are wielded together into one rational and expedient mechanism. He will fare best who follows its directions, subordinating his spontaneity to the anonymous wisdom which ordered everything for him." CF. Marcuse, H., *Technology, War and Fascism, Collected Papers of Herbert Marcuse,* p. 46.

95 Marcuse, H. The New Forms of Control, in Technology and Values: Essential Readings, C., Hanks, (ed.), West Sussex, UK: Wiley-Blackwell Edition, 2010, p. 160.

society is a system of domination which operates already in the concept and construction of techniques [...]. As a technological universe, advanced industrial society is a political universe, the latest stage in the realization of a specific historical project—namely, the experience, transformation, and organization of nature as the mere stuff of domination [...]. As this project unfolds, it shapes the entire universe of discourse and action, intellectual and material culture. In the medium of technology, culture, politics, and the economy merge into an omnipresent system which swallows up or repulses all alternatives. The productivity and growth potential of this system stabilize the society and contain technical progress within the framework of domination. Technological rationality has become political rationality.[96]

Because the society is subjugated by such a technological domination, the result, according to Marcuse is that "technological rationality homogenizes nature and people into neutral objects of manipulation. The result is a 'one-dimensional' universe;"[97] one in which technology becomes a new form of social control in industrialized and capitalist society.

c) *Jürgen Habermas (1929)*

Jürgen Habermas "inherited from the first generation of critical theorists a concern with the dominance of instrumental reason... and has relatively little to say about science and technology as such"[98] instead, he was rather more concerned with "technological knowledge and ideas about technology."[99] He criticized the overriding role of scientific-technological rationality and the culture of experts or specialists. In an early essay, he argued that science cannot help us decide between functionally equivalent technologies, but that values must intervene.[100] Even in *Technology and Science as 'Ideology'* Habermas recognizes that "social interest still determine the direction, function, and pace of technical progress"

The quasi-autonomous progress of science and technology then appears as an independent variable on which the most important single system variable, namely, economic growth, depends. Thus arises a perspective in which the development of a social system *seems* to be determined by the logic of scientific- technical progress [...] the culturally defined self- understanding of a social lifeworld is replaced by [...] categories of purposive-rational action and adaptive behaviour [...]. The manifest domination of the

96 Marcuse, H., *One-Dimentional Man*, pp. xv–xvi.
97 Marcuse, H. *The New Forms of Control, in Technology and Values*, p. 160.
98 Outhwaite, W., *Habermas: A Critical Introduction*, 2nd Edition, Cambridge: Polity Press, 2009, p. 19.
99 Habermas, J., *The Philosophical Discourse of Modernity: Twelve Lectures*, Cambridge, Mass: MIT Press, 1987, p. 228.
100 Habermas, J., *Theory and Practice*, Boston: Beacon Press, 1973, pp. 270–271.

authoritarian state gives way to the manipulative compulsions of technical operational administration.[101]

The implication for this is that technical development is not an independent affair but from the political and social point of view, it is dependent on the cultural structure where it imposes its influence under the appearance of modernization. The question for Habermas is then "how can the power of technical control be brought within the range of the consensus of acting and transacting citizens?"[102]

Habermas is one thinker whose theory of communicative action relates squarely to democratizing technology. The goal of his theory of communicative action then, is that of "clarifying the presuppositions of the rationality of processes of reaching understanding, which may be presumed to be universal because they are unavoidable"[103] In his *Philosophical Discourse of Modernity*, Habermas develops his intersubjective approach to modernity using the concept of 'communicative rationality'.

> The communicative rationality recalls older ideas of logos, in as much as it brings along with it the connotations of a non coercively unifying, consensus-building force of a discourse in which the participants overcome their at first subjectively based views in favour of a rationally motivated agreement.[104]

A realm which Habermas refers to as instrumental action takes place within a rational system, such as the state, economy, or any organizational device. For Habermas, the key difficulty with the lack of public regulation or control over technology comes about because of the interference of such instrumental action into the lifeworld. The lifeworld provides a contextual framework from which communication, and therefore communicative action, can draw from in order to reach a consensus. Communicative action subsists in the lifeworld because it is in it that socialization and discourse develop. When instrumental action intrudes into the lifeworld, it

---

101 HABERMAS, J., "Technology and Science as Ideology", in *Towards a Rational Society*, trans. J. SHAPIRO, Boston: Beacon Press, 1970, p. 105.

102 HABERMAS, J., *Toward a Rational Society: Student Protest, Science and Politics*, Boston: Beacon Press, 1971, p. 57.

103 HABERMAS, J. "Questions and Counterquestions" in *Habermas and Modernity*, R,J. BERNSTEIN, ed., Cambridge, Mass.: MIT Press, 1985, p. 196.

104 HABERMAS, J., *The Philosophical Discourse of Modernity*, pp. 294, 315. Although Habermas sees communicative rationality as being threatened by actual modern society, he nevertheless argues that the core of communicative rationality, "the unconstrained, unifying, consensus-bringing force of argumentative speech," is a "central experience" in the life of a human being." Cf. HABERMAS, J., *Theory of Communicative Action, vol. I*, Cambridge, Mass.: MIT Press, 1983, p. 10.

replaces such discourse with methods of manipulation and subjugation of the public rather than encouraging their participation. According to Habermas' renowned explanation, therefore, "the problem with technological modernization has little to do with technology itself, but with the way in which the influence, scope, and success of technology in modern life have tended to authorize only "purposive" of instrumental and strategic notions of rationality, and to delegitimate (as unresolvable and subjective) genuinely practical or political questions."[105]

d) Andrew Feenberg (1943)

Andrew Feenberg, one of the leading voices in philosophy of technology in his "*Critical Theory of Technology*," is interested in reconceptualizing the relation of technology, rationality and democracy with the aim of a truly radical reform of industrial society.[106] Feenberg calls for a democratic transformation of technology because "there will generally be ways of rationalizing society that democratize rather than centralize control"[107] According to him:

> Theories of technology fall into one of two major categories: the instrumental theory, and the substantive theory. The instrumental theory, offers the most widely accepted view of technology. It is based on the commonsense idea that technologies are 'tools' standing ready to serve the purposes of their users. Technology is deemed 'neutral,' without valuative content of its own.[108]

---

105 Pippin, R., *Idealism as Modernism (Hegelian Variation),* Cambridge: Cambridge University Press, 1997, p. 201. Habermas indicates impatience with Critical Theory for abolishing the problem he describes simply by proclaiming a premature victory for technocracy. He argues that the "thesis of the autonomous character of technical development is not correct." For him, "as little as we can accept the optimistic convergence of technology and democracy, the pessimistic assertion that technology excludes democracy is just as untenable." Cf. *Toward a Rational Society*, pp. 59/60.

106 Feenberg, A., *Transforming Technology: A Critical Theory Revisited*, New York: Oxford University Press, 2002, p. 3.

107 Feenberg, A., *Questioning Technology*, London & New York: Routledge, 1999, p. 76.

108 Feenberg, A., *Critical Theory of Technology*, New York: Oxford University Press, 1991, p. 5. According to Feenberg, the concept of "neutrality" of technology usually implies at least four points: 1. The neutrality of technology is merely a special case of the neutrality of the instrumental means, which are only contingently related to the substantive values they serve. 2. Technology also appears to be indifferent with respect to politics, at least in the modern world, and especially with respect to capitalism and socialism. 3. The socio-political neutrality of technology is usually attributed to its "rational" character, the universality of the truth it embodies. 4. Technology is neutral because it

Feenberg stresses further that environmentalist's struggles with technology represent "the single most important domain of democratic intervention into technology. Environmentalists want to reduce harmful and costly side effects of technology to protect nature and human health."[109] Feenberg is here concerned with how one lives a good life given one's technological environment. In particular, he is interested in how we might "intervene in the design process in the defense of the conditions of a meaningful life and a livable environment."[110] According to him:

> Technology is not inherently good or bad, and can be used to whatever political or social ends desired by the person or institution in control. Technology is a 'rational entity' and universally applicable [...].One may make exceptions on moral grounds, but one must also understand that the "price for the achievement of environmental, ethical, or religious goals [...] is reduced efficiently.[111]

Feenberg is of the view that technology is subject to conscious, sensible social control and existing technology is not the result of simply rational decision about the most competent and efficient way to do things but depends on social choices between alternative paths with different environmental consequences. Hence, regulations, according to him, can lead to technological changes that enhance economic activity rather than obstruct it.[112]

e) Langdon Winner (1944)

Langdon Winner, in *Autonomous Technology*, is equally of the view that "technology and its various manifestations have become virtual obsessions in discussions about politics and society on a wide variety of fronts." According to him, people from different sections and professional backgrounds in the society "are now united in the conclusion that something we call 'technology' lies at the core of what is most troublesome in the condition of the world."[113] Winner laments the overall effects which new technologies has on civil society by destabilizing the vitality of citizens in such a way that people see themselves not as citizens, not as public

---

stands essentially under the very same norms of efficiency in any and every context. Cf. FEENBERG, A., *Transforming Technology: A Critical Theory Revisited*, pp. 5–6.

109 FEENBERG, A., *Questioning Technology*, p. 93.

110 *Ibid.*, Preface, p. xiv.

111 FEENBERG, A. *Critical Theory of Technology*, p. 6.

112 FEENBERG, A., *Between Reason and Expereince: Essays in Technology and Modernity*, Cambridge, Massachusetts: The MIT Press, 2010, p. 3.

113 WINNER, L., *Autonomous Technology: Technics-Out-Of-Control as a Theme in Political Thought*, Cambridge, Massachusetts: The MIT Press, 1978, p. 2.

persons in any expressive sense, but as human constituents in current processes of technology-centered production.

> Our whole way of life is increasingly technologically mediated. You can't separate society from technology any more, or politics from technology, or culture from technology. The life around us is very largely influenced by choices in design and shape of technology.[114]

Langdon advocates for a "political theory of technology" because technology is now a primary means of constructing our society. He maintaines that there should be that recognition that "every political philosophy in a given time implies a technology…and…that every technology of significance to us implies a set of political commitments that can be identified if one looks carefully enough."[115] He concludes that we must think long and hard about the systems we choose because they can be either constraining or flexible in terms of future choices. Technology, he argued, should be motivated by and serve political aims.

f) From Technology to Technoscience: Gilbert Hottois (1946)

Recently, the term "technoscience" coined by Belgian philosopher Gilbert Hottois[116] in the late 1970s has found favour in the writings of some, if not all, philosophers of technology as well as historians of technology. Advocates of the term maintain that the practices, objects, and theories of sciences and technology, which even if they once were separate professional communities, have blurred to a point where they share many important features, without a doubt, to a point where their similarities outweigh their differences. With a tone of restraint, Barry Barnes writes of "near consensus on the predominance of technoscience as something characteristic particularly of recent times."[117] Latour in *Science in Action* defined and in fact popularized "technoscience":

114 WINNER, L., *New Options*, 67, May 28, 1990.

115 WINNER, L., "Technē and Politeia," in *The Whale and the Reactor: A Search for Limits in an Age of High Technology*, Chicago: University of Chicago Press, 1986, p. 52.

116 Gilbert Hottois is a Belgian professor of philosophy at the Free University of Brussels who specialises in Bioethics. His works on technosciences includes: HOTTOIS G., *Philosophies des sciences, philosophies des techniques, sous l'égide du college de France*, Paris: Odile Jacob, 2004; HOTTOIS G., *La Philosophie des technosciences*, Abidjan: Presses des Universités de Cote d'Ivoire, 1997; HOTTOIS, G. *La Science entre valeurs modernes et postmodernité*, Paris: Vrin, 2005.

117 BARNES, B., "Elusive Memories of Technoscience," in *Perspectives on Science*, 13, no. 2, 2005, (142–165), p. 155.

> To remind us of this important distinction I will use the word technoscience from now on, to describe all the elements tied to the scientific contents no matter how dirty, unexpected or foreign they seem, and the expression 'science and technology', in quotation marks, to designate *what is kept of technoscience* once all the trials of responsibility have been settled. The more 'science and technology' has an esoteric content the further they extend outside. The more 'science and technology' is only a sub-set which seems to take precedence only because of an optical illusion.[118]

Science, as we understand it today, is fundamentally different from the *doctrina* and *scientia* of the Middle Ages, and especially from the Greek *epistemi.*[119] Technology is more than equipment employed as a means to verify scientific hypotheses.

Though not exhaustive, this is as far as we will go in examining the meaning and the question of technology and science as is conceived by different contemporary schools of thought. We have chosen these authors in particular because some of their insights will help us to understand the critical environment within which Heidegger examines the question of technology as well as open an avenue for responsible and democratic steps as alternative approach to technology in the last part of our study. In what follows, we shall be examining the question of technology in Heidegger and here, we will see how his interpretation of technology is influenced by his quest for the real meaning of Being.

4.4. Heidegger and the Question of Science and Technology

Martin Heidegger is often regarded, and rightly too, as the philosopher who has so strongly shaken the foundations of modern science and technology. He criticizes not the symptoms of decay exhibited by science nor its instrumental use but rather the very scientific-technological vista of our present life.

The primary question perhaps could be: what motivates one to study Heidegger's philosophy of technology? Or to put it in another way, what calls for philosophical reflection on the topic or idea of technology itself? There is something within the nature of technology that allows it to be questioned and sought after as idea. That a philosophy of technology is not only possible, but also inseparable from modernity is one of the key points of Heidegger's works on the subject. A good analysis would therefore turn to his works; *The Question Concerning Technology* and *The Turning,* where Heidegger spelt out expressively his philosophy of technology.

---

118 LATOUR, B., *Science in Action: How to Follow Scientists and Engineers through Society*, Cambridge: Harvard University Press, 1987, pp. 174–175.

119 ACKERMANN, R., *Data, Instruments, and Theory: A Dialectical Approach to Understanding Science,* Princeton: Princeton University Press, 1985, p. 125.

Heidegger recognizes technology as an act of revealing instead of simply a human accomplishment. In the modern epoch, technology symbolizes the way in which *Homo sapiens* stand with and interrelate with the world. The danger of this era is the degree to which the whole lot has become available, accessible, and disposable to human manipulation, practically without limit.

Like other technology thinkers both existential and critical, Heidegger argues, as we shall see below, that technology has grown beyond control and warned of a technological understanding of being. Instead of saying "yes" or "no" to technology, he offered a new idea of letting-be or 'releasement' (*Gelassenheit*); his answer was both "yes" and "no" to technology. This will make our relationship with technology somehow simple but complex because we let the technical devices enter our daily life as necessity, and simultaneously, leave them outside, that is to say, let them alone, as insignificant, things which are nothing absolute but remain reliant on something higher.

What we will see from this part of our study is that Heidegger's philosophy of technology stems directly from his ontology. That is, the root of his philosophy of technology lies in his ontology, and in the question of being. Specifically that his notion of technology, as the enframing destining spirit of this age, is a revelation of being itself as technology in this age. The metaphysical objectification of Being is profoundly challenging, which obtains its most menacing expression in the age of technology or the world picture (*Gestell*). Heidegger's philosophy of technology therefore proclaims not only the danger inherent in the technical, but also how this danger can be surmounted.

Trish Glazebrook in *Heidegger's Philosophy of Science* positions science as the 'essence' of modernity in Heidegger's work. According to her, Heidegger more often refers to the technological than the scientific. Technology enframes our world view but, she argues, it is the scientific that is the 'essence' of modernity. "For Heidegger, science is not just one phenomenon among several: it is the determination of the metaphysics of modernity."[120] In this section we shall first examine Heidegger's teaching on Science and calculative thinking which according to him, gives rise to modern technology.

120 Glazebrook, T., *Heidegger's Philosophy of Science*, New York: Fordham University Press, 2000, p. 208.

4.4.1. The Question of Science

Heidegger made two seemingly simple but related statements about science.

> The essence of what we today call science is research.[121]
> Science is the theory of the *real*.[122]

These seemingly quite common place statements appear to be well within the competence of everyone. Yet that simplicity is deceptive as any student of Heidegger might have guessed.

What, then, is science? Science, too, is a way "in which all that is presents itself to us." But, like technology, Heidegger affirms, science is no mere "fabrication of man." In modern science Heidegger sees a destining beyond simply man's quest to know. In the end, he states that "[modern] science is the theory of the real" and that the present age of science has been thought through since Greek time. Likewise, science invites a contrast between Greek and East Asian approaches to the real. Whereas science remains something which begins in Greek thinking, it is clear that Heidegger sees it, today, as something that has moved away from the essential character previously thought.

He listed three attempts to define modern science and to distinguish it from ancient and medieval science. It has been assumed that modern science deals with facts while the older science dealt with speculative propositions: "This is true in a certain respect. But it is equally undeniable that the medieval and ancient sciences also observed the facts and that modern science also works with universal propositions and concepts."[123]

A second suggestion is put forward which distinguishes the varied epochs of science on the basis of experimentation. The establishment of knowledge experimentally however is not the exclusive characteristics of modern science. Experiment was quite familiar to medieval man. "This kind of experience lies at the basis of all contact with things in the crafts and in the use of tools."[124]

A third position argued that modern science was distinct by virtue of its calculations and measurements. But science is always calculating and measuring. The ancient science no less than the modern science sought to measure than to calculate.

---

121 HEIDEGGER, M. "Age of the World Picture," in *QCT*, p. 118.

122 HEIDEGGER, M. "Science and Reflection," in *QCT*, p. 157.

123 HEIDEGGER, M. "Modern Science, Metaphysics, and Mathematics," in *BW*, p. 271.

124 *Ibid.*, p. 272.

Setting aside these positions, Heidegger proposed the distinctive characteristics of modern science. "From Kant comes the oft-quoted but still little understood sentence, however, I maintain that in any particular doctrine of nature only so much *genuine* science can be found as there is mathematics to be found in it."[125]

The meaning of modern science then is derived from the meaning of mathematics. "Ta mathemata means for the Greeks that which man knows in advance in his observation of whatever is."[126] To come to the knowledge of a vegetable requires a prior knowing of vegetableness. "...the body as the bodily, the plant-like of the plant, the animal-like of the animal, the thingness of the thing and so on."[127] The mathematical then "is the fundamental presupposition of the knowledge of thing." It is out of and within this mental presupposition that things can be learned at all. As Foltz, interpreting Heidegger, explains:

> Modern science is not mathematical due to its use of mathematics, but rather the converse: it necessarily employs mathematics—and indeed, employs a peculiar kind of mathematics that was unknown to the ancient Greek mathematicians—because it is essentially mathematical, because it is founded upon ta mathemata, that which can and must already be known in advance of any actual knowledge of things.[128]

The mathematical in this Greek meaning signifies "the fundamental condition for the proper possibility of knowing."[129] It is because the mathematical means learning in this way that numbering is itself something mathematical.

Heidegger explained numeration as mathematical in a simple example. To see three chairs in a room is a commonplace event. "What 'three' is the three chairs do not tell us, or three apples, three cats, or any other three things."[130] We count chairs as already knowing three, recognize it as something already attained and possessed.

> This means that number is something mathematical. Only because numbers represent, as it were, the most striking of always-already-knowns, and thus offers the most familiar instance of the mathematical, is "mathematical" promptly reserved as a name for the numerical. In no way, however, is the essence of the mathematical defined by numberness.[131]

---

125 *Ibid.*, p. 273.

126 "The Age of the World Picture,"*QCT*, p. 118.

127 *Ibid.*

128 Foltz, V. B., *Inhabiting the Earth*, p. 66.

129 "Modern Science, Metaphysics, and Mathematics," in *BW*, p. 278.

130 *Ibid.*, p. 276.

131 "The Age of the World Picture," *QCT*, p. 119. Numeration and numbering is the manner of mathematical and the mathematical means a knowing for which "something is stipulated in advance as what is already known." *Ibid.*

To elaborate the meaning of the mathematized science, Heidegger instituted a contrast between Aristotle and exponents of modern science concerning the common question of motion. Aristotle began from the motion "which shows itself in beings." In this, Aristotle began from "the unimpeachable evidence of the senses as to each fact."[132] Bodies which belong to nature are "in themselves movable with respect to location." "How a body moves, i.e., how it relates to place and to which place it relates—all this has its basis in the body itself."[133]

This natural body moves according to its nature. The earthly body moves downward. The fiery body moves upward. Each body has its place and it strives toward that place. According to Aristotle then, the foundation for natural movement lies in the nature of the body itself in its most proper being. The earthly body falling towards the centre, the fiery body moving from the centre, each moves in a straight line. But the heavenly bodies move the center in a circular motion. In these distinctive motions, the circular motion is of the higher order.

> In circular motion the body has its place in the motion itself; for this reason such motion is perpetual and truly in being. In rectilinear motion the place lies only in one direction, away from another place, so that motion comes to an end over there.[134]

For Aristotle, motion has a definite relation to place. Circular motion which derives from the being of the thing is most complete and dependent because place is within the motion itself. Because it arises out of the very being of the body in question, the motion is independent of anything outside itself.

> With Aristotle, however, this "force," dynamis, the capacity for its motion, lies in the nature of the body itself. The kind of motion of the body and its relation to its place depends upon the nature of the body. The velocity of natural motion increases the nearer the body comes to its place; that is, increase and decrease of velocity and cessation of motion depend upon the nature of the body.[135]

In 1638, Galileo set forth a law of motion which was an anticipation of Newton's law. Galileo was concerned with the body on a horizontal plane with every obstacle removed. "…The motion of the body over this plane would be uniform and perpetual if the plane were extended infinitely."[136] This law of Galileo considered a motion which did not exist anywhere. For Aristotle, bodies move according to their nature. For Galileo, all bodies are alike and motion is uniform and in a straight line

132 HEIDEGGER, M., "Modern Science, Metaphysics, and Mathematics," *BW*, p. 282.
133 *Ibid.*, p. 284.
134 *Ibid.*
135 *Ibid.*, p. 285.
136 *Ibid.*, p. 290.

and all bodies fall equally fast. In the presence of his intellectual opponents, Galileo dropped bodies of unequal weight from the leaning tower of Pisa:

> Bodies of different weights did not arrive at precisely the same time after having fallen from the tower, but the difference in time was slight. In spite of these differences and therefore really against the evidence of experience, Galileo upheld his proposition.[137]

His opponents disagreed. The Aristotelians correctly saw that the falling bodies did not reach the ground at the same time and they argued from the evidence of their eyes that motion was not uniform. Galileo argued that the time difference was slight enough to be ignored and argued against the evidence that motion was uniform. Galileo did this because he already thought and continued to think mathematically regarding the essence of a body and the nature of its motion. Thinking mathematically, Galileo came to the evidence with *a priori* knowledge, with *a priori* conception of what ought to happen. Galileo says: "I think in my mind of something movable that is left entirely in itself,"[138] and this mathematical thinking in his mind provided the measure for "the laying-out of the realm which is in the future."

In 1686–87, Newton published *Mathematical Principles of Natural Philosophy*. In this work, Newton stated the first law of motion of the principle of inertia: "Every body continues in its state of rest, or uniform motion in a straight line, unless it is compelled to change that state by force impressed upon it."[139]

This law of motion contradicted Aristotle on every form. The law referred to everybody and thus the Aristotelian distinction between earthly and heavenly bodies is set aside. The Aristotelian priority of circular motion over straight -line motion is rejected. Indeed in Newton's physics it is circular motion which requires explanation. Place is no longer where a body belongs according to its nature, it is simply a position in relation to other positions. Nature is no longer the *inner* principle over which the motion of the body follows. Newton's law referred to a body which is not molested by any force, a body left to itself.

For Newton then no less than for Galileo, the law of motion reveals the nature of the mathematical. Newton's law of inertia is mathematical because it has determined in advance spheres where investigations can be pursued. He does this

137 *Ibid.*

138 *Ibid.*

139 Heidegger, M., "Modern Science, Metaphyiscs and Mathematics," in *BW*, pp. 279–280. Cf. Issac Newton, *Mathematical Principle of Natural Philosophy and His Sytem of the World*, trans. Andrew Motte, 1729; revised translation, Florian Cajori, Berkeley: University of California Press, 1946, p. 13.

because it has posed in advance conditions "to which nature must answer in one way or the other." Within the arising of this law of inertia, things of nature can show themselves.

> Upon the basis of the mathematical, the experiential becomes the modern experiment. Modern science is experimental because of the mathematical project. The experimenting urge to the facts is a necessary consequence of the proceeding mathematical skipping of all facts.[140]

Things no longer appear as they are in themselves, but only as they are allowed to appear in the axiomatically determined sphere of the mathematical.

It is not possible to return to the original statement on research as the essence of science. Modern science is mathematical. Therefore the characteristic procedure of the mathematical will enter into the understanding of science as research. Because of the determining role of the mathematical, science as research will be initially established as research "through the projected plan and through the securing of that plan in the rigor of procedure."[141] The science of physics is "the knowledge of material corporeality in its motion." But modern physics is also mathematical and this means that in modern physics, "something is stipulated in advance as what is already-known."[142] It is only within a ground plan arising out of the mathematical nature of science that "an event in nature becomes visible as such an event."

> The projection sketches out in advance the manner in which the knowing procedure must bind itself and adhere to the sphere opened up [...]. Through the projecting of the ground plan [...] procedure makes secure for itself its sphere of objects within the realm of Being.[143]

Projection of the ground plan, and the rigor with which it is adhered to, become properly themselves in methodology and this is the second requirement of research as the essence of science. Methodology brings the sphere that is projected into rule and law and objectivity. "Methodology...has the character...of explanation. Explanation takes place in investigation. In the physical sciences investigation takes place by means of experiment."[144]

It is because modern science is mathematical that it is necessary for it to be experimental and research is but the extension of the mathematical bases of modern thinking into new spheres of knowing. The medieval "doctrina" and the Greek

---

140 Heidegger, M., "Modern Science, Metaphysics, and Mathematics," in *BW*, p. 292.
141 Heidegger, M., "The Age of the World Picture," in *QCT*, p. 120.
142 *Ibid.*, p. 119.
143 *Ibid.*, p. 118.
144 *Ibid.*, p. 121.

"episteme," even when they utilized measure and number were not experimental in the modern meaning of the word. Even the Medieval Roger Bacon who demanded the *experimentum* was firmly a medieval thinker. He demanded that the *argumentum ex-re* replace the *argumentum ex-verbo.*[145] He sought to replace the discussion of doctrine with a careful observation of things. Yet he was not a forerunner of modern science.

> For in all this, that which is decisive about the experiment is completely missing. Experiment begins with the laying down of a law as a basis. To set up an experiment means to represent or conceive (vorstellen) the conditions under which a specific series of motions can be made susceptible of being followed in its necessary progression, i.e., of being controlled in advance by calculation.[146]

Even the careful observations of Aristotle were not experiment in this methodological way. They merely observed things as they presented themselves to them. They did not delimit in advance the frames of reference within which things could be observed. They did not like the observation nor the measure, what they like was the mathematics which forms the possibility of approaching reality within the parameters of acquired knowing.

Heidegger mentioned a third characteristic of science as research. Every science involves "the projection of a circumscribed object sphere."[147] Science then is necessarily knowledge of individualized character. The findings of experiments reveal further data, which creates the need for further experiment, and creates the demand for specialization. Such specialization is not necessarily about things. "Is rather an essential necessity of science as research, specialization is not the consequence but the foundation of the progress of all research."[148]

Research culminates in on-going activity. The method of science becomes circumscribed by means of its own results. In its on-going activity, varied scientific discipline expands and multiplied. In this expansion, a certain institutional character of science is consolidated. Conforming types of method are able to work together to facilitate checking of results. This institutional aspect of research as

---

145 Heidegger, M., "The Age of the World Picture," in *QCT*, p. 122. *Argumentum ex-re* means *argument from the things* while *argumentum ex-verbo* means *argument from the word.*

146 *Ibid.*, p. 121. Heidegger defines Experiment as "that methodology which, in its planning and execution, is supported and guided on the basis of the fundamental law laid down, in order to adduce the facts that either verify and confirm the law or deny it confirmation." *Ibid.*, p. 122.

147 *Ibid.*, p. 123.

148 *Ibid.*

on-going activity indicates "that modern science is beginning to enter under decisive phase of its history." This development of the institutional character of science also affects the scientists themselves.

> The scholar disappears. He is succeeded by the researchman who is engaged in research projects. These, rather than the cultivating of erudition, lend to his work its atmosphere of incisiveness. The research man no longer needs a library at home. Moreover, he is constantly on the move. He negotiates at meetings and collects information at congresses. He contracts for commissions with publishers. The latter now determine alone with him which books must be written.[149]

At this point, the second statement of Heidegger concerning science as theory of the real comes into view. The content and meaning of the term 'real' requires to be determined. For the Greeks, *Physis* signifies "things in so far as they originate and come forth from themselves."[150] *Physis* "is the process of arising, of emerging from the hidden whereby the hidden is first made to stand." *Thesis* for the Greeks meant to place, to position. "*Physis* is thesis: from out of itself to lay something before, to place it here, to bring it hither and forth (*her-und vor-bringen*), that is, into presencing."[151]

This laying before, this placing here involves work. Reality then meant for the early Greeks a working which brought something to presence of itself. Reality meant for the Greeks the presencing of self-bringing-forth. And it was this working of self-bringing-forth that Aristotle translated as *energia*. In this rendering, Aristotle was largely faithful to the pre-Socratic understanding. *Energia* signified "that something comes to stand and to lie in unconcealment."

The Latin translated *energia* as '*actio*' and '*actus*'. The context of this '*actus*' is that which is brought hither and forth and follows as a result of action. It follows out and follows upon: "The real is now that which has followed as consequence. The consequence is brought about by the circumstance (*Sache*) that precedes it, i.e., by the cause (*Ursache*) (*causa*)."[152] *Energia* as action has come to be as doing and that which follows from such doing is the factual. The factual is "something that presence which sets itself forth from out of itself."[153] The real now shows itself as an object, as that which stands over against (*Gegen-Stand*).

---

149 *Ibid.*, p. 125.

150 Heidegger, M., "Modern Science, Metaphysics, and Mathematics," in *BW*, p. 274.

151 Heidegger, M., Science and Reflection, in *QCT*, p. 159; See also, *Introduction to Metaphysics*, p. 16.

152 *Ibid.*, p. 161.

153 *Ibid.*, p. 162.

It is now time once again to take up Heidegger's second statement concerning science as the theory of the real. The Greek verb "*theorein*" meant to look attentively "on the outward appearance wherein what presences becomes visible and, through such sight—seeing—to linger with it."[154] The life which receives its existential determination from such seeing and lingering is the life of beholding. The beholder is that one "who looks upon the pure shining-forth of that which presences."[155]

In Greek term "theory" means the beholding of the shining forth of that which presences, of that which gives itself standing presence. The Latin translation of *theorein* into *contemplare* and *theoria* into contemplation made what was essential in the Greek words to vanish at a stroke. The *Templum* which is at the centre of the Latin word is a place set apart from the profane of everyday life. The *Templum* is the marked-out spot which can be seen from all points and from which all points can be seen. From this marked out place, the Latin diviners could make their observations about the future. In the Latin contemplation, it is no longer question of a radiating presence into a simple beholding which was untainted by ulterior purposes. In the Latin words *contemplare* and *contemplation*, there is question of seeing which has set things apart. In the Latin language, contemplation was a seeing which encroached upon the real by compartmentalizing.[156]

The modern scientific community speaks of the theory of relativity, of the quantum theory, of the theory of evolution. In modern scientific understanding, the theory is an observation of the real. Such a scientific observation is also encroaching upon the real so that the thing being observed will present itself "in a surveyable series of related causes."[157] The theory in the modern science "sets upon the real." The theoretical nature of modern science then does not mean for Heidegger that

154 *Ibid.*, p. 163.

155 *Ibid.*, p. 164. At the root of the noun 'theoria' there is 'thea'. This thea is also at the root of aletheia, that is, "the unconcealment from out of which and in which that which presences, presences." Theoria then is "paying heed to the unconcealement of what presences." *Ibid.*

156 *Ibid.*, p. 166. Contemplation in the Latin language signified "a looking-at that sunders and compartmentalizes." The Greek 'theoria'which was a pure relationship to the outward appearance to that which presences was lost in the Latin contemplation. Theorein which was the beholding of "those appearances that in their radiance...brings the presence of God to shine forth" was radically uttered in the Latin contemplare. In another text, Heidegger states: "In the Greek language, one is not speaking about the action of seeing, about videre, but about that which gleams and radiates. But it can only radiate if openness has already been granted." Cf. "Task of Thinking," in *On Time and Being*, p. 66.

157 Heidegger, M., "Science and Reflection,"in *QCT*, p. 168.

science is a disinterested search for the real. The sciences are theoretical because they set upon the real and mark out 'in advance the possibilities for the posing of question'.

Science whose essence is research understands that which is present as the factual and science which is theoretical sets upon that factual and traps it and sets up the conditions under which it can be known. Science makes secure at any given time a region of reality as its object area. The theoretical nature of science determines in advance the way in which the thing can be observed. The real is that aspect of the factual which theory can impose upon, set upon and entrap it in its observation. Greek and medieval thought would find this approach to reality to be incomprehensible.

For all the capacity of the sciences to observe and set upon the real, there are questions which fall outside of the predetermined possibilities imposed by the sciences as theoretical

> Physics as physics can make no assertions about physics. All the assertions of physics speak after the manner of physics. Physics itself is not a possible object of a physical experiment. The same holds for philology. As the theory of language and literature philology is never a possible object of philological observation. This is equally the case for every science.[158]

Nature for physics, language for philology, history for historiography, all reveal that despite the success of the sciences, there is something "which is not to be gotten around."[159] Something is moving in the sciences which the sciences cannot subdue or get around. This something is inconspicuous and withdrawn yet it holds sway in the sciences. That which the sciences cannot get around, that which remains unthought in the sciences directs man "on a way that brings us before that which is worthy of questioning."[160] But that which is worthy of questioning; that which calls for thinking is "the presence of what is present, the being of Beings."[161] Science which is the theory of the real and an enframing of being leaves unthought that which cannot be gotten around. Of its nature "science itself does not think and cannot think". The thinking involved in the sciences does not think Being. Scientific thinking is a calculative thinking for which nothing comes "save what can be calculated."[162] Science does not think Being rather it enframes Being. For this reason, it must be overcome.

---

158 *Ibid.*,p. 176.

159 *Ibid.*, p. 175.

160 *Ibid.*, p. 179.

161 HEIDEGGER, M., *What is Called Thinking*, p. 244.

162 HEIDEGGER, M., "What is Metaphysics" in *Existence and Being*, p. 387.

4.4.2. *Calculative Thinking*

This transition from theoretical science to ontology, according to Heidegger, demands a movement from Dwelling Thinking to Called Thinking. To wrest Being out of hiddenness, there is required the event of Called Thinking

Thus, Heidegger presents the philosophical dialogue with a drastic distinction of two modes of thinking: *calculative thinking and rational meditative thinking.*[163] The calculative mode predominates in modern secular scientific and technological age. It is the most precarious feature of our scientific age captivated with the superficial thinking that confuses us from profound thinking. Though Heidegger is not disparaging its ability to organize our world, he is rather warning against its power to absorb completely our energy and attention. It is based on willfulness and the desire both to objectify everything and to dominate the object of thought. It is concerned with the ontic level of the being of man, and the being of worldly things.

In a work titled *What is Metaphysics?* we discover possibly Heidegger's strongest interpretation of the difference between calculative thinking and meditative thinking which he sometimes calls *essential thinking* as we saw in earlier part of our study. Without doubt, Heidegger refers back over and over again to this work. To start with, with respect to calculative thinking, Heidegger states:

> All calculation makes the calculable "come out" in the sum so as to use the sum for the next count. Nothing counts for calculation save for what can be calculated. Any particular thing is only what it "adds up to", and any count ensures the further progress of counting. This process is continually using up numbers and is itself a continual self consumption. The "coming out" of the calculation with the help of what-is counts as the explanation of the latter's Being. Calculation uses every-thing that "is" as units of computation, in advance, and, in the computation, uses up its stock of units. This consumption of what-is reveals the consuming nature of calculation. Only because number can be multiplied indefinitely [...] is it possible for the consuming nature of calculation to hide behind its "products" and give calculative thought the appearance of "productivity" [...]. Calculative thought places itself under compulsion to master everything in the logical terms of its procedure.[164]

Calculative thinking is not simply a synonym for the method of experimental science but exemplifies any thinking technique or procedure that plans to control and influence circumstances or surroundings. "It [calculative thinking] has no notion that in calculation everything calculable is already a whole before it starts working out its sums and products, a whole whose unity naturally belongs to the

---

163 Heidegger, M., *Discourse on Thinking*, Introduction, p. 24.

164 Heidegger, M., "What is Metaphysics," in *Existence and Being*, pp. 387–388.

incalculable which, with its mystery, ever eludes the clutches of calculation. That which, however, is always and everywhere closed at the outset to the demands of calculation and, despite that, is always closer to man in its enigmatic unknowableness than anything that 'is', than anything he may arrange and plan, this can sometimes put the essential man in touch with a thinking whose truth no "logic" can grasp."[165]

Heidegger states further that calculative thinking "calculates," i.e., it "plans and investigates"[166]; it sets target and wants to achieve them. It serves particular function; it reflects on and works out countless fresh and always different potentialities to expand. As Rojcewicz puts it:

> Calculative thinking [...] is the attitude that beings are there simply for what we can get out of them that the world is there for us to exploit. In order words, what Heidegger here calls calculative thinking is precisely the attitude of modern technology, the impositional attitude [...]. It is our contemporary attitude, by which we approach the things of nature with our gas-guzzling self-interest [...]. It sees all things as there to be ravished and motivates their actual ravishment.[167]

In spite of this efficiency and impositional mentality of thinking; regardless of our modern tremendous advances in research proceedings and investigations in various quarters, Heidegger in *Memorial Address*, insists that a "growing thoughtlessness" [*Gedankenlosigkeit*], which hinges on the idea that modern "man today is *in flight from thinking*"[168] should be dealt with. According to Heidegger:

> Let us not fool ourselves. All of us, including those who think professionally, as it were, are often enough thought-poor; we all are far too easily thought-less. Thoughtlessness is an uncanny visitor who comes and goes everywhere in today's world. For nowadays we take in everything in the quickest and cheapest way, only to forget it just as quickly, instantly. Thus one gathering follows on the heels of another. Commemorative celebrations grow poorer and poorer in thought. Commemoration and thoughtlessness are found side by side.[169]

Calculative thinking computes. It's constantly computing, increasingly showing potentials and at the same time more cost-effective potentials. "Calculative

---

165 *Ibid.*, p. 388.

166 Heidegger, M., "Memorial Address," in *Discourse on Thinking*, p. 64.

167 Rojcewicz, R., *The Gods and Technology: A Reading of Heidegger*, Albany: State University of New York Press, 2006, pp. 215–216.

168 Heidegger, M., "Memorial Address" in *Discourse on Thinking*, p. 45.

169 *Ibid.*

thinking races from one prospect to the next. Calculative thinking never stops, never collects itself".[170]

Calculative thinking was concerned with things, objects, abstracted from live experience. It positioned a thing in the "single exclusive reified context",[171] of thing experience; exclusive because "my 'I'"[172] was removed."[173] The result of this "Iness" is that "calculative thinking sees nothing other than itself and therefore considers nothing other than itself. In calculative thinking humans act like machines, albeit possibly extremely smart machines devoid of emotion, or more precisely, devoid of the experience of emotion… Calculative thinking might *think* that it is acting wisely, but it has no capacity for recognising genuine meditative thinking (or wisdom)."[174] In Heidegger's ultimate mental picture, modern man have lost the original wholesomeness and holy embeddedness in Being, and have become lost in the material world of things, of human projects, of human pig-headedness, what he calls "fallenness" and have forgotten the awareness of Being.

Heidegger therefore makes it clear that "calculative thinking is not meditative thinking, not thinking which contemplates the meaning which reigns in everything that is."[175] Meditative thinking engaged with meaning as it included me, the meaning of a situation for me at a particular time, and the contemplation of such. It involved thinking the whole of experience, experiencing and experienced together; not just experienced alone as in abstract, calculative thinking, or as Heidegger called it elsewhere, "technical interpretation of thinking" which goes "back to Plato and Aristotle."[176]

---

170 *Ibid.*, p. 46. To say that calculative thinking 'computes' does not necessarily mean that it is computational. It does not require calculators or computers. It is not necessarily scientific or sophisticated. It should be understood in the sense in which a person is said to be "calculating", which means that he is designing; he uses others for his own selfish interest. Cf. Rojcewicz, R., *The Gods and Technology*, p. 215.

171 Heidegger, M., *Towards the Definition of Philosophy*, trans. T. Sandler, London: The Athlone Press, 2000, p. 58.

172 *Ibid.*, p. 62.

173 *Ibid.*, p. 72.

174 Haynes, J. D., *Perspectival Thinking for Inquiring Organisations*, p. 46.

175 Heidegger, M., "Memorial Address," in *Discourse on Thinking*, p. 46.

176 "Letter on Humanism" in *BW*, p. 218. The present supremacy of calculative thought, as evidenced in the modern triumph of science and technology, was preceded by a long history of metaphysical exploration that born a similar ultilitarian focus. Cf. Leslie P. Thiele, *Timely Meditations: Martin Heidegger and Postmodern Politics*, Priceton, New Jersey: Princeton University Press, 1995, p. 107.

Meditative thinking is a sort of thinking associated with man, it is constituent of his character; but all the same, it is a method of thinking that requires being stimulated or agitated. When Heidegger asserts above that modern man is "in flight from thinking" he was actually talking of man being enmeshed in calculative thinking and flight from meditative thinking. What differentiate calculative thinking from meditative thinking is that meditative thinking detect, examine, contemplate, and stimulate a consciousness of what is essentially happening close to us and within us. As Rojcewicz states: "It does not have a practical interest, it does not view things as means to an end, but instead dwells on things for the sake of disclosing that which makes them be what they are."[177]

Meditative thinking, according to Heidegger, "does not just happen by itself any more than does calculative thinking." It entail effort, commitment, willpower, concern, practice, but all together "it must be able to bide its time, to await as does the farmer, whether the seed will come up and ripen."[178] During the "*Memorial Address*," Heidegger himself made an effort to execute Meditative thinking when he attempts to take the listeners from a condition where they are docile, inactive listeners to a situation wherein they effectively meditate and think in relation to what is happening, beyond the mere occasion of commemoration. His speech below gives us an illustration of what the practice of Meditative thinking is all about:

> What does this celebration suggest to us, in case we are ready to meditate? Then we notice that a work of art has flowered in the ground of our homeland. As we hold this simple fact in mind, we cannot help remembering at once that during the last two centuries great poets and thinkers have been brought forth from the Swabianland. Thinking about it further makes clear at once that Central Germany is likewise such a land, and so are East Prussia, Silesia, and Bohemia. We grow thoughtful and ask: does not the flourishing of any genuine work depend upon its roots in a native soil? Johann Peter Hebel once wrote: "We are plants which—whether we like to admit it to ourselves or not—must with our roots rise out of the earth in order to bloom in the ether and to bear fruit." [...]. The poet means to say: For a truly joyous and salutary human work to flourish, man must be able to mount from the depth of his home ground up into ether. Ether here means the free air of the high heavens, the open realm of the spirit. We grow more thoughtful and ask: does this claim of Johann Peter Hebel hold today? Does man still dwell calmly between heaven and earth? Does a meditative spirit still reign over the land? Is there still a life-giving homeland in whose ground man may stand rooted, that is, be autochthonic?[179]

---

177 Rojcewicz, R., *The Gods and Technology*, p. 216.

178 Heidegger, M., "Memorial Address," in *Discourse on Thinking*, p. 46–47.

179 "Memorial Address" in *Discourse on Thinking*, pp. 47–48.

What Heidegger is making clear to his countrymen here is that, although man, over and above the animals, is a thinking being, yet we need to coach ourselves in the aptitude to think meditatively, to meet reality head-on and in doing so, face our live meditatively.

The danger for man is to be dislodged not only from his existential world of reality, but he is also alienated from himself. The danger in the age of modern technology is precisely a threat to the essence of man. But if man thinks meditatively, he will become conscious of the risk concealed in modern technological and its value, and hence, he can take action on it. Heidegger stresses in his speech that the greatest danger is not that technological things such as atom bombs, might get out of hand and annihilate human existence. The impending danger is that calculative thinking, the impositional attitude, might overthrow meditative thinking. As he puts it further:

> What great danger then might move upon us? Then there might go hand in hand with the greatest ingenuity in calculative planning and inventing indifference toward meditative thinking, total thoughtlessness. And then? Then man would have denied and thrown away his own special nature—that he is a meditative being. Therefore, the issue is the saving of man's essential nature. Therefore, the issue is keeping meditative thinking alive.[180]

To facilitate an understanding of what this means, Heidegger recommends that we reflect on the attitude or mind-set we have towards modern technological appliances. There is no doubt that technological appliances are indispensable in today's mechanized and computerized world. Today, our live is structured in such a way that virtually, "we depend on technical devices."[181] Calculative thinking allow us to be slaves to these devices and they even challenge us to think more or less on how to develop more new sophisticated devices to enhance greater efficiency. In technological society, calculative thinking is promoted over the more fundamental 'meditative thinking'. Calculative thinking is ordained toward the management and control of things and people through rigid frameworks of organized thought. Calculative thinking is an abstract and practical process confined to organizing, managing and controlling. It is a form of thinking which does not consider meaning and yet has the "power to absorb completely our energy and attention." Thus as Gavin Rae observed, "the entire point of Heidegger's critique of technology is to show how it creates a particular form of human understanding that, in turn,

180 *Ibid.*, p. 56.
181 *Ibid.*, p. 53.

creates a particular society defined by quantifiable, calculative thinking."[182] Let us now turn to Heidegger's question concerning technology and his call on man to look for the essence of technology.

4.4.3. *The Question of Technology*

One can say without fear of contradiction that we are becoming increasingly immersed and captivated within a technological culture with each passing day. We appreciate and welcome the way computers are getting faster and faster and at the same time cheaper and cheaper, without knowing how we will use the amazingly flexible computing power they give us. From computers to cell phones, microwaves to Handheld Ultrasound, fax machines to the 3-D camera, technology permeates virtually every part of our lives. Within a little space of time, we decoded the human genome and virtually everyone got a cell phone, an iPod, a GPS and a DVD. Even fast food chains that give us cheap and instant service at any time of day or night stand out as technological triumphs of proficiency, effectiveness and adaptability. In military technology, we live today in a Dronocratic Hypermodernity in which everything is moving and changing fast with computerized precision; where you are recognized by how constructive or destructive your little gadget can be. Ours is a generation on a technological fast lane; from Super Drones to 3D Printers, Smarter appliances and virtual reality; from next generation robotics to emergent artificial intelligence; from Neuromorphic technology to Digital genome as well as precise genetic engineering techniques etc.

We use technology at work, to communicate, to travel, entertainment etc. Practically everything we do involves some sort of technology. In fact, most of us would be completely lost without it. Modern technology has made our lives easier, well-organized, more comfortable and generally more pleasurable. While enormous technological advances in computing, communications, transportation, health, genetics, biotechnology, military technology and even nanotechnology are widely expected, Heidegger, among others, was not particularly enthusiastic about the prospects of technology.

If we consider all that modern technology has made possible, we quickly recognize that along with all the conveniences come just as many dangers, with instruments of war and mass destruction, environmental degradation and global warming being the most notable, though certainly not the only, examples. Interestingly enough, however, Heidegger does not primarily identify the danger

182 RAE, G., *Ontology in Heidegger and Delueze: A Comparative Analysis*, Palgrave Macmillan, 2014, p. 51.

of modern technology with what it does, or at least can, produce. Rather, he sees the danger of modern technology at work in its very essence. Suffice it to say here that Heidegger was not against technology. His aim in *The Question Concerning Technology* was to disclose to man the essence of technology which:

> In no way confines us to a stultified compulsion to push on blindly with technology or, what comes to the same thing, to rebel helplessly against it and curse it as the work of the devil. Quite to the contrary, when we once open ourselves expressly to the essence of technology, we find ourselves unexpectedly taken into a freeing claim".[183]

Heidegger states at the beginning of this text that his concern is the essence of technology. "The essence of technology is by no means anything technological."[184] The text contains his critique of modern technology in addition to his views concerning the possibility for an authentic philosophy of technology. Even though Heidegger distinctively addresses the true nature of technology within this essay, his ideas pertaining to a philosophy of technology are anything but predictable. He is mainly interested in how we, as *Dasein*, can have a free relationship to technology:

> We shall be questioning concerning *technology*, and in so doing we should like to prepare a free relationship to it. The relationship will be free if it opens our human existence to the essence of technology. When we can respond to this essence, we shall be able to experience the technological within its own bounds.[185]

By "essence of technology", Heidegger does not mean the "nature" of the "concept" of technology, or the definition of technology. By "essence of technology", he means how the "phenomenon" of technology "comes to presence." The expression of *Dasein* choosing something freely does not satisfactorily define a free relationship to technology, for it involves much more than a simple act of the will. Furthermore, Heidegger envisions a free relationship with modern technology as one that will entail an absolutely different way of being in the world. If *Dasein* is able to have a free relationship to technology, then Heidegger tells us that it will be one that is entrenched in an entirely new "attitude." He calls this new attitude or distinctively different way of being in the world, "releasement" [*Gelassenheit*].

Heidegger made the distinction between technology and the technological. To understand this distinction, two renowned definitions of technology abound: Technology is a means to an end; it is an "instrument" to meet our needs. Second, Technology is a human activity." These two statements about technology are interconnected "for to posit ends and procure and utilize the means to them is

183 *QCT.*, pp. 25–26.
184 *Ibid.*, p. 4.
185 *Ibid.*, pp. 4–5.

a human activity."[186] These two current conceptions of technology are called instrumental and anthropological definitions of technology. According to Frede, Heidegger proceeds on two levels;

> He distinguishes between (a) the "ontic" level of the factual (for human existence Heidegger introduces the special term "existenti*ell*") that is open to observation, the level of field studies for the phenomenologist, and (b) the "ontological" level, the phenomenological description of the deep structures that underlie and explain the ontic (for the structure of human existence Heidegger introduced the term "existential").[187]

Drawing upon man's traditions rooted deep within Western thinking, Heidegger delves deep into an investigation regarding the essence of technology. To understand what is meant by the essence of something, Heidegger offers the example of a tree. When the word tree is uttered, the idea of a specific tree or a species of tree may strike to mind immediately. Yet there is a fundamental, general quality that subsumes all such images or concepts of trees. It is this basic, entrenched sense of tree, a sense that pervades and suffuses our understanding of every tree that serves as the essence of trees, that is, the "treeness" of the tree, it's "quiddity," its "whatness" that is common to all trees.[188] In another instance, modern technology may remove distances (*Entfernungen, Abstände*), but it does not grasp the essence of the nearness (*Nähe*) and remoteness (*Ferne*) of things because the latter never arises as a question for it.[189]

On the one hand, to arrive at the essence of technology does not mean reverting from the present advanced complexities to the primitive simplicity. In 1955, in an address to the Black-forest peasants, Heidegger declared: "The world now appears as an object open to the attacks of calculative thought, attacks that nothing

---

186 *Ibid.*, p. 4.

187 Frede, D. "The Question of Being," In Guignon, C. ed. *The Cambridge Companion to Heidegger*, Cambridge, UK: Cambridge University Press, 1993, (pp. 42–69), p. 55. Outwardly, this ontology of technology seems to bear some resemblance with the Platonic notion of the Ideal form yet, as Guignon pointed out, a major difference can be drawn between Heidegger's "substance ontology" and the traditional notion of the "metaphysics of presence." According to Guignon, Heidegger's approach challenges the idea that "reality must be thought of in terms of the idea of substance at all." *Ibid.* p. 4 In this way, it is possible for Heidegger to claim that the essence of technology existence prior to the industrial and scientific revolutions of the seventeenth and eighteenth centuries.

188 *QCT*, p. 4.

189 See *GA* 79: 3–4= *Bremer und Freiburger Vorträge*, Ed. Petra Jaeger, Frankfurt: Klostermann, 1994.

is believed able any longer to resist. Nature becomes a gigantic gasoline station, an energy source for modern technology and industry."[190] Such statement makes it appear as if Heidegger is a Luddite who would like to revert from the present commercialism, the negligence and exploitation of natural resources, and information technology, to the primordial world of the pre-Socratics or old traditional Black Forest peasant's way of life.

On the other hand, while Heidegger does not contest that technology has its harmful effects, as his philosophy progresses, he arrives at the startling and interesting conclusion that dwelling on its failure and devastation is more or less technological. "All attempts to reckon existing reality … in terms of decline and loss, in terms of fate, catastrophe, and destruction, are merely technological behaviour."[191] Viewing our condition as creating a crisis that must be unravelled by suitable and courteously proper action is also technological. As Heidegger states;

> That is why the Instrumental conception of technology conditions every attempt to bring man into the right relation to technology. Everything depends on our manipulating technology in the proper manner as a means […]. The will to mastery becomes all the more urgent the more technology threatens to slip from human control.[192]

Heidegger made it plain that being more technological in order to solve technological problems will not help because we are not prepared for the transformation as well as our inability to confront the situation meditatively. Therefore, "[N]o single man, no group of men, no commission of prominent statesmen, scientists, and technicians, no conference of leaders of commerce and industry, can brake or direct the progress of history in the atomic age. No merely human organization is capable of gaining domination over it."[193]

In this section of our study, Heidegger contends that the conventional ways of perceiving technology are deficient. Both the anthropological and instrumental views that technology is one system of human activity among others, and that technology is a neutral instrument, a means to an end, respectively, can be subjugated and manipulated by conscious human understanding of the essence of modern technology. For Heidegger, technology is a unique mode of revealing or connecting to Reality. Intrinsically it is never purely under conscious human regulation,

---

190 Heidegger, M. *Discourse on Thinking*, p. 50. In this address he laments the springing up of television antennaes on the peasant's dwelling, and lunched an attack on the media and its power as we discussed in Poetic Dwelling in section one of our study.

191 *QCT*, p. 48.

192 *QCT*, p. 5.

193 *Discourse on Thinking*, p. 52.

because technology's unique mode of revealing Reality continuously stands before every intended act taken on the basis of what is already revealed. And technology cannot be just one among many neutral kinds of behaviour in which man engages. As a mode of revealing of Being, it is the most essential behaviour of man. In what follows, we shall examine these terms (*Destining, Revealing, Enframing, Gelassenheit* etc) to discover their interrelatedness with regard to Heidegger's interpretation of modern technology.

4.4.4. The Destining of Technological Being

In the writings following *Being and Time*, Heidegger started to speak of destining or sending (*Schicksal/Geschick*) of being.[194] In *Letter on Humanism*, he states: "Man does not decide whether and how beings appear, whether and how God and the gods or history and nature come forward into clearing of Being, come to presence and depart. The advent of beings lies in the destiny of being."[195] This means that it is not Dasein who determines, regulates or controls how or the form in which beings present themselves to humans, but rather being itself does. Further in the text, Heidegger developed his view of technology:

> For such dialogue it is certainly also necessary to free oneself from naïve notions about materialism, as well as from the cheap refutations that are supposed to counter it. The essence of materialism does not consist in the assertion that everything is simply matter but rather a metaphysical determination according to which every being appears as the material of labour.... The essence of materialism is concealed in the essence of technology, about which much has been written but little has been thought. Technology is in its essence a destiny within the history of Being and the truth of Being, a truth that lies in oblivion. For technology does not go back to the *technē* of the Greeks in name only but derives historically and essentially from *technē* as a mode of *alētheuein*, a mode, that is, of rendering beings manifest (*Offenbarmachen*). As a form of truth technology is grounded in the history of metaphysics, which is itself a distinctive and up to now the only perceptible phase of the history of being.[196]

This passage suggests Heidegger's overall view of technology and highlights the central themes around which his discussion on technology will revolve. Thus, there is, according to Heidegger, certain inevitability with regard to technological being. Once modern science with its technological undertone set it into motion, it could not apparently be stopped. Such a concept illustrates that, for Heidegger, modern technology is not only a challenging-forth; it is a *destining* as well. As a

194 *QCT*, pp. 3–35; 36–49; *Principles of Reason*, pp. 54–83, *Pathmark*, p. 252.
195 Letter on Humanism, in *BW*, p. 234.
196 *Ibid.*, pp. 243–244.

destining, modern technology sends man on to an experience of being which apparently leaves room for nothing other than challenging-forth which enframes Being. Once Enframing (*Ge-stell)* takes hold, man no longer see nature explicitly as other. Rather, he conceives of nature as being for himself, thereby having no other recourse than to order it as standing reserve (*Bestand*). How, then, are we to recover other forms of revealing seeing as we are already deeply immersed within an age of technological being? In the hope of offering a possible solution, Heidegger looked back to the beginnings of modernity, to a time when the destining had not yet taken hold. In doing so, he believed we can locate a certain setting-forth of the destining of *Ge-stell* and thus recover a human element at work in technological being which in no way exists today. If the problem of modernity is technology's monopoly on revealing, any possible solution must somehow get outside of a technological mindset. The result of this destining is that Dasein must cultivate a sort of receptiveness or openness to the inevitability that being transmits, which Heidegger refers to as *Gelassenheit*.

Albert Borgmann clarifies destining in uncomplicated language when he states; "Destiny is neither an inevitable fate that descends on humanity, as Heidegger claims, nor the result of human willing. Disclosure of destiny and human freedom are one and the same."[197] And Heidegger defines this freedom as "that which conceals in a way that opens to light, in whose clearing shimmers the veil that covers what comes to presence of all truth and lets the veil appear as what veils. Freedom is the realm of the destining that at any given time starts a revealing on its way."[198]

From this part of our study, we shall discover how Heidegger's interpretation of modern technology revolves around the idea of destining, because as we shall see, all revealing and enframing is a destining of man in his quest to be technological.

4.4.4.a. Technology as Revealing

The instrumental definition of technology as noted earlier (i.e., as a means to an end) provide an accurate description of what technology is. Yet it does not shed much light on what the essence of technology is. To arrive at this, Heidegger turns to a classical line of investigation that uses a four-fold schema of causality. The schema of four causes goes back to Aristotle. First philosophy for Aristotle as the ultimate wisdom is the knowledge of things in their first causes. Therefore, one hopeful route to the essence of technology is through an appraisal of causation.

---

197 Borgmann, A.,"Technology" in *A Companion to Heidegger*, eds. Hubert L. Dreyfus and Mark A. Wrathall, Malden, MA: Blackwell Publishing, 2005, p. 429.

198 *QCT*, p. 25.

The next question about such wisdom is then is to know something about Aristotle's understanding of cause.

Aristotle did not explicitly say how he arrived at the system of causes, but he suggests that each cause corresponds to a particular question about a thing or evidence which requires explanation. It seems moreover that the making of anything demands a fourfold series of causes. It is equally evident that the philosophic explanation of change demands the matter of change, the form attained in change, the agent of change and the purpose or finality of change. In the *Metaphysics,* Aristotle did not long delay on the elaboration of these four causes: "We have studied these causes sufficiently in our work on nature."[199]

In the first instance then, reference must be made to the *Physics* to establish its fourfold schema of causes. There is first the *material cause* "...that out of which a thing comes to be and which persists, is called cause."[200] The material out of which a thing is made is an intrinsic cause in the making of and explanation of a thing. A further intrinsic cause is the *formal cause* "...the form or archtype i.e., the statement of the essence...are called cause."[201] The form actuates the matter and it is this actuation which constitutes the causality of the formal cause. Aristotle's use of the term archtype is not typical of him. In fact he did not avail himself at all of the examplar cause of Plato. He mentioned further of the *efficient cause* which is the primary source of change. "...the primary sources of change are coming to rest. Example, the man who gives advice is the cause. The father is the cause of the child..."[202] Aristotle emphasized efficient cause as against Plato who never refered to it. A further cause is that of *final cause* and is concerned with the purpose of the thing or the action "...in the sense of end or that for the sake of which a thing is

199 Aristotle, *Metaphysics,* Bk. 1, Ch. 3, 983b, 31 Aristotle draws from his earlier work, The Physics, were he had listed and analyzed the causes as four in number. His objective here is to trace and see if more than the four known causes are discoverable among the thinkers before him, and if any of the ancient philosophers discovered all four together. The history of philosophy indicates that the early philosophers enunciated different causes for reality. For Thales, the material cause was water, Anaximanes and Diogenes posited air, Heraclitus posited fire, and Empedocles enunciated the four elements including earth. The Pythagoreans posited number while for Anaxagoras the principle is infinite in number and when he evokes Reason, a type of efficient cause, it is a sort of *Deus ex machina*. Parmenides held it to be the One in love and strife (therefore a formal and final cause). Plato posited the forms as essence and as material cause. See. Iroegbu, P., *Metaphysics: The Kpim of Philosophy*, p. 145.

200 Aristotle, *ThePhysics*, Bk. 2, Ch. 2, 194b, 25.

201 *Ibid.*, Bk. 2, Ch. 2, 194b, 25.

202 *Physics*, Bk. 2, Ch. 3, 194b, 30.

done, e.g., health is the cause of walking about. Why is he walking about? We say to be healthy and having said that we think we have assigned to cause."[203]

Although the conventional interpretation of Aristotle is inclined to recognize every one of these aspects independently and overlooks their mutual affiliation, Heidegger emphasized that the fundamental nature, the real meaning of causation should rest on what brings together the four. "The four causes are the ways, all belonging at once to each other, of being responsible for something else."[204] Heidegger see "causality" as inevitably linked to "instrumentality" (i.e., employment of means of attaining a certain end) as both means and the end are mutually causes of each other.

The term *causa* can be seen to have a least two meanings within this context. For the ancient Greeks, *causa* is a term linked to the verb *cadere*, which means "to fall." And so as the cause prompts something to happen, a result will fall out of this occurrence. In other words, a cause-and-effect relationship is created. For later thinkers, such as the Romans, the word *causa* is linked to an idea of indebtedness, where the result is seen as indebted to the occurrence of the cause, semantically a subtle but useful differentiation from mere cause and effect.

According to Heidegger, what the Greeks intended by the use of the word cause was "that to which something else is indebted (*das, was ein anderes verschuldet*)."[205] He explained this indebtedness with the example of a silver chalice:

> Silver is that out of which the silver chalice is made [...] it is co-responsible for the chalice. The chalice is indebted to, i.e., owns thanks to, the silver for that out of which it consists. But the sacrificial vessel is indebted not only to the silver [...] that which is indebted to the silver appears in the aspect of a chalice and not in that of a brooch or a ring.[206]

The chalice can be described as owing a debt to the silver for its existence. It could not exist without this material to constitute its body and mass. However, the chalice is not exclusively indebted to the silver, because it is also not a silver ornament or fork but a chalice; thus it is further indebted to the shape or the "aspect" of the chalice. Sequentially, this form owns something to the context of its intended end use for which that specific shape is suitable. In this case, the intended use for the chalice is as an instrument used in a sacrificial rite. Heidegger states further:

> But there remains yet a third [...] it is that which in advance confines the chalice within the realm of consecration and bestowal [...] that which gives bounds, that which

---

203 *Ibid*, Bk. 2, Ch. 3, 194b, 32.
204 *QCT.*, p. 7.
205 *QCT.*, p. 7.
206 *Ibid.*

completes, in this sense is called in Greek *telos*, which is all often translated as "aim" or "purpose.' The silversmith considers carefully and gathers together the three aforementioned ways of being responsible and indebted.[207]

The context of this use gives bounds to existence of the chalice, and once made, this precise shape makes sense given its intended ritualistic use. These issues of material, form and end use are fully well thought-out by the maker of the chalice, which, in turn, comes to exist using the method and techniques this maker uses to fashion it. Therefore, the chalice does not come to exist as simply a resultant effect due to one single cause; rather, it owes something to the occurrence of each of these causes and to the interplay between them. Thus, these four ways of being's indebtedness, gather together, to bring something into appearance. "They let it come forth into presencing (*An-wesen*). They set it free to that place and so start it on its way, namely into its complete arrival."[208] It is in the sense of starting something on its way into arrival that being responsible and indebted is an occasioning. "…'To occasion'… is the name for the essence of causality thought as the Greeks thought it."[209] The gathering together of the modes of responsibility and indebtedness constitute a revealing, a bringing forth of what is not yet present into presencing.

Indeed, the idea of revealing is embedded in the meaning of the word technology. Heidegger writes that technology stems from the Greek *Technikon* and "*technikon* means that which belongs to *technē*."[210] That which is decisive about *technē* is not that it is a technique for making or manipulating but that it is a revealing, a bringing forth. It is this meaning of *technē* as revealing that lies at the root of the word technology. According to Heidegger, it is only by focusing on technology as a mode of revealing that the essence of modern technology will show itself to us. The question then is to determine the way in which modern technology is revealing. To answer the question, Heidegger differentiates various phases to arrive at his conclusion.

It too, is a revealing. Only when we allow our attention to rest on this fundamental characteristic does that which is new in modern technology show itself to us. And yet,

---

207 *Ibid.*, p. 8.

208 *Ibid.*, p. 9. "By writing *An-wesen*, Heidegger stresses the composition of the verb *an-wessen*, translated as "to presence." The verb consists of *wesen* (literally, to continue or endure) with the prepositional prefix *an-* (at, to, towards). It is man who must receive presencing, man to whom it comes as enduring." Cf. footnote, no. 7; *On Time and Being*, p. 12.

209 *QCT*, p. 10.

210 *QCT*, p. 12.

> the revealing that holds sway throughout modern technology does not unfold into a bringing-forth in the sense of poiēsis. The revealing that rule modern technology is a challenging (*Herausfordern*), which puts to nature the unreasonable demand that it supply energy that can be extracted and stored as such.[211]

The revealing of the primitive technology is essentially contained with the precincts of nature bestowal, for example, the case of an old windmill that is completely left to the manipulation of the wind and does not extract energy from the wind with the aim of hoarding it for future use. Heidegger asks whether this unfashionable and obsolete windmill did not make the same demand upon the energies of nature: "But does this not hold true for the old windmill as well? No. Its sails do indeed turn in the wind; they are left entirely to the wind's blowing. But the windmill does not unlock energy from the air currents in order to store it."[212] The dam, on the other hand, requires the adjustment of nature in modern technology. The river is transformed through the construction of the dam, and the potential energy of the stored water made available as a type of standing pool to be used as needed. In trapping the water, the river loses something of its essence, becoming known more singularly as a source of power and less essentially as a river that might fulfill a variety of roles.

> The hydroelectric plant is set into the current of the Rhine. It sets the Rhine to supplying its hydraulic pressure, which then sets the turbines turning. This turning sets those machines in motion whose thrust sets going the electric current for which the long-distance power station and its network of cables are setup to dispatch electricity. In the context of the interlocking processes pertaining to the orderly disposition of electrical energy, even the Rhine itself appears as something at our command.[213]

The damming of the river is still considered a revealing, but it is a revealing where nature is captured, ordered, and transformed, challenging it to unlock and release its stored energy. "The hydroelectric plant is not built into the Rhine River as was

---

211 *QCT*, p. 14. This *Herausfordern* is derived from a verb root *fordern* (which means to provoke, to demand, to challenge) as well as the prefixes: *her*-(hither) and *aus*-(out). Cf. *Langenscheidt- Redaktion Wörterbücher*, Berlin: Cornelsen Verlag GmbH & Co., 2002.

212 *QCT*, p. 14. For more explication on the difference between the old windmill and the modern one, see Rojcewicz, R., *The Gods and Technology*, pp. 72–75.

213 *QCT*, p. 16: „Das Wasserkraftwerk ist in den Rheinstrom gestellt. Es stellt ihn auf seinen Wasserdruck, der die Turbinen daraufhin stellt, sich zu drehen, welche Drehung diejenige Maschine umtreibt, deren Getriebe den elektrischen Strom herstellt, für den die Überlandzentrale und ihr Stromnetz zur Strombeförderung bestellt sind. Im Bereich dieser ineinandergreifenden Folgen der Bestellung elektrischer Energie erscheint auch der Rheinstrom als etwas Bestelltes." „Die Frage nach der Technik", *in VA* p. 15.

the old wooden bridge that joined bank with bank for hundreds of years. Rather the river is dammed up into the power plant. What the river is now, namely, a water power supplier derives from out of the essence of the power station."[214] The river thus loses something of its character and essence to this singular application of purpose. It is a revealing that occurs within the driving context and mandate of obtaining power to the exclusion of all else. The river remained a river, perhaps coming to light for the first time as what we understand a river to be with its currents, banks, and floods, and people acknowledged it as such in their practices. In its old role, the bridge was a "thing thinging." Today even when the river is regarded as part of a pleasing landscape rather than untapped power, it is so, says Heidegger, only "as an object on call for inspection by a tour group ordered there by the vacation industry."[215]

The older technology of farming depended upon the soil and its fertility, upon the influence of weather and the cultivation of the farmer. But in the new technology, the fields are challenged to bring forth vast quantities of grain or food which are stockpiled for future use. Modern technology not only challenges the real to bring forth, it stores that which is brought forth. As Heidegger states further:

> The revealing that rules throughout modern technology has the character of a setting-upon, in the sense of a challenging forth. That challenging happens in that the energy concealed in nature is unlocked, what is unlocked is transformed, what is transformed is stored up, what is stored up is, in turn, distributed, and what is distributed is switched about ever anew [...]. What kind of unconealment is it, then, that is peculiar to that which comes to stand forth through this setting-upon that challenges? Everywhere everything is ordered to stand by, to be immediately at hand, indeed to stand there just so that it may be on call for a further ordering.Whatever is ordered about in this way has its own standing. We call it the standing-reserve [*Bestand*].[216]

---

214 *QCT*, p. 16.

215 *QCT*, p. 16.

216 *QCT*, pp. 16–17. As Zimmerman explains; "everyday life is determined according to the demand of the economic system. In this hectic world, no longer understand death, pain or love. We are uprooted and alienated [...] rivers and streams become sewers; the air is poisoned; forests are annihilated; mountains are flattened for their ore, or to make room for highways; farms become 'agri-business' operations which degrades the soil with the imposition of artificial fertilizers and pesticides; homes become high-rise apartment complexes; work becomes repetitive, simplified and boring, biochemists study how to manipulate man's genetic structure; and all of this happens under the aegis of self-development, self – emancipation, and progress." Cf. ZIMMERMAN, M. *Eclipse of the Self: The Development of Heidegger's Concept of Authenticity*, p. 223.

The essence of modern technology is a revealing of events, often completely detached from the normal events of life, wherein human infringement on nature, human challenging-forth of nature for personal satisfaction, not simply dislocates nature but deeply alters it."Nature, as the totality of what is deemed 'knowable' by modern science is thereby conceived from the outset in terms of equipment, instrumentality, becoming what Heidegger calls '*Bestand*' usually translated as "standing reserve."[217] Machine-powered technology is no longer the simple bringing forth of the Greeks. It is different from all other technologies because it is based on modern physics as an exert science. Modern physics is not experimental physics because it applies machinery to the questioning of nature. Rather the reverse is true. "Because physics, indeed already as pure theory, sets nature up to exhibit itself as coherence of forces calculable in advance."[218] Modern science is a setting-upon the real. Consequently, modern technology based on modern science is a setting-upon reality; a challenging of the real. As Ihde points out; "Technological revealing takes its particular shape from its field of possibilities, its framework. And its framework is a particular form of the human taking up a relation to a world through some existential intentionality. There is thus some particular presumed shape to world and some particular activity that responds to that shape of the world."[219]

---

217 PATTISON, P., *The Later Heidegger*, p. 54. "In ordinary German, *Bestand* means "stock-in-hand" or "inventory". The root of the word is *stehen*, which means "to stand," and Heidegger certainly does play on that root meaning. *Stehen* and *Stellen*, standing and posing, or imposing, are interwined." In general, "the *stell*-words apply to the things of modern technology. So, *Bestand* and *stehen* conveys the general idea of things as stockpiles resources." Cf. ROJCEWICZ, R., *The Gods and Technology*, pp. 83–84. *Bestand* is also a term which characterizes the mode of "unconcealment" peculiar to the challenging sort of "seting upon". In his work titled *Heidegger*, W. Biemel cites a relevant passage from the "erste Ausarbeitung" of the essay "Die Frage nach der Technik": "Der Bestand bestellt. Er besteht, sofern er auf ein Bestellen gestellt ist. In das Bestellen gewendet, ist er in das Verwenden gestellt. Das Verwenden stellt jegliches im vorhinein so, daß das Gestellte dem folgt, was erfolgt. So gestellt, ist alles: in Folge von. Die Folge aber wird zum Voraus als Erfolg bestellt. Der Erfolg ist jene Art von Folge, die selbs auf das Ergebnis weiterer Folgen abgestellt bleibt. Der Bestand besteht durch ein eigentümliches Stellen." Cf. Biemel, W., *Martin Heidegger*, Reinbek bei Hamburg: Rowohlt, 1973, p. 115. See also, LOSCERBO, J., *Being and Technology: A Study in the Philosophy of Martin Heidegger*, The Hague: Martinus Nijhoff Publishers, 1981, pp. 136–139.

218 *QCT*, p. 21.

219 IHDE, D., Heidegger's Technologies: Postphenomenological Perspectives, New York: Fordham University Press, 2010, p. 34.

Therefore, the revelation which is at the basis of technology is no longer a bringing forth. It is a setting-upon which challenges the real to unlock its secret rules and to store them in standing reserve for future use. An aircraft packed upon the apron reveals itself as a standing reserve "in as much as it is ordered to endure the possibility of transportation."[220] Even unemployed people become "human resources" footing in line at the local unemployment administrative centre. Thinking itself in the modern age is mostly dedicated to preparation, exploring, and organizing on the basis of specified conditions and precise intentions. Not only the physical sciences but the sciences of social organization look for exact results if not in the short-term future, then in the long run. In all, the reality of the thing itself is concealed and is revealed only as a standing reserve "...even the object disappears into the objectlessness of standing reserve."[221] Because the essence of modern technology lies in placing the object in the objectlessness of standing reserve, it is evident that the essence of modern technology lies in the enframing or unconcealment.

4.4.4.b. Technology as Enframing (Ge-stell)

We have seen how through modern technology, man challenges nature into objectlessness of standing reserve for future use. Yet, Heidegger states that further than this challenging, man himself is equally challenged outside his control to challenge nature, a situation which he refered to as *Ge-stell* (*enframing*). "Modern technology, as an ordering revealing is, then, no merely human doing. Therefore we must take the challenging that sets upon man to order the real as standing-reserve in accordance with the way in which it shows itself. That challenging gathers man into ordering. This gathering concentrates man upon ordering the actual as standing-reserve."[222] While technology as Revealing challenges nature, technology as Enframing challenges man to challenge nature.

Etymologically, *Gestell* is derived from two roots "*Gebirg*" and "*Gemüt*."

> That which primordially unfolds the mountains into mountain ranges and courses through them in their folded togetherness is the gathering that we call "*Ge*birg" (mountain chain). That original gathering, from which unfold the ways in which we have feelings of one kind or another, we name "*Ge*müt" (disposition).[223]

220 *QCT*, p. 17.

221 *QCT*, p. 19.

222 *QCT*, p. 19.

223 *QCT*, p. 19. Heidegger is invoking here the distinction in German between *der Berg* ("mountain") and das *Gebirg* ("the mountains as a whole," "the highlands") and between

Thus, "analogously, by the *Gestell* or enframing Heidegger means to refer to what collects the setting-upon (*stellen*) into a whole."[224] Heidegger gives a precise definition:

> Enframing means the gathering together of that setting-upon which sets upon man, i.e., challenges him forth, to reveal the real, in the mode of ordering, as standing-reserve. Enframing means that way of revealing which holds sway in the essence of modern technology and which is itself nothing technological.[225]

From this interpretation, it is clear that Heidegger sees *Gestell* as the most appropriate name for the present event in the history of Being, i.e., the appropriate name for Being in our modern technological era.

As the essence of technology, enframing is not a common essential characteristic in all technology. It is a way in which being reveals itself through a particular ordering of things. The ordering of *Gestell* is obviously not a physical ordering, but a positioning of one's thinking such that it includes one's surroundings as a realm of possible manipulation. From Heidegger's perspective "enframing is the way in which truth reveals itself as standing-reserve."[226] As standing-reserve, things become potentialities at the disposal of human beings. At its inception, the modern world becomes a collection of usable objects ready at hand for the subject. As Otto Pöggeler puts it:

> Where beings are mere reserve and where revealing is a setting upon and ordering of the reserve, there unconcealment holds sway as enframing (*Ge-stell*) [...]. In the enframing of technology the Being of beings is presentability and deliverability, the disposability of there serve; what presences in Being, the respective occurring of the unconcealedness which is not at one's disposal, remain forgotten.[227]

---

der Mut ("courage," "cheer") and *das Gemüt* ("temperament," "general temper"). Cf. Richard Rojcewicz, *The Gods and Technology*, p. 102.

224 Verbeek, P-P., *What Things do: Philosophical Reflections on Technology, Agency, and Design*, trans. Robert P. Crease, Pennsylvania: The Pennsylvania State University Press, 2005, p. 55.

225 *QCT*, p. 20.

226 Godzinksi, R., Jr. "(En)Framing Heidegger's Philosophy of Technology" *Essays in Philosophy*, 6(1), article 9 2005. see especially, p. 3.

227 Pöggeler, O., *Martin Heidegger's Path of Thinking*, Atlantic Highlands, N.J.: Humanities Press, 1991, p. 198. The nature of truth was elaborated in a lecture which Heidegger gave in the 1920s. The lecture was entitled "Being-Here and Being-True. Theodore Kisiel elaborates on this Heideggerian notion of truth as unconcealment while commenting on that essay: "If truth is a disclosive letting-see-and-be-seen, we then have a threefold sense of truth here: a) disclosure of a being by way of prevalent views of it which includes something already seen in them; b) disclosive entry into hitherto

The question of the oblivion or forgetfulness of Being was a constant and persistent theme of Heidegger. Over many years, he gave a variety of explanations for the oblivion. The oblivion did not arise only from a failure of philosophy. That failure is real. Modern philosophy has become anthropology and in post Kantian philosophy the rejection of metaphysics is axiomatic. The failure of philosophy then to raise the question of being is one simple explanation for the oblivion of being. But more deeply still, it is being itself that is hidden and that hides.

> "Being"—that is not God and not a cosmic ground. Being is farther than all beings and is yet nearer to man than every being, be it a rock, a beast, a work of art, a machine, be it an angel or God. Being is the nearest yet the near remains farther from man. Man at first clings always and only to beings[228]

Being as unconcealed issues from concealment and "inclines back towards it" at the same time. This inclination of being to return to concealment makes the metaphysical question difficult to rise. But the added factor of the darkening of the world makes the question virtually incomprehensible. "The essential happenings of this darkening are: the flight of the gods, the destruction of the earth, the reduction of human beings into a mass, the pre-eminence of the mediocre."[229]

In the darkening of the world, the life of man slides into superficiality. Man lacks the inner resources and depth to respond to the call which calls man to thinking. Hence, the realm of unconcealment is one within which human beings always already find themselves:

> Wherever man opens his eyes and ears, unlocks his heart, and gives himself over to meditating and striving, shaping and working, entreating and thanking, he finds himself everywhere already brought into the unconcealed. The unconcealment of the unconcealed has already come to pass whenever it calls man forth into the mode of revealing allotted to him.[230]

---

unknown domains of being; c) constant struggle with the chatter which gives itself out to be disclosive and knowing. In all three cases, disclosure is a human affair, a basic comportment of Dasein." Cf. KISIEL, THEODORE, *The Genesis of Heidegger's Being and Time*, Los Angeles: University of California Press, 1993, p. 285.

228 HEIDEGGER, M., "Letter to Humanism",in, *BW.*, p. 234.

229 HEIDEGGER, M., *Introduction to Metaphysics*, p. 47.

230 *QCT*, pp. 18–19. „Wo immer der Mensch sein Auge und Ohr öffnet, sein Herz aufschließt, sich in das Sinnen und Trachten, Bilden und Werken, Bitten und Danken freigibt, findet er sich überall schon ins Unverborgene gebracht. Dessen Unverborgenheit hat sich schon ereignet, so oft sie den Menschen in die ihm zugemessenen Weisen des Entbergens hervorruft." „Die Frage nach der Technik", in *VA*, p. 22.

Thus it is unquestionable that human beings in their way reveal the actual, and in so doing play a part in such a revealing to the extent that they conceive, form and operate things they consider real. But it remains uncertain if their taking part in revealing the actual by way of conceiving, forming and operating takes place within a realm that is absolutely under their control.

Furthermore, as an ordering that "sets upon man", enframing is not a willed activity in which humans engages. As Heidegger explains; "Man can indeed conceive, fashion, and carry through this or that in one way or another. But man does not have control over unconcealment itself, in which at any given time the real shows itself or withdraws."[231] Again, as unwilled activity, enframing becomes a destining and "[A]s a destining, it banishes man into that kind of revealing which is an ordering. Where this ordering holds sway, it drives out every other possibility of revealing. Above all, Enframing conceals that revealing which, in the sense of *poiēsis,* lets what presences come forth into appearance."[232]

Given that the essence of modern technology lies in *Enframing* or revealing as challenging and ordering, it "starts man upon the way of that revealing through which the real everywhere more or less distinctly, becomes standing-reserve."[233] This 'starting upon the way' means also "to send" in ordinary usage, and Heidegger calls this the "sending-that-gathers (*versammelde Schicken*) which first starts man upon a way of revealing, *destining* (*Geschick*)."[234] Thus Heidegger concludes that "Enframing as a challenging-forth into ordering, sends into a way of revealing. Enframing is an ordaining of destining, as is every way of revealing. Bringing-forth, as is every way of revealing. Bringing-forth, *poiēsis*, is also a destining in this sense."[235] As the sentinel of destining, "man might be admitted more, sooner and ever more primally to the essence of that which is unconcealed and to its unconcealment, in order that he might experience as his essence his needed belonging to revealing."[236] Above and beyond, as soon as *Enframing* holds sway or dominates, it intrinsically obstructs revealing, and therefore ensures severe damage to man's freedom. As Heidegger clarifies:

---

231 *QCT*, p. 18.

232 *QCT*, p. 27.

233 QCT, p. 24.

234 *Ibid.*

235 *QCT*, pp. 24–25. The unconcealment of any particular entity as standing reserve reveals that entity as useful and purposeful is in the words of Carmen Taylor *Zuhandenheit*. Cf. TAYLOR, C., *Heidegger Analytic: Interpretation, Discourse and Authenticity in Being and Time*, New York: Cambridge Unviersity Press, 2003, p. 15.

236 *QCT*, p. 26.

....the destining of revealing holds complete sway over man. But that destining is never a fate that compels. For man becomes truly free only insofar as he belongs to the realm of destining and so becomes one who listens and hears [*Hörender*], and not one who is simply constrained to obey [*Höriger*].[237]

This situation is even made clearer when Jean Ladriere argues that the technological scheme or logos "becomes an exterior power" imposing "its own law on humanity."[238] Affected by our culture of technique, innovation and modernism, we are anything but masters, we are the inventors or users of technology, but in this structured relation to technology, humanity are subordinated to technology. Therefore, when destining reigns, according to Heidegger, humanity finds itself in greatest danger.

4.4.4.c. Technology as the Danger and Saving Power

What does Heidegger mean here by danger? He states categorically that "in whatever way the destining of revealing may hold sway, the unconcealment in which everything that is shows itself at any given time habours the danger that man may quail at the unconcealed and may misinterpret it."[239] The threat of misinterpretation of the being of an entity, due to the notion that one is taken up with the destining of the age, has always been the threat in any mode of revealing. Heidegger perceived both American capitalism and Soviet communism as symptomatic of destining—two sides of the same technological coin. It is far too simple to let oneself be carried away in the spirit of an age, and in so thinking, to be absorbed into that spirit. Heidegger does not see any need to continue with technological advancement in his society. In fact, his description of the everyday world reverberated with conservative criticisms of modern commercial society. In *An Introduction to Metaphysics*, he identified the presence of a full-blown spiritual crisis in the West. He speaks to the nature of this crisis in the following passage:

This Europe, in its unholy blindness always on the point of cutting its own throat, lies today in a great pincers, between Russia on the one side and America on the other. Russia and America, seen metaphysically, are both the same: the same hopeless frenzy of

---

237 *QCT*, p. 25.

238 LADRIERE, J., *Les Enjeux de la rationalité*, Paris: Aubier-Montaigne, 1977, p. 71. Cited by DOMINIQUE JANICAUD, *Power of the Rational*, trans., Peg and Elizabeth Birmingham, Indianapolis: Indiana University Press, 1994, p. 58.

239 *QCT*, p. 26. In welcher Weise auch immer das Geschick der Entbergung walten mag, die Unverborgenheit, in der alles, was ist, sich jeweils zeigt, birgt die Gefahr, daß der Mensch sich am Unverborgenen versieht und es mißdeutet." („*Die Frage nach der Technik*," in *VA*, 30).

unchained technology and of the rootless organization of the average man. When the farthest corner of the globe has been conquered technologically and can be exploited economically; when any incident you like, in any place you like, at any time you like, becomes accessible, as fast as you like; when you can simultaneously "experience" an assassination attempt against the king in France and a symphony concert in Tokyo; when time is nothing but speed, instantaneity, and simultaneity, and time as history has vanished from the lives of all Dasein of all peoples; when a boxer counts as a great man of a people; when the tallies of millions at mass meetings are a triumph; then, yes then, there still looms like a specter over all this unproar the question: What for? – Where to? – And what then?[240]

On Heidegger's description, Europe stands floating on the brink of cultural suicide. Russia and America threatens to transform Europe into a place characterized by technological "progress" and collective standardization. This does not necessarily mean that Russia and America are the ultimate *sources* of danger. Rather, Russia and America are simply symptomatic of an age characterized by the growth of global technology and mass communications, economic exploitation, pragmatic excess, historical forgetfulness, and collective exhibition. Heidegger continues further:

The spiritual decline of the earth has progressed so far that people are in danger of losing their last spiritual strength, the strength that makes it possible even to see the decline (which is meant in relation to the fate of "Being"), and to appraise it as such. This simple observation has nothing to do with cultural pessimism—nor with any optimism either, of course; for the darkening of the world, the flight of the gods, the destruction of the earth, the reduction of human beings to a mass, the hatred and mistrust of everything creative and free has already reached such proportions throughout the whole earth that such childish categories as pessimism and optimism have long become laughable.[241]

According to this passage, *all* of the nations in the world are in danger of submitting to nihilism to such an extent that they are barely able to recognize that they are threatened at all. Hence, the worldly disillusionment, environmental destruction, escalation of mass culture, conformism that threatens to extinguish the final fragment of spiritual vitality in Europe also threatens the world all together.

Given contemporary views of technology, we might too swiftly and easily conclude that for Heidegger, too, the "danger" of technology lies in our inability to control our own machines and devices and their side-effects. Nuclear power plants disruption, acidic rain, and "the greenhouse effect" are all good instances of the threats posed by our modern techniques. The extent of the danger and risk that

240 Heidegger, M., *Introduction to Metaphysics*, p. 40.
241 Heidegger, M., *Introduction to Metaphysics*, p. 40–41.

arises from misinterpretation could be such that within a way of revealing that sees everything in terms of cause and effect coherence, "even God, for representational thinking, lose all that is exalted and holy, the mysteriousness of his distance."[242] Representational or Calculative by no means stops to contemplate the Being of other beings it thus controls. In such thinking, beings disclose themselves in the same way as technological objects and cannot reveal their essential Being because of self-concealing nature of enframing. This is where Heidegger sees the danger which is not just any danger.

> This danger attests itself to us in two ways. As soon as what is unconcealed no longer concerns man even as object, but does so, rather, exclusively as standing-reserve, and man in the midsts of objectlessness is nothing but the order of the standing-reserve, then he comes to the very brink of a precipitous fall; that is, he comes to the point where he himself will have to be taken as standing-reserve. Meanwhile, man, precisely as the one so threatened, exalts himself to the posture of lord of the earth. In this way the impression comes to prevail that everything man encounters exists only insofar as it is his construct.[243]

The more successful we are at turning things around us into standing reserve, the more we come to regard ourselves as just one more resource to be put to use for the ends we pursue; hence the more we lose track of ourselves as Dasein, the being which is open to a revelation of Being. We model ourselves on the dominant machine and become computers, programmed for tasks, taking input and giving output.

In the light of the relationship between man and destiny, Heidegger points out another happening that, man as the one most endangered in particular, applauds himself and poses as lord of the earth. In order words, it is of the essence of modern technology that man seems to exalt himself to become the lord of the earth. This result in self-deception because whatever man comes across subsists merely as his own concept, which is but the ultimate illusion: "It seems as though man everywhere and always encounters only himself."[244]

As standing-reserve, being is ordered in and through the means of technology. As the object of science, the subject understands itself in its objectness as it grasps other objects. Yet as subject, part of modern being is hidden from the lens of science and technology. Technology, therefore addresses a facet of human existence, but not being as such. It is this twofold nature of modern human being as both

---

242 *QCT*, p. 26. „So kann, wo alles Anwesende sich im Lichte des Ursache-Wirkung-Zusammenhangs darstellt, sogar Gott für das Vorstellen alles Heilige und Hohe, das Geheimnisvolle seiner Ferne verlieren" (*VA*, 30).

243 *QCT*, p. 26–27.

244 *QCT*, p. 27.

subject and object that sheds light on the role of technological achievement in the modern age.

> Intrinsic to technology is an urgent proceeding always toward some confidently foreseen goal. Preoccupied with itself and supremely confident in itself as worthy of perpetuation, technology advances from undertaking to undertaking [...]. Transpiring thus, as ever in train toward some envisioned goal, technology focuses its attention on usefulness.[245]

From this perspective, Lovitt & Lovitt expresses the nature of technology as a *usefulness* applied towards an end. On the sphere of the modern era, technology offers itself as a solution to the world's problems. More well-organized transportation, more effective, qualitative and quantitative food production, and the possible cure of disease place technological advancement as the very mechanism through which progress happens. The concept of progress attained through the means of technology, must, nevertheless be understood beyond the scope of the modern scientific program if the implications of technology are to be realized in terms of being as such.

Technological progress does not enable society to become more human. On the contrary, the infusion and mixture of technology in every aspect of human lives brings human existence closer to an automated bond of order. Technology, therefore, has not brought to presence the unconcealment of being as much as it has grounded itself as a self-perpetuating promise of improvement. Lovitt grounds the inadequacy and shortfall of technology on its very essence as usefulness that is predicated on a conception of 'what is' as standing-reserve.

> In the domain of technology, everything – whether natural feature of sophisticated artifact, whether person or art work or idea – is viewed solely in terms of the place it occupies in a continually proliferating complex. Everything possesses significance only insofar as it is seen and is taken charge of as something useful for the serving of an end beyond itself. Technological using is a derogating employment that, arising out of intense purposefulness, leaves no place for true self-presenting.[246]

Heidegger describes this false deception that badly affects human being under the sway of enframing in this way:

> Man stands so decisively in attendance on the challenging-forth of Enframing that he does not apprehend Enframing as a claim, that he fails to see himself as the one spoken

245 Lovitt, W. & Lovitt H. Brundage, *Modern Technology in the Heideggerian Perspective: Problems in Comtemporary Philosophy*, Vol. I, Lewiston, New York: The Edwin Mellen Press, 1995, p. 225.

246 Lovitt, W. & Lovitt, H. Brundage, *Modern Technology in Heideggerian Pespective*, p. 229.

> to, and hence also fails in every way to hear in what respect he ek-sists, from out of his essence, in the realm of an exhortation or address, and thus *can never* encounter only himself.[247]

This is taken as a disadvantage by Heidegger for the true "advent of beings lies in the destiny of Being."[248] But if we are able to see enframing for what it is, that is, as a destining that reveals, the indispensable way through which Being manifest itself, then, even in the midst of the extreme danger, it can reveal to us the real meaning of Being.

> It is precisely in enframing, which threatens to sweep man away into ordering as the supposed single way of revealing, and so thrusts man into the danger of the surrender of his free essence –it is precisely in this extreme danger that the innermost indestructible belongingness of man within granting may come to light, provided that we, for our part, begin to pay heed to the essence of technology. Thus the coming to presence of technology harbours in itself what we least suspect, the possible arising of the saving power.[249]

The Saving Power: First, we have to call to mind that *technē* in Heidegger's interpretation belongs to *poiēsis* and is accordingly something poetic. "Once there was a time when the bringing-forth of the true into the beautiful was called *technē*. And the *poiēsis* of the fine arts was also called *technē*."[250] Second, we must also bear in mind something essential that Heidegger has already hinted at previously as regards the word *stellen*. We recall that the word *stellen* within the term *Ge-stell* preserves the resonance of another *stellen*, which itself belongs to *poiēsis*. Hence, what could save us from the supreme danger that enframing is, is something that is already hinted at by what is echoed within enframing itself. For this reason, we can understand why, after saying that "where enframing reigns, there is danger in

---

247 *QCT*, p. 27 "Der Mensch steht so entschieden im Gefolge der Herausforderung des Ge-stells, daß er dieses nicht als einen Anspruch vernimmt, daß er sich selber als den Angesprochenen übersieht und damit auch jede Weise überhört, inwiefern era us seinem Wesen her im Bereich eines Zuspruches ek-sistiert und darum *niemals* nur sich sleber begegnen *kann*."(*VA* 31).

248 Heidegger, M., "Letter on Humanism" in *BW*, p. 234.

249 *QCT*, p. 32. "Gerade im Ge-stell, das den Menschen in Bestellen als die vermeintlich einzige Weeise der Entbergung fortzureißen droht und so den Menschen in die Gefahr der Preisgabe seines freien Wesens stößt, gerade in dieser äußerten Gefahr kommt die innigste, unzerstöbare Zugehörigkeit des Menschen in das Gewährende zum Vorschein, gesetzt, daß wir an unserem Teil beginnen, auf das Wesen der Technik zu achten…." (*VA* 36).

250 *QCT*, p. 34.

the highest sense," Heidegger immediately follows it up with the words of Friedrich Hölderlin, the poet:

> *But where the danger is, grows*
> *The saving power also.*[251]

In the face of danger, Heidegger alludes to hope in the form of a saving power. He finds in these lines from the poem "Patmos" a formulation of the paradox he wants to describe: within the "supreme danger" of humanity's enframing orientation to the world lies the potential of a rescue from that very danger. Where the danger of totalitarian thinking within the essence of technology holds sway, there is also present a clue as to what we are to do.

This two-sided interplay of danger and saving power reflects the two faces of being itself. Being presents itself as a present-at-hand façade, but also withdraws into enigmatic bottomless depth. Technology is not a regrettable human deed of pre-historic times or the Modern Industrial Revolution, but an inescapable facet of being itself. Hence, humans cannot force a change in the essence of technology to occur and must passively wait.

It is only in "*The Turning,*" a parallel philosophical piece to "*The Question Concerning Technology,*" that Heidegger gets the opportunity to examine and completely unload the contents of this proposition. The danger and the saving power are not two diverse things. Rather Heidegger argues that "the selfsame danger is, when it is *as* the danger, the saving power."[252] Whereas both essays deal with parallel subjects, the crucial point where they come together can be recognized in Heidegger's dealing of enframing [*Ge-stell*]. In his phenomenological interpretation of enframing, he points out that there is a peculiar phenomenon that takes place within the movement of enframing or *Gestell*. More explicitly, he maintains that a twofold movement of concealment can be observed within the overall movement of enframing. Heidegger says that there is a concealment that is innate to the very nature of Being itself. Being conceals itself in order to presence. Likewise, nothing would be able to come to presence without this concealment. In the process of presencing or coming to be, things unavoidably conceal themselves.

> The coming to presence of Enframing is the danger. As the danger, Being turns about into the oblivion of its coming to presence, turns away from this coming to presence,

---

251 *QCT*, p. 28 "*Wo aber Gefahr ist, wächst Das Rettende auch." VA 32.* From "Patmos." Cf. *Friedrich Hölderlin Poems and Fragments*, trans. Michael Hamburger, Ann Arbor: The University of Michigan Press, 1966, pp. 462–63.

252 "The Turning" in *QCT*, p. 42.

> and in that way simultaneously turns counter to the truth of its coming to presence. In the danger there holds sway this turning about not yet thought on.[253]

The danger and the saving power are both contained within enframing. Both are two different moments of the same process. Since the danger is the essence of technology as a way of revealing, the saving power, too, must come from the very essence of this way of revealing. The distinguishing trait of the unconcealing of technology is that it drives out all other ways of revealing and gets us increasingly firm in its clutch. How can this also save us? Heidegger points out that the essence of technology has an unclear meaning. On the one hand, it is the challenging which barricades other possibilities of Being, but on the other it is a granting that lets us endure as the being that discloses Being.

It requires to be understood that any given enframing of Being is not necessarily final and irrevocable. "In the coming to presence of the danger, there comes to presence and dwells a favour, namely, the favour of the turning-about of the oblivion of Being into the truth of being."[254] The turning of Being into oblivion can turn again to the presencing of being. This turning about of the oblivion comes to pass suddenly in the lightening flash of revelation and lightening. Being turns about, "brings itself into its own brightness."[255] The truth of Being flashes, the essence of Being clears and lights itself up. This turning about of oblivion, this breaking free of enframing, Heidegger calls in-turning. The turning-about of Being into its self-denying hiddenness in-turns into being; into the lightening-up; into the ontological difference. Now is the epoch of being, the era of being.

> The coming to presence of technology is Enframing. In-turning, as the bringing to pass of the turning about of oblivion, turns in into that which now is the epoch of Being. That which genuinely is, is in no way this or that particular being. What genuinely is, i.e., what expressly dwells and endures as present in the "is," is uniquely Being. Only Being "is," only in Being and as Being does that which the "is," names bring itself to pass; that which is, is Being from out of its essence.[256]

He concludes that the saving power also subsist in Being's aptitude to send itself in another way. In order for this to become a genuine possibility, Heidegger thinks that we must act similar to "shepherds" and vigorously organize the way for a new sending.

253 *Ibid.*, p. 41.
254 "The Turning" in *QCT,* p. 44.
255 *Ibid.*
256 "The Turning" in *QCT*, p. 44.

Perhaps we stand already in the shadow cast ahead by the advent of *this* turning. When and how it will come to pass after the manner of a destining no one knows. Nor is it necessary that we know. A knowledge of this kind would even be most ruinous for man, because his essence is to be the one who waits, the one who attends upon the coming to presence of Being in that in thinking he guards it. Only when man, as the shepherd of Being, attends upon the truth of Being can he expect an arrival of a destining of Being and not sink to the level of a mere wanting to know.[257]

As a destining which man has no control over, Heidegger states therefore that having a free relationship with technology allows us to see Being as a destiny while living with technicity. This free relationship is what he called "Releasement" (*Gelassenheit*).

4.4.4.d. Releasement (Gelassenheit)

Gelassenheit means "releasement" or "letting go." Heidegger initially made use of this word under the influence of Meister Eckhart (1260–1327);[258] but more specifically as a determination of the proper composure vis-à-vis technology, and concretely in the relation to things. Eckhart's concern was not the thinking of Being or the belonging together of Being and man but a meditating on the mystical union of God and soul. In this sense, it is used in the sense of "letting-go" of the things of this world and clinging to the things of God. But "Releasemcnt" toward things releases, that is, frees both things and human beings. It lets beings be, even as it frees the human being herself in her Being. As Heidegger states:

We can use technical devices, and yet with the proper use also keep ourselves so free of them, that we may let go of them at any time [...]. We can affirm the unavoidable use of technical devices, and also deny them the right to dominate us, and so to warp, confuse,

257 *Ibid.*, pp. 41–42.

258 For a discussion of *Gelassenheit* in Heidegger and Eckhart, see SCHÜRMANN REINER, *Wandering Joy: Meister Eckhart's Mystical Philosophy,* Massachussetts: Lindisfarne Books, 2001, pp. 81–82, 206–209; SCHÜRMANN REINER, "Heidegger and Meister Eckhart on Releasement" in H. DREYFUS AND M. WRATHALL, (eds.), *Heidegger Reexamined,* New York and London: Routledge, 2003, Vol. III, pp. 295–319. John Caputo emphasizes of the difference between Heidegger and Eckhart in J., CAPUTO, *The Mystical Elements in Heidegger's Thought*, 2000. But most recently, Barbara Dalle Pezze has engaged critically with and goes against Caputo's stance by pointing out that there is in fact a similarity between Heidegger's and Eckhart's Gelassenheit. Cf. DALLE PEZZE, BARBARA, *Martin Heidegger and Meister Eckhart: A Path Towards Gelassenheit*, The Edwin Mellen Press, 2009.

> and lay waste our nature [...]. I would call this comportment toward technology which expresses "yes" and at the same time "no," by an old word, *releasement toward things*.[259]

Here Heidegger envisions a free relationship with modern technology as one that will require a completely different way of being in the world. If *Dasein* is able to have a free relationship to technology, then Heidegger tells us that it will be one that is rooted in an entirely new "attitude," "the attitude toward modern technology that might serve to hasten the advent and foster the growth of that which might save."[260]

We cannot attack science and technology blindly, and try to do away with them. This too, would be self-destructive. In Heidegger's view, we can use scientific and technological devices properly and keep ourselves free from them in such a way that we may let go of them any time: "We let technical devices enter our daily life, and at the same time leave them outside, that is, let them alone, as things which are nothing absolute but remain dependent upon something higher."[261] If we situate this within the Biblical parlance, releasement would mean to be *in* the technological world but not *of* that world, availing oneself of technological things but bequeathing one's heart and soul elsewhere. This will involve "being in the technological world and yet simultaneously withdrawing from it."[262] It is a mystery how something may offer itself and withdraw while doing so; be closest, yet the closest remains farthest. But this is precisely how Being presents itself, and as a result the mystery is Being.

According to Heidegger, man is becoming "encircled ever more tightly by the forces of technology. These forces, which everywhere and every minute claim, enchain, drag along, press and impose upon man under the form of some technical contrivance or other—these forces, since man has not made them, have moved

---

259 Heidegger, M., "Memorial Address," in *Discourse on Thinking*, p. 54. *Gelassenheit*, p. 23. This was a commemorative speech in which Heidegger introduced the concept of Releasement. Heidegger entitled the speech *Gelassenheit*. However, he was for the most part, occupied with the distinction between two ways of thinking and Releasement was surprisingly rarely mentioned. This is probably why the English translation of the text came under the title *Discourse on Thinking*. The speech itself was given the title "Memorial Address" while the dialogue part was called "Conversation on a Country Path about Thinking." These English titles no doubt laid emphasis on the question of thinking but missed Heidegger's own focus on the attitude embodied in the thinking, namely, the attitude of *Gelassenheit*, releasement from technological things.

260 Rojcewicz, R., *The Gods and Technology*, p. 214.

261 "Memorial Address," in *Discourse on Thinking*, p. 54; *Gelassenheit*, p. 23.

262 Rojcewicz, R., *The Gods and Technology*, p. 214.

long since beyond his will and have outgrown his capacity for decision."[263] One is already in a relationship with technology, so it is not a matter of whether or not we will respond to it. Rather, it is a matter of how we will respond to it. More importantly, our response to the challenge that enframing emits, is neither completely predetermined nor free. It is only when the essence of something is disclosed to our existence that we can establish something like a "free relation" to it. "We can affirm the unavoidable use of these (technological) devices and at the same time deny them the right to dominate us and lay waste our very own Being."[264] In other words we should not satisfy ourselves with looking for the appropriate philosophical theory about technology and the technical, but we must somehow accomplish a practical transformation of thinking itself, so as to make it attentive to the essence of the technical. This hydra-headed and ambivalent stance with regard to modern technology, with its inclination toward yes and no at the same time corresponds to the two modes of thinking we have referred to earlier. Through calculative thinking, we can use our resources efficiently; through meditative thinking, we can make sure that technicity will not overpower us. It will rather make it possible for us to come to a freedom with respect to the things that lets beings be (*Gelassenheit*), by maintaining an openness to the mystery that is concealed in modern technology.

> That which shows itself and at the same time withdraws [i.e., our understanding of being] is the essential trait of what we call the mystery. I call the comportment which enables us to keep open to the meaning hidden in technology, *openness to the mystery*. Releasement toward things and openness to the mystery belong together. They grant us the possibility of dwelling in the world in a totally different way. They promise us a new ground and foundation upon which we can stand and endure in the world of technology without being imperiled by it.[265]

As Heidegger indicates here, releasement with regard to beings and openness to the mystery belong together. This releasement makes possible "openness to the mystery" of technology, an attentive stance which reveals the hidden side of technological activity. Nevertheless, such acceptance of the mystery of the gift of an understanding of being cannot be Heidegger's whole account about how to overcome technological nihilism. By means of them, thought can grant us in principle the possibility of dwelling in the world in a new way. For as Heidegger added immediately: "Releasement toward things and openness to the mystery give us a

263 "Memorial Address," in *Discourse on Thinking*, p. 51; *Gelassenheit*, p. 22.

264 Kockelmans, J. J., *Heidegger and Science*, p. 254.

265 "Memorial Address" in *Discourse on Thinking*, p. 55; *Gelassenheit*, p. 26.

vision of a new autochthony (rootedness) which someday even might be fit to recapture the old and now rapidly disappearing autochthony in a changed form."[266]

However for the moment, man finds himself still in a perilous and dangerously precarious situation. This is because, according to Dreyfus "mere openness to technology leaves out much that Heidegger finds indispensable to overcoming nihilism: rootedness in nature, nearness or localness, and new mutual meaningful differences."[267] As a result releasement, despite the fact that it gives us a free relation to technology and guarding our nature from being skewed and upset, cannot of its own accord provide us one of these. The danger remains that the impending wave of the technological revolution will fascinate and mesmerize us, because calculative thinking is still quite universally accepted and practiced as the only way of thinking. But Heidegger argues that mankind's only hope against technology is the cultivation of meditative thinking; only this type of thinking will establish the distance from technology which is essential for "releasement." For Heidegger, at this juncture, there are two fundamental issues and one of them is very obvious: "The issue is the saving of man's essential nature. Therefore, the issue is keeping meditative thinking alive."[268] Keeping meditative thinking alive entails overcoming thoughtlessness and avoiding giving up and throwing away that which genuinely makes man human. "If releasement toward things and openness to the mystery awaken within us, then we should arrive at a path that will lead to a new ground and foundation."[269]

While the *Memorial Address* focuses most simply and directly to the task of meditative thinking, the second work, *Conversation on a Country Path about Thinking* deals with the apparent difficulty in the movement of this mode of thought. Heidegger here articulates the need for a non-representational mode of thinking, which does not focus on objects per se but upon the horizon or "region" in which the objects appear. In fact, Heidegger utilized a distinctive set of terms such as horizon, region (*Horizont, Gegend, Gegnet*), to speak of what he habitually calls 'Being' (*Sein*). The horizon is something that is open: and it is 'openness' which is not the result or upshot of our looking or representing, that lets the horizon be. This primordial openness is what Heidegger refers to with the term 'Gegnet'—an

---

266 "Memorial Address" in *Discourse on Thinking*, p. 55; *Gelassenheit*, p. 26.

267 Dreyfus, H., "Heidegger on Gaining a Free Relation to Technology" in *Readings in the Philosophy of Technology*, D. M. Kaplan, (ed.), Lanham, U.S.A: Rowman & Littlefield Publishers, Inc., 2004, p. 59. Originally published in *Technology and the Politics of knowledge*, ed. Feenberg and Hannay, Indiana University Press, 1995, pp. 97–107.

268 "Memorial Address" in *Discourse on Thinking*, p. 56; *Gelassenheit*, p. 27.

269 "Memorial Address" in *Discourse on Thinking*, pp. 56–57; *Gelassenheit*, p. 28.

old German term for 'Gegend' (region).[270] Hence, that-which-regions ('Gegnet') is not just a region, nor the abstracted content of all regions, but "the region of all regions" ("*die Gegend aller Gegenden*"). Presented in a dialogue form between the scientist, teacher and scholar, Heidegger states:

> Regioning is a gathering and re-sheltering for an expanded resting in an abiding [...]. So the region itself is at once an expanse and an abiding. It abides into the expanse of resting. It expands into the abiding of what has freely turned towards itself [...]. That-which-regions is an abiding expanse which, gathering all, opens itself, so that in it openness is halted and held, letting everything merge in its own resting.[271]

By introducing these two terms, namely, expanse and abiding, Heidegger tries to completely bring out the meaning of 'Gegnet' as spatio-temporal. As R. Guilead states; "This unique 'Gegnet', synonym of openness, has not only a spatial aspect, but a temporal aspect as well: it is at the same time the expanse (die Weite) and the abiding (die Weile)."[272] Things do not any more have the nature or character of 'objects'; they no longer 'stand against' our representing. Everything 'rests' in the regioning of the 'abiding expanse': the foundation and origin, from where everything blossoms.

However, Heidegger recounts the complexity of this approach.

> *Scientist*: I must confess that I can't quite re-present in my mind all that you say about region....

---

270 The German word for region is Gegend. Heidegger prefers to use the old term 'Gegnet' because it is not tied to the pejorative meaning that the term 'Gegend' easily falls to. What is in question here, for Heidegger, is not region in general, but as he said, "the region of all regions" ("*die Gegend aller Gegenden*"). So 'Gegnet', though an old term for 'Gegend', carries a richer meaning for Heidegger: an expanse (Weite) and an abiding (Weile); and it has etymological affinity to 'Gegnen', 'Vergegnis' (regioning), 'Bedingnis'(bethinging) and Èreignis' (Appropriation)—all of which are central to the Heideggerian thinking. (See Conversation on a Country Path, translators note, in *Discourse on Thinking*, p. 66.

271 "Gegnen ist das versammelnde Zurückbergen zum weiten Beruhen in der Weile... Demnach ist die Gegend selbst zumal die Weite und die Weile. Sie verweilt in die Weite des Beruhens. Sie weitet in die Weile des frei In-sich-gekehrten...Die Gegnet ist die verweilende Weite, die, alles versammelnd, sich öffnet so daß in ihr das Offene gehalten und angehalten ist, jegliches aufgehen zu lassen in seinem Beruhen." "Conversation on a Country Path" in *Discourse on Thinking*, p. 66; *Gelassenheit*, p. 40.

272 "Cet 'Gegnet'unique, synonyme de l' ouverture, n'a pas seulment un aspect spatial, mais aussi un aspect temporal: elle est a la fois l'amleur (die Weite) et le séjour (die Weile)." Guilead, R., *Etre et Liberté: Une Etude sur le Dernier Heidegger*, Louvain: Nauwelaerts, 1965, p. 144.

Scholar:	Probably it can't be re-presented at all, in so far as in re-presenting everything has become an object that stands opposite us within a horizon.
Scientist:	Then we can't really describe what we have named?
Teacher:	No. Any description would reify it.
Scholar:	Nevertheless it lets itself be named, and being named it can be thought about....[273]

According to Heidegger, when one is released from calculative thinking, one is open within responsive thought to "that-which-regions," which is no other than Being. Through such praxis something like a freer relation to technology should be possible, not by radicalizing technology, not through a willing acceptance and control of the technical, but rather by becoming attuned to the technical as a sort of destiny to which we always already belong. The premise for the whole argument is thus that we, in our customary way of calculative thinking, are already chained to technology, besides when we believe ourselves to be most free from it. Releasement towards things and openness to the mystery belong together and together they assist in the overcoming of technology. Thus, in the danger of being lies the possibility of a turn (*Kehre*) away from the forgetting of being into the truth of being itself.

Nevertheless, releasement, so to speak, is merely a step, a sort of positive feature we can take part in anticipation of a new understanding of being which would give a shared content to our openness, a new autochthony or rootedness. Is therefore not surprising that Heidegger goes a step further to say that "only when man, in the disclosing coming-to-pass of the insight by which he himself is beheld, renounces human self-will ... may, as the mortal, look out toward the divine."[274] This is reverberated in Heidegger's renowned statement in his famous 1966 interview with *Der Spiegel*: "Only a god can save us now."

4.4.4.e. *The Work of Art and the Call for a God*

What seems to be a new direction in Heidegger's abstract creativity began with the publication of a lecture on *Hölderlin and the Essence of Poetry* (1936). Rather than taking his text from the rationalist philosophers, he now turned to the romantic poets. Instead of restricting himself to an analysis of human existence as projected in a world of technical relations, he considered man in a wider natural setting, in a close interaction with apparently supernatural power.

---

273 Heidegger, M., "Conversation on a Country Path," in *Discourse on Thinking*, p. 67. *Gelassenheit*, p. 41.

274 *QCT*, p. 47.

If modernity is defined by its preoccupation with the subjectness of the subject, the man of antiquity and the Christian medieval era was preoccupied with being in its transcendental reality and ultimately with God. To these metaphysical realities, he gave his watchful attention and unending marvelling. Ancient and medieval man was full of a cosmic and divine enthusiasm. The world itself was understood as an epiphany of the divine. Poets, Seers and Philosophers participate by divine power or fervour in seeing the truth which gleams and radiates, that is, which is good and beautiful. Metaphysics is about knowing achieved in the speculative intellect. The speculative intellect directs that which it apprehends to the consideration of truth. The intellect as speculative knows and knows only for the sake of knowing truth. But the issue for art is to make something and art is a matter of the intellect as practical. The intellect as speculative and the intellect as practical do not connote a division between two separate powers. It is rather a distinction between two ways in which the intellect operates. The intellect as practical directs what it apprehends towards operation; the intellect as practical knows for the sake of action.

As a form of making, art does not necessarily begin from a concern for the beautiful as such. The question of making originated in the human order in the fabrication of tools and useful instruments. In time, as the requirements of utility was satisfied, man made things because of the delight in making them. Tools even though useful, do not necessarily exclude the beautiful. The ornamentation of the Samurai sword does not enhance its utility as a weapon of war. It merely enhances the beauty of that particular weapon. An Aeroplane in flight can be a beautiful thing, but neither the sword nor the Aeroplane was made for their beauty. They were made precisely for utilitarian purposes. But because the utilitarian is not the ultimate factor of the human spirit, man can go beyond the utilitarial and the industrial into the fine arts in order to make things for the simple reason of beauty. Because art is a making, it initially includes all making. The carpenter, the silversmith possesses a perfection of the practical intellect. Such a making can develop into the making of things particular to the fine arts into the making of beautiful things. The fine arts are concerned with the good of the artist himself. The virtue of art causes a man to act in the right way with regard to a particular thing to be made. The particular being of the virtue of art does not make a man good. It seeks only the good of the object of art.

> …the good of those things depends not on man's appetative faculty being affected in this way or that but on the goodness of the work done. For a craftsman, as such, is commendable not for the will with which he does a work, but for the quality of the work.[275]

Heidegger generates a tripartite relation between art, artist and work in an encompassing hermeneutic. As Joseph Kockelmans has observed, "if art is the origin of the work of art, art lets those who intimately belong together in regard to the work, namely the one who artistically produces it and those who try to preserve it artistically, each in his own essence, be what they are."[276]

Heidegger opens his treatise, *The Origin of the Work of Art* with very similar claim as in *The Question Concerning Technology*. He is after the *origin* of the art, which here means "the source of its essence or nature."[277] Artistic inspiration emerges into consciousness and culminates in the operation of making. In that operation of making, the artist must discover some medium which is sufficiently durable to survive the work of making and sufficiently plastic. Heidegger is not particularly concerned with the properties of artistic objects or with tools, techniques, and artistic innovation. For him, the essence of the work of art concerns its truth and its role at the center of a cultural paradigm. The painter for instance, works in the medium of colour, of paint or oil, of canvas, of perspective and drawing. Of each of these, an artist requires a technical mastery and competence. Without such mastery, the artistic inspiration and creativity will attain only disorder. The reality which the painter seeks to paint under the influence of inspiration and under the instrumentality of his acquired technique is the visible corporeal thing. The painter grasps some aspect of the secret depth of the corporeal thing. The work of art foregrounds the *thinghood* of the thing, its objectivity, to a higher status than the genius of its creator, thus saving us from the type of violent subjectivity that technology reveals. What Heidegger is after is a *phenomenology of the thing*, its place in a world of human events, social processes, and cultural horizons. He took as an example the painting of a pair of farming clogs (shoes) by Van Gogh. In reality, the farmer simply wears the working clogs and does not advert to them, one way or the other. The clogs are things that open up the world of the peasant woman. But in the painting of the clogs, Van Gogh says many things.

> From the dark opening of the worn insides of the shoes the toilsome tread of the worker stares forth. In the stiffly rugged heaviness of the shoes there is the accumulated tenacity

---

275 Aquinas, T., *Summa Theologia*, Ia-IIae, q. 57, a.3, C.

276 Kockelmans, J. J., *Heidegger on Art and Artworks,* Dordrecht and Boston: Martinus Nijhoff, 1985, p. 209.

277 Heidegger, M., "The Origin of the Work of Art," in *PLT*, p. 17.

of her slow trudge through the far-spreading and ever uniform furrows of the field swept by a raw wind. On the leather lie the dampness and the richness of the soil.Under the soles slides the loneliness of the field path as evening falls. In the shoes vibrates the silent call of the earth, its quiet gift of the ripening grain and its unexplained self-refusal in the fallow desolation of the wintry field. This equipment is pervaded by uncomplaining anxiety as to the certainty of bread, the wordless joy of having once more withstood want, the trembling before the impending childbed and shivering at the surrounding menace of death.This equipment belongs to the *earth*, and it is protected in the *world* of the peasant woman.[278]

The nature or essence of the peasant woman's shoes as a work of art does not lie in their usefulness; we do not experience them in that way. We cannot help but experience the shoes in their relation to the being of peasant natural life. According to Heidegger, the existential condition of the shoes is what shines forth at the inception. This mode of revealing is not enframing (*Gestell*) but bringing-forth (*poiēsis*). The shoes let peasant life, at its genuineness, to display its most authentic existent. The clearing created by the shoes is open; it opens into the fullness of peasant life.

In Van Gogh painting also, Heidegger recognized the whole world emerging, he could see in the particularity of the shoe, certain boundlessness, a certain metaphysical aptitude. Heidegger speaks of earth and world; clues to what this means are given in another passage:

The world is the self-disclosing openness of the broad paths of the simple and essential decision in the destiny of a historical people. The earth is the spontaneous forthcoming of that which is continually self-secluding and to that extent sheltering and concealing. World and earth are essentially different from one another and yet are never separated. The world grounds itself on the earth, and earth juts through the world.[279]

There is a symbiotic relationship between earth and world. Earth is our basic, indispensable existential condition, as well as our material condition. Earth is sea and sand, tree and forest, and birth and death, but what's more tools and technology. World is what we make of these conditions. World is the way we treat earth; the way it becomes present to us. Art opens up the relationship between earth and world at the same time as it ties them together in a logical and articulate

278 *Ibid.*, p. 33. As Michael J. Golee explains; "For Heidegger, the artwork does not have its origins in van Gogh or the shoes themselves; the painting is not a representation of shoes there to be recorded by the artist. Rather, the artwork originates in the system-framing of *Gestell*. Cf. GOLEE, J. M., "Heidegger's 'From the Dark Opening...' in Boetzkes, A & Vinegar, A (eds), *Heidegger and the Work of Art History*, Surrey, England: Ashgate Publishing Limited, 2014, p. 110.

279 *Ibid.* p. 47.

fashion for a people. Let us use architecture to illustrate the manner in which man is artistically fashioning and moulding his world.

Man is not only a spatial being; his historical existence is always located in a given place. This localizing of things is a matter of grave moment. Man is continually marking off and making places for himself and this is the beginning of architecture. Stones arranged in various shapes can locate the holy and serve to mark it distinct from the profane. There are locations for the sitting of political power. Within his own places man creates other places for actions and operations which are significant and important for human life. Man not only receives his place, he creates his places in the virtual place that is the building. Architecture has rarely expended artistic effort and ideas on the human need for shelter and actual living space. The great artistic effort of architecture has been devoted to the creation of a virtual religious space: The great pre-christian Temples of Greece and Rome, the great Christian Churches and Mosques of Islam; it is in such buildings that architecture artistically fashioned a virtual place where men might meet the deity. A whole religious world and environment is expressed and given shape in these virtual places. The great tombs of the Egyptian Pharoahs architecturally create the illusions of the underworld. A virtual human place is created which symbolizes the kingdom of the dead. Architecture creates a public virtual place which at its best can convey a mode which expresses an entire world-view. Heidegger places a great artistic work as the means of manifesting a people's comprehension of being, offering them an articulate concentration and comportment for their lives. He uses the example of a Greek temple to express what he means.

> It is the temple-work that first fits together and at the same time gathers itself the unity of those paths and relations in which birth and death, disaster and blessing, victory and disgrace endurance and decline acquire the shape of destiny for human being.[280]

Here, the temple is a focal point for a people, and, as such, it *worlds*. It creates a sense of identity for people by revealing their fundamental condition as beings-in-the-world. "People whose practices were manifested and focused by the temple had guidelines for leading good lives and avoiding bad one."[281] The temple sets up a world at the same time as it reveals it to a people. However, unlike the mode of revealing that modern technology sets up, artistic works have many paths; they are more open in their revealing than technology, which narrows the earth by enframing it: "The temple-work, standing there, opens up a world and at the same

280 *Ibid.*, p. 41.

281 Dreyfus, H., "Heidegger on Gaining a Free Relation to Technology" in *Readings in the Philosophy of Technology*, p. 59.

time sets this world back again on earth, which itself only thus emerges as native ground [...]. The temple, in its standing there, first gives to things their look and to men their outlook on themselves."[282]

Technology's enframing demands full disclosure and accounting of entities as standing reserve. In doing so, it forgets the guiding question of being. Nature is not there solely for man to subjugate it to his own advantage, but nature needs man for its proper development.[283] The challenge, then, is to find marginal practices and works of art that allow a common meaning for us—something we can all participate in and look to as focusing and manifesting a moral space, how to live and die, what matters, etc. Although Heidegger does not explicitly state this, art can be seen as one example of a god that might temporarily save us from the enframing of the world by technological thinking.

In allusion to the clout of the enframing, as well as the vulnerability of our era, the era of technology, Heidegger made a renowned declaration in an interview with *Der Spiegel*:

> ....only a god can save us. I think the only possibility of salvation left to us is to prepare readiness through thinking and poetry, for the appearance of the god or for the absence of the god during the decline; so that we do not, simply put, die meaningless deaths, but that when we decline, we decline in the face of the absent god.[284]

The tasks facing our era—the age of absence of God—are: the recovery of the meaning of Being and the arousal of the sense of the holy. "But the holy [...] comes to radiate only when Being itself beforehand and after extensive preparation has

---

282 "The Origin of the Work of Art," in *PLT*, pp. 41–42.

283 Dastur, F., La Question de la Technique: Le Dialogue Entre Patočka et Heidegger. In Frogneux, N., editor, Jan Patočka : Liberté, Existence et Monde Commun, Le Cercle Herméneutique Editeur, Argenteuil, 2012, p. 60. La nature n'est pas sous la domination de l'homme qui ne peut jamais la soumettre à son pouvoir, mais elle a néanmoins besoin de l'homme, auquel elle donne la mission de favoriser et de compléter son proper développment.

284 Heidegger, M., "Nur noch ein Gott kann uns retten," Der Spiegel, 31 May 1976, pp. 193–219. The interview with Rudolf Augstein and Georg Wolff took place on 23 September 1966 in Heidegger's Freiburg residence. Translation by Maria Alter & John D. Caputo, "Only a God can Save Us: Der Spiegel's Interview with Martin Heidegger," in *Philosophy Today* 20 (Winter 1976): 267–284. Cf. also, Gunther, Neske & Emile, Kettering (eds.), *Martin Heidegger and National Socialism*, New York: Paragon House, 1990, pp. 41–66; Richard Wolin, (ed.), *The Heidegger Controversy, A Critical Reader*, Cambridge, Massachusetts: MIT Press Edition 1991, 1993, pp. 91–116.

been illuminated and is experienced in its truth."[285] Heidegger assigned the first objective to the philosopher and the second to the poet.

> Since poetry and thinking are most purely alike in their care of the word, the two things are at the same time at opposite poles in their essence. The thinker utters Being. The poet names what is holy.[286]

In the poem "*Homecoming*," Hölderlin spoke of the serenity and of the serene and the serenification of the serene. "The gods are the serenifiers who in the serenification announce the greeting which the Serene sends."[287] For the poet, the serene is the holy. "It is the holy. For the poet, the "highest" and the "holy" are one and the same: the serene."[288] The poet comes to the homeland of the serene and in this nearness to the serene; he is empowered to name the holy.

Our age is a time of need in a dual way. There is a lack of both the holy and Being itself. To Heidegger, Hölderlin appears as the herald of the contemporary man. Hölderlin has set a pattern on the artistic level for human existence in a transitional period when the old gods have departed and the god of the future has not yet come.

> It is the time of the gods that have fled and of the gods that is coming. It is the time of need, because it lies under a double lack and a double Not: the No-more of the gods that have fled and the Not-yet of the god that is coming.[289]

In the observation of Julian Young, "the gods are absent because we live in the age of 'metaphysics', of dis-enchantment, the age of the victory of the prosaic over the poetic."[290]

Heidegger was of the view that there was nothing people could 'do' to launch a new age. In actual fact 'doing' something further would only magnify and intensify the grasp of the technological system since 'doing' for the benefit of controlling and calculating is fundamental to productionist metaphysics of modernity.[291] To bring out the change of attitude which will pave the way for the arrival of the new gods, Heidegger urged people to reflect on the poetry of Hölderlin which according to

---

285 "Letter on Humanism," in *BW*, p. 242.

286 Heidegger, M.,"What is Metaphysics?" *Postscript*, in *Existence and Being*, p. 391

287 Heidegger, M., "Remembrance of the Poet" in *Existence and Being*, p. 273.

288 *Ibid.* p. 271.

289 Heidegger, M., "Hölderlin and the Essence of Poetry" in *Existence and Being*, p. 313.

290 Young, J., *Heidegger's Later Philosophy*, p. 98.

291 Zimmerman, M. E., *Heidegger's Confrontation with Modernity: Technology, Politics, Art*, Bloomington and Indianapolis: Indiana University Press, 1990, p. 236.

him was an appropriate way of "*Schritt zurück*", a "step back" from the technological inclination of modernity.

The Heideggerian god, if there is such a thing, is the result of a continual event of appropriation that itself reflects the evolution of human understanding. Dreyfus thinks that Heidegger's god is a new human practice that will impart meaning to the world. He refers to such practices as "cultural paradigms."[292] Furthermore, Heidegger speaks of god not in some strict religious sense, but only to illustrate a need to appreciate or acknowledge the unknown. In fact, as far as Christian God is concerned, Heidegger's thought was wrapped in enigma and he even called for silence towards the question of God: "Someone who has experienced theology in his own roots, both the theology of the Christian faith and that of philosophy, would today rather remain silent about God when he is speaking in the realm of thinking."[293] Since the danger for Heidegger is a state of human being in which calculative thinking is the only possible mode of thought, he is saying that "work of art" will represent a radical alternative (the new gods), which will become apparent to us upon reflection (meditative thinking) and so from the danger itself a saving power will come. Consequently, the answer to the problem of technology comes in the form of quiet reflection. Releasement from human self-will means to be released from the need for action and into the mystery of the revealing truth that comes from thinking. Iain Thomson echoes this sentiment in his own writings:

> Heidegger's resacralization of the simple 'thing' reminds us that the conditioned has roots in the unconditioned, the secular in the sacred, and thus suggest that we should

---

292 DREYFUS, H., "Heidegger on Gaining a Free Relationship to Technology," in *Readings in the Philosophy of Technology*, p. 60 According to Dreyfus, "a cultural paradigm focuses and collects the scattered practices of a culture, unifies them into coherent possibilities for action, and holds them up to the people who can then act and relate to each other in terms of the shared exemplar." He refers to the need for us to appreciate the marginal practices—what Heidegger calls the saving power of insignificant things—practices such as friendship, back—packing into the wilderness, and drinking the local wine with friends. All these practices are marginal precisely because they are not efficient. The expanding technological efficiency is the greatest danger. But these saving practices could come together in a new cultural paradigm that held up to us a new way of doing things, thereby focusing on a world in which formerly marginal practices were central and efficiency marginal. Such a new object or event that grounded a new understanding of reality Heidegger would call a new god.

293 HEIDEGGER, M., *Identity and Difference*, pp. 54–55. "Wie die Theologie, sowohl diejenige des christlichen Glaubens als auch diejenige der Philosophie, aus gewachsener Herkunft erfahren hat, zieht es heute vor, im Bereich des Denkens von Gott zu schweigen." *ID.* p. 121.

adopt a very different attitude toward our world, a *Grundstimmung* much more reflective and thankful than the thorough-going instrumental reasoning characteristic of our technological mode of revealing.[294]

We are all called to be poets, says Heidegger, not necessarily in the sense of turning out literature but rather of "dwelling poetically on the earth." While the major part of human existence is taken up with the habitual transactions of work and family life, to be authentically human something more is needed. This "poetic dimension", is a state of mind in which things are seen in the light of their Being, much as Van Gogh described, and Heidegger understood, an entire world of human involvement in a simple pair of farming boots. We are each of us called to "name the holy" for ourselves and thus create spaces for God's return.

Certainly there are many understanding which a Christian might share with Heidegger on the nature of technology and its danger for our world. However, Heidegger's answer is not an explicitly Christian one, and Heidegger does not seem to be campaigning or agitating for a turn to religion. The indifference of modern man toward the experiencing of the divine is a necessary outcome of his deafness to the voice of Being and not as a result of irreligiousity. In BH Heidegger states in clear terms how the question of God is to be situated in the thinking of Being:

Only from the truth of Being can the essence of the holy be thought. Only from the essence of the holy is the essence of divinity to be thought. Only in the light of the essence of divinity can it be thought or said what the word "God" is to signify.[295]

Here, Heidegger speaks principally of the primacy of the question of Being, and he stresses that the problem of God has to be seen in the light of the truth of Being. Nietzsche had already proclaimed God dead and toward the end of *The Turning*, Heidegger declares:

Whether the god lives or remains dead is not decided by the religiousity of men and even less by the theological aspirations of philosophy and natural science. Whether or not God is God comes disclosingly to pass from out of and within the constellation of

294 Iain, Thomson, "What's Wrong with Being a Technological Essentialist? A Response to Feenberg, *Inquiry, 43*(4), 2000, pp. 439–440, (429–444).

295 Heidegger, M., "Letter on Humanism" in *BW*, p. 253. "Erst aus der Wahrheit des Seins läßt sich das Wesen des Heiligen denken. Erst aus dem Wesen des Heiligen ist das Wesen von Gottheit zu denken. Erst im Licht des Wesens von Gottheit kann gedacht und gesagt werden, was das Wort 'Gott' nennen soll." BH, in *Wegmarken*, pp. 347–348.

> Being. So long as we do not, through thinking, experience what is, we can never belong to what will be.[296]

So, the experiencing of the situation of godlessness should be understood within the backdrop of the diverse history of Being. "The meaning of godlessness consists not in a mere incredulity (*Ungläubigkeit*) or in a moral incapacity (*Unfähigkeit*) of man, but in something more fundamental, namely in the coming to pass of the history of Being itself, in a destiny (*mittence*) of Being (*ereignete Geschichte des Seins selbst*)."[297] Heidegger is said to have declared to a group of professors and students in Zürich in 1951:

> I would never attempt to think the unfolding of God, thanks to Being [...]. I think that Being cannot at all be thought as the ground and unfolding of God—however, the experience of God and his manifestation (insofar as it is encountered by man) comes to pass in the dimension of Being.[298]

One is tempted to take Being to be God or at least that Being has taken the place of God in the philosophy of Heidegger. In a text on *Heidegger and Theology*, John Caputo demonstrates, in no small terms, the contrast between Being and God.

> Thinking is directed toward being, not God. Being is not God but the event of manifestness, the happening of the truth of being, the coming to pass of the history of the epochal manifestations of being, from the early Greeks to the will to power.[299]

The God that appeared in Heidegger, according to Caputo, is a poetic god, manifested in the fourfold, as derived from Heidegger's interpretation of Hölderlin's

296 "The Turning," in *QCT*, p. 49. "Ob der Gott lebt order tot bleibt, entscheidet sich nicht durch die Religiosität der Menschen und noch weniger durch theologische Aspirationen der Philosophie und der Naturwissenschaft. Ob Gott Gott ist, ereignet sich aus der Konstellation des Seins und innerhalb ihrer...." TK, p. 46.

297 Kovacs, G., *The Question of God in Heideger's Phenomenology*, Evanston, Illinois: Northwestern University Press, 1990, p. 24. Otto Pöggeler reiterates this by saying that godlessness for Heidegger is an experience of the history of Being, a way of understanding an epochal mittence in the destiny of Being (*seinsgeschichtlich erfahrene Gott-losigkeit*). Cf. Pöggeler, O., *Der Denkweg Martin Heidegger*, p. 217.

298 Cited by François Fédier, "Heidegger et Dieu" in Kearney, R. & O'Leary J., (eds.), *Heidegger et la Question de Dieu*, Paris: Bernard Grasset, 1980, p. 38. "Jamais je ne tenterais de penser le deployment de Dieu gràce a l'être...Je crois quel'être ne peut pas du tout être pensé comme fond et deployment de Dieu—et cependant l'expérience de Dieu et de sa manifesté (pour autant qu'elle est rencontrée par l'homme) parvient a soi-même dan la dimension de l'être".

299 Caputo, J. D., "Heidegger and Theology," in *The Cambridge Companion to Heidegger*, p. 283.

text.[300] The Hölderlin interpretations speak profusely of the experience of the absence of God as the main distinctive character of the present age, age of the forgetfulness of Being. The withdrawal or flight of God is closely linked to that of Being. "We are too late for the gods and too early for Being."[301] The current technological culture that the modern man is in, is distinct by the forgetfulness of Being, so is it bereft of all spiritual energy. Thus, only by taking a reflective "*Schritt zurück*," a "step back" can we step out of our modern technological predicament.

Concluding Remarks

What we have done in this part of our study is the breakdown of the main tenets of Heidegger's question concerning technology. We want to make it a maxim worthy of our retention that Heidegger's philosophy of technology rests principally on his fundamental ontology, the question of Being. The reality of being's oblivion derives from being itself. Being has no equal whatever. Precisely because Being has no equal, Being is of the highest rank. Because it is of the highest rank, Being does not show itself as one pleases[302]. Being maintains its rank and withholds itself. The very moment of the revelation of being is also the moment of hiddenness. This means that the history of being begins with the oblivion of being since being keeps to itself. At the very moment of its disclosure, Being is endangered with closure and hiddenness. The danger beyond any danger is the danger that being in its unconcealment would be entrapped by concealment and denied 'its own coming to presence'.[303]

The ultimate ontological endangerment of being is enframing "which entraps the truth of its coming into presence with oblivion"[304]. The modern positivist sciences together with their attendant technologies are a manner of enframing being. They are enframing because the sciences are a manner of being's revelation and unconcealment. But because these sciences are exact and the calculating method of approach to reality, they perceive the incalculable as something unknown. It is only that which can be repeated experimentally, repeatedly calculated and measured which can be accepted as accurate and valid. At the very moment when the sciences are revealing being, they simultaneously hide being and entrap

300 *Ibid.*

301 Heidegger, M., "The Thinker As Poet," in *PLT*, p. 4.

302 Heidegger, M., *Introduction to Metaphysics*, p. 141.

303 "The Turning," in *QCT*, p. 40.

304 *Ibid.*, p. 36.

its coming into presence in oblivion. The world of modern science does not allow being to hold the highest rank. All trace of the ontological difference is wiped out.

Henceforth, it is man whose being it is to think being, who is claimed by being for the task of thinking being, who dedicates the Dasein he has won to preservation of the dignity of being, who allows being to slip back into concealment. He pursues science and technology in a manner of thinking wherein "the very relation between presencing and what is present remained unthought"[305]

Heidegger begins his discourse of technology by making a significant distinction between mere technology and the essence of technology. Mere technology includes the tools and techniques of modern science, in conjunction with all the gadgets and devices created to add material comfort to human life. However, the essence of technology is, literally, the being of technology and it is this essence that Heidegger claims we must comprehend if we are to gain a free relation to technology.

Expounding the nature of a free relationship to technology is much more difficult for Heidegger than explaining why the prevalent relationship to technology is unfree. The relationship to technology as something technical is unfree because we are dominated by enframing. However, the opposite is not the case in a free relationship: It consist neither in *human mastery over* enframing nor in *freedom from* enframing. Heidegger's notion of a free relationship to technology is difficult to grasp because he obliterates and stamps out the graphic precincts found in the established concepts of freedom. He is not suggesting that we should reflect and reflect, gather insight and knowledge, and that the more we know about technology the freer we will be in our relationship with it. We must allow technology to show itself.

We can break here to recapitulate the major points of the argument and conclusions of Heidegger's teaching on technology in this form:

1. We are inclined to conceive technology as a mechanism, a way of accomplishing something. This description, still, is an oversight to the genuine essence of technology, and inclines towards making us believe that we will dominant technology and unravel the difficulties that go alongside with it by refining and intensifying our technological expertise.
2. Modern man's application of technology and his approach toward the world leads to "enframing," that is to say; a situation in which it is seen as "standing reserve," a source of raw materials to be stockpiled for future use. But then, in

305 HEIDEGGER, M. *Early Greek Thinking*, (Anaximanders Fragments), p. 50.

this enframing, lies the possibility for two additional demeanors with different scenarios:

i. It is a *danger*: "On the one hand, enframing challenges forth into the frenziedness of ordering that blocks every view into the coming-to-pass of revealing and so radically endangers the relation to the essence of truth."[306]
ii. It is also a "*saving power*" and *prospect or chance:* Due to the prospect of enframing, there is need for modern man to take responsibility and care for the world because man "is needed and used for the safekeeping of the coming to presence of truth."[307]

In *The Discourse on Thinking*, Heidegger states that we must say *yes* and *no* to technology, we must practice "releasement" toward technology. The term releasement refers to our being gathered by things rather than controlled by them. This requires a completely new cultural paradigm, or as Heidegger has it, a new god, which can only 'arrive' when we dwell poetically on earth.

Heidegger is not suggesting that we all should become artists or poets; rather we should assimilate more of the artistic and poetic vision into our dealing with the world. In so doing, we can resist the dangers of enframing, and enter into a "free", persistently critical, relentlessly questioning our affinity with the technology that is habitually making new intrusions into our lives. Hence Heidegger concludes his text with these words:

> Thus questioning, we bear witness to the crisis that in our sheer preoccupation with technology we do not yet experience the coming to presence of technology, that in our sheer aesthetic-mindedness we no longer guard and preserve the coming to presence of art. Yet the more questioningly we ponder the essence of technology, the more mysterious the essence of art becomes. The closer we come to the danger, the more brightly do the ways into the saving power begin to shine and the more questioning we become. For questioning is the piety of thought."[308]

When Heidegger was asked if he knew of a possible solution to the potential threat that modern technology presented, he responded by saying; "I know of no paths to the immediate transformation of the present situation of the world, assuming that such a thing is humanly possible at all."[309]

In terms of everyday, concrete experience, Heidegger's idea of technology appears to remain abstract or hypothetical. Could this be due, to some extent, to

306 *QCT*, p. 33.
307 *Ibid.*
308 *QCT*, p. 35.
309 Heidegger, M. Der Spiegel, in *Philosophy Today*, p. 280.

his emphases on the ontological aspects of technology over and against the ontic ones? Possibly the ontological characteristics that are engraved to the movement of modern technology deserve to be related more to a concrete experience with the aim of making Heidegger's concept more meaningful, especially in terms of its relation with political, social, and environmental realities. In view of this, just as we did at the end of the previous part of our study, we will turn to Hans Jonas, one of Heidegger's Jewish follower to go beyond Heidegger's interpretation and seek to emancipate Heidegger from the shackles of ontology into techno- existential ethics.

Section Three
Toward a Responsible and Reflexive Technology

Chapter V: The Question of Modern Technology and Responsibility

Introduction

This book has the unflinching aim of looking beyond Heidegger's interpretation of modern technology. In the preceeding chapters, we have shown how his reflections on technology proceed rationally from the question of Being, by his offering a scientifically informed account of the nature of man (Dasein) as being-in-the-world, that recognizes and makes use of instruments which culminated in technological know-how.

But today more than ever, modern technology is beginning to infiltrate and direct all of culture and way of life. The ultimate result will be a "technological culture" in which technology puts a stamp on nearly everyone and almost everything becomes reliant on technology. As soon as this reliance is connected with the economy, culture tends to become one-dimensional and simple. Environmental problems arise, and the degradation of nature ensues. In the same way, human development becomes one-sided, narrow and unexciting; hence, society begins to fall apart. Modern technology is dynamic and has expanded tremendously, probably beyond Heidegger's anticipation. It has left its mark on human culture and has become a world-encompassing system. In modern technology, everything is connected to everything else, somehow making everybody to be connected to somebody. The result is a technological environment. But with the increase in industrial development, the adverse effects on our environment are also increasing. In the words of Keith Smith; "The rising technology of the rich countries is normally seen as helping to prevent disaster through better forecasting systems and safer construction techniques. However, the more a society becomes dependent on advanced technology, the greater is the potential for disaster if the technology fails."[1] The cities which were once refreshing for living are now becoming focal points of polluted air and soil.

The scale of some environmental problems, such as pollution, climate change and human overpopulation, go beyond any one national frontier and requires harmonized governance policies and a shift in the culture of modernity. People discuss the threat of nuclear weapons or radioactive waste from nuclear power

1 Smith, K., *Environmental Hazards: Assessing Risk and Reducing Disaster*, 3rd Edition, New York: Routledge, 2001, p. 47.

plants. The depletion of raw materials, the question of biodiversity, extinction of many plant and animal species, deforestation, desertification, the depletion of the ozone layer and the increase of exhaust fumes with their far-reaching consequences for life and climate, the harzadous health and pollution problems from e-waste, the scale of urban sprawl are all severe problems. Then there is the growing threat of overestimating genetic modification techniques and underestimating the repercussions of cloning and human gene therapy in biotechnology. Finally, the latest Information and Communications Technology (ITC) promise ample information and communication. All the same, there is actually less face-to-face contact between people than ever before, leading to mutual alienation, loneliness, and social disintegration. Globalisation has given individuals the power to influence the lives of other people all over the world, and the scientific and technological progress has given the human species power to destroy the environment and kill masses of people. In a word, our society has aptly been called a risk society (Beck, 1995). While in earlier times people feared mostly the capriciousness or unpredictability of nature as a threat to their existence, today they fear risks attached to technological interventions and the occasional catastrophic consequences of unsuspected interconnections in the humanly-constructed technological environment. The universal character of modern technology confronts us with comparatively new questions and dimensions of human responsibility. Jonas formulated the new ethical principle of responsibility in this context, calling for individual as well as collective action to prevent disaster. Yet fundamental questions search for answers:

What is the added risk from the newer technological advancements? What is the normative structure for technology? Who is responsible and for what? What are the norms for this responsibility, and how can this responsibility be summoned? Ulrich Beck has observed that modern society has become a laboratory in which no one is responsible for the outcome of the experiment. He even ventures to talk about controlled irresponsibility. Politicians, for instance, dismiss all responsibility, saying they do not produce the technology and can at best manipulate or influence its expansion only indirectly. At the same time, scientists and technologists maintain their only task is to participate in solid research and create new technological possibilities thereby multiplying options for the world to choose. They repudiate responsibility for what is done with their outcomes. Business chiefs and Industrialists, on the other hand, also maintain that it is not they who at the end of the day determine what happens and what does not. They count on the market to decide that. The end user, they argue, has the last say in deciding what is preferred.[2]

Heidegger suggests that refocusing on the primary questions that make it meaningful to be human—the question of Being—could create the means for alternative discourse that both challenge and circumvent the attempt for total supervision and total control.

This part of our study will be divided into two chapters: In the first chapter, we shall lay out the practical implications of Heidegger's concerns, which are so to speak, brought down from the realm of metaphysics by Hans Jonas to address the practical question of survival in a technological age. Moreover, many basic ethical notions such as harm, risk, action, and responsibility appear in need of reassessment in order to be meaningful in a modern technological world. In the next chapter, we shall be less metaphysical and more sociological, i.e., from the question of modernity to a reflexive modernization and democratic technology in which different actors ought to be allowed to participate in decision making about technological developments as it affects our society. In this way, we shall be looking into alternative existential possibilities that Heidegger shut down and probably never envisaged in his solution to the problem of technology.

5. Hans Jonas: Toward an Ethics of Technological Responsibility

Heidegger set the stage for Hans Jonas' critique of technology. What Jonas bewails about technology is how, in its modern scientific form, it has amplified the power of human accomplishment beyond our capacity to contain it. In an ambiance that offers hope while announcing pessimism, Jonas delineates an ontologically based, social and political ethics that responds to the political and scientifically technological theaters performed on the stage of modernity. His work stirred a lively

---

2 DEDEURWAERDERE, T., "La dimension éthique dans la sphère de la rationalité technologique: normes, contextes et arrière-plans," *Les Carnets du Centre de Philosophie du Droit*, N° 82, 2000, 21pp. D'une part, les politiciens rejettent toute responsabilité, en disant qu'ils ne produisent pas les technologies et qu'ils peuvent au mieux influencer leur développement de façon indirecte. Mais en même temps, les scientifiques et les ingénieurs affirment que leur seule tâche est de s'engager dans la recherche fondamentale ou de créer de nouvelles possibilités technologiques. Ils dénient toute responsabilité quant à ce qui est fait avec les résultats de leurs travaux. Enfin, les chefs d'entreprise affirment qu'eux non plus ne déterminent pas le développement technologique. Ils dépendent du marché qui est le principal déterminant: c'est le consommateur qui décide de ce qui doit être préféré.

public debate on the risks of technological development and the ethics needed to counter these risks effectively, especially in Germany.

Hans Jonas studied under Martin Heidegger and Rudolf Bultmann in the 1920s and his mentors included Edmund Husserl.[3] His work can be divided into three main periods. (i) In the beginning he studied *gnosis* – which is one of the most representative forms of western dualism – in its own negativity. The most important publication of this time is *Gnosis und spätantiker Geist.*[4]

(ii) In his later life, Jonas dealt with the philosophical problems of life. As Davis Levy explains; "In his introduction to his 1974 anthology of essays, Jonas speaks tellingly of the motivations of his work during the "five years of soldiering in the British Army" in which, "cut off from books and all the paraphernalia of research" he had to cease work on the study of Gnosticism."[5] This topic made it possible for

---

3 Hans, J., *Philosophical Essays: From Ancient Creed to Technological Man*, Chicago: University of Chicago Press, 1980, p. xi. In his own words, Jonas maintained that he was educationally "formed during the 1920s in the school of such teachers as Husserl, Heidegger and Bultmann." In 1933, after Hitler came to power, he fled Germany for England to escape the Nazi regime, and from England he moved to Palestine. In 1940 he returned to Europe to join the Bristish Army that had been arranging a special brigade for German Jews wanting to fight against Hitler. Immediately after the war, he returned to Mönchengladbach to search for his mother, but found that she had been sent to the gas chambers in Auschwitz concentration camp. Having heard this, he refused to live in Germany again. So he returned to Palestine and took part in Isreal's war of independence in 1948. Jonas taught briefly at the Hebrew University of Jerusalem before moving to North America. In 1950, he left for Canada to teach at Carleton University, Ottawa; five years later, he moved to New York City in 1955 where he lived for the rest of his life. He was Alvin Johnson Professor of Philosophy at the New School of Social Research 1955–1976 and died in New York, 5 Febrary 1993. Cf. Carl Mitcham, *Forward* to 2010 Reprint edition; see http://en.wikipedia.org/wiki/Hans_Jonas.

4 Hans, J., *Gnosis und spätantiker GeistI -II* (1934 and 1954), Göttingen: Vandenhoeck & Ruprecht, 1964–1988, 1993. An English abridgment of the two volumes, under the title *The Gnostic Religion: The Message of the Alien God and the Beginnings of Chiristianity*, Boston: Beacon Press, was published, again to great critical acclaim, in 1958. See David, J. Levy, *Hans Jonas: The Integrity of Thinking*, Columbia & London: University of Missouri Press, 2002, p. 6.

5 David, J. L., *Hans Jonas: The Integrity of Thinking*, p. 5 "The apocalyptic state of things, the threatening collapse of a world, the climatic crisis of civilization, the proximity of death, the stark nakedness to which all issues of life were stripped, all these were ground enough to take a new look at the very foundations of our being and to review the principles by which we guide our thinking on them. Thus, thrown back on my own resources, I was thrown back on the philosopher's basic duty and his native business–thinking. And, while living in tents and barracks, being on the move or in position,

him to transcend dualism positively. The most important work of this period is *The Phenomenon of Life: Towards a Philosophical Biology.*[6]

(iii) Finally, he applied his results to the challenges of the modern era focusing on the problems brought about by modern technology. The most significant study of this era is *Das Prinzip Verantwortung. Versuch einer Ethik für die technologische Zivilisation*.[7] This book, which will serve as our priniciple text on Jonas, can be considered as a response to Heidegger's essay, "*The Question Concerning Technology*." Just like Heidegger, Jonas is less concerned with specific technologies than with the overall phenomenon of modern technology in an abstract sense. However, much of Jonas's philosophy of this period responds to contemporary historical and political challenges: mass culture, dictatorship or totalitarianism, the events of the Holocaust, nuclear fears, environmental devastation (Chernobyl), and, later, the risk against genetic or biological engineering (e.g. cloning). Heideggerian concerns are, as it were, put into concrete balance by Jonas. In fact, Jonas aims to re-unite metaphysics with ethics, and drive the "objective imperatives for man in the scheme of things." This project has to surmount the Heideggerian ontology by bridging "the alleged chasm between scientifically ascertainable "is" and morally binding "ought.""[8] Or as John Okoro explains, "Hans Jonas' philosophy based on nature and responsibility is a giant effort to analyse the course of bringing the often-problamatic issues of ontology and ethics to a single shelter."[9] In other words, the "good" or "value" needs to be gounded in being and thus, a theory of responsibility.[10] In the *Preface* to the English Edition, *The Imperative of Responsibility* Jonas states unequivocally in the opening lines:

---

tending the guns or firing them, all the reductive primitivism and ordered waste of the soldier's life in a long war are most unfavourable to scholarly work, they do not prevent, are even preeminently conductive to thinking–and thinking to the point–when there is a will to it" Hans J., *Philosophical Essays: From Ancient Creed to Technological Man*, p. xii.

6 Hans, J., *The Phenomenology of Life: Toward a Philosophical Biology*, Chicago & London: The University of Chicago Press, 1966.

7 Hans, J., *Das Prinzip Verantwortung: Versuch einer Ethik für die technologische Zivilisation*, Frankurt am Main: Insel Verlag, 1979. The English edition came under the title, Hans, J., *The Imperative of Responsibility: In Search of an Ethics for the Technological Age*, (See our Introduction).

8 Hans, J., *The Imperative of Responsibility*, Preface, p. x.

9 OKORO, J. I., *The Earth as a Living Superorganism: From Scientific Gaia (Hypothesis) to the Metaphysics of Nature*, Frankfurt am Main: Peter Lang, 2005, p. 311.

10 Hans, J., *The Imperative of Responsibility*, p. 79.

> Modern technology, informed by an ever deeper penetration of nature and propelled by the forces of market and politics, has enhanced human power beyond anything known or even dreamed of before. It is a power over matter, over life on earth, and over man himself; and it keeps growing at an accelerating pace. Its unfettered exercise for about two centuries has raised the material estate of its wielders and main beneficiaries, the industrial "West," to heights equally unknown in the history of mankind [...]. But lately, the other side of the triumphal advance has begun to show its face, disturbing the euphoria of success with threats that are as novel as its welcomed fruits [...] the peaceful and constructive use of world wide technological power, a use in which all of us collaborate as captive beneficiaries through rising production, consumption and sheer population growth—that poses threats much harder to counter. The net total of these threats is the overtaxing of nature, environmental and (perhaps) human as well. Thresholds may be reached in one direction or another, points of no return, where processes initiated by us will run away from us on their own momentum—and towards disaster.[11]

With the above depiction of modern technology, it is not surprising therefore that Jonas builds up his ethics of technological responsibility along five key hypotheses in the *Preface* to his text:

i. "The altered, always enlarged nature of human action, with the magnitude and novelty of its works and their impact on man's global future, raises new moral issues for which past ethics...has left us unprepared. A new reflection on ethical principles... is required for coping with those issues."
ii. "The lengthened reach of our deeds move *responsibility*...into the centre of the ethical stage." Responsibility must be commensurate with the scope and exercise of technology. This means that "for its discharge today...we need lengthened foresight, that is, a scientific futurology"
iii. Even at its best, certainty and completeness predictions "will fall short of the causal pregnancy of technological deeds". Consequently, we must apply "an imaginative 'heuristic of fear,' replacing the former projections of hope" which must "tell us what is *possibly* at stake and what we must beware of."
iv. "What we must avoid at all costs is determined by what we must preserve at all costs." Since the teachings of revealed religion that gave us the "image of man" is "in eclipse today" by secular reasoning, "a philosophy of nature is to bridge the alleged chasm between scientifically ascertainable 'is' and morally binding 'ought'"

11 *Ibid.*, p. ix.

v. This will lead us to "maximize the inherent technological dangers of overstraining nature, the more modest and fitting goal is set to save the survival of humanity of man from the excesses of his own power."[12]

In Jonas' interpretation, the human domination of ethical question is broken especially with their gaining an almost-dominerring authority over every aspect of their existence.

> Modern technology has introduced actions of such a novel scale, objects, and consequences that the framework of former ethics can no longer contain them. No previous ethics has to consider the global condition of human life and the far-off future, even existence, of the race.These now being an issue, demand a new conception of duties and rights, for which previous ethics and metaphysics provide not even the principles, let alone a ready doctrine.[13]

As we shall see from this text, Jonas questioned technoscience with radicalism that leads to a new ethical reflection with a critical uncertainty. As Hottois, lending voice to to the issue of ethics states:

> Contrary to what the adherents of any religion may believe or ideology which awaits the salvation of the world of compliance and the imposition of universal morality, what is really at stake in the contemporary ethical business is the ethical, the value of ethics, its conditions of possibility and validity of its features, its limitations, its temptations, illusions, its ambivalences, its ambitions and excesses.[14]

In this part of our study, contrary to Heidegger's ontological interpretation, we shall see how Jonas reinstates moral obligation and responsibility to the center of ethical stage for the modern human race in the quest for progress, an obligation, not just for the here and now, but for the future generation; a responsibility, not only to humanity, but to all animate things on earth.

---

12 Hans, J., *The Imperative of Responsibility*, Preface, pp. ix, x.

13 *Ibid.*, pp. 6 and 8.

14 Hottois, G., *Aux Fondements d'une Èthique Contemporaine: H. Jonas et H.T. Engelhardt en Perspective*, Paris: J. Vrin, 1993, p. 13. "contrairement á ce que peuvent croire les adeptes de telle ou telle religion ou idéologie qui attendant le salut du monde de l'observance et de l'imposition universelles de leur morale, ce qui est réellement en jeu dans l'affairement éthique contemporain, c'est l'éthicité, la valeur de l'éthique, ses conditions de possibilité et de validité, ses fonctions, ses limites, ses tentations, ses illusions, ses ambivalences, ses ambitions, et ses outrances." (Translation, mine)

5.1. Modern Man-Nature Relations: Jonas' Critique of Heidegger

It will not be an overstatement to say that Jonas's concept of existential thought owes much to Heidegger. In fact, Oliver Depré observed that Jonas's phenomenology would be impossible without Heidegger's *Being and Time*.[15] As a student of Heidegger, Jonas shares with his master the view that technology is "the focal fact of modern life." First he introduces the practical and moral-ethical components of the ancient man-nature relations and the practical and ethical components of the modern man-nature relations, and then he points out that the latter circumstance can best be characterized as an ethical vacuum. Until his death, Hans Jonas drew our attention to this "ethical vacuum" at the core of our culture: a vacuum caused by both traditional ethics and modern natural science.

Technology today has changed the very nature of human action by giving us free-rein to affect nature, both outside ourselves and within, in ways that are long-range, cumulative, irreversible, and planetary in scale. But traditional ethics has assumed that the effects of our actions are rather restricted. Ethical import belonged to relations between humans, not between us and nature. And while the moral good or evil of our actions lay close at hand, the long unforeseen circumstances are left to chance, fatality of destiny, or divine intervention as in medieval mentality. But all of these have changed with the advent of modern technology, and traditional ethics leaves us unprepared or ill-equipped to account for our responsibilities when the very future of humanity is at stake.

Jonas thus invites us to reflect on technology from three perspectives: the "formal dynamics," the "material content," and the "importance of assessment." The first is a contemplation of technology according to its internal logic, a consideration of technology as a whole. The second is technology in use, and includes not only the history of particular artefacts but also the phenomenology of everyday interactions with technology. The third aspect is where we take a decisive detachment on our tools and lives and ask where we are going and why. This is the feature of understanding technology that most clearly and unswervingly draws on the tradition of Western Philosophy that started with Socrates. Sophocles, about 2500 years ago, composed *Antigone* analyzing the renowned refrain, which as it were, strikes a technological note: "*Many the wonders but nothing more wondrous than man [...]. Clever beyond all dreams the inventive craft that he has made which*

15 Depré, O., *Hans Jonas*, Paris: Ellipses Édition, 2003, p. 8. "...il est indiscutable que la phenomenology jonassienne du vivant eut été impossible sans *Sein und Zeit*: car sa "biologie philosophique" est largement conditionnée par une application au vivant des existentiaux qui serévelent dans l'analytique du *Dasein* chez Heidegger".

may drive him one time or another to well or ill."[16] This tells of man's disruptive violation of celestial order which serves either for good or for vice. Jonas points out that "making free with the denizens of land and sea and air, he yet leaves the encompassing nature of those elements unchanged, and their generative powers undiminished. He cannot harm them by carving out his little kingdom from theirs."[17] The "little kingdom" is a small world, a human commune where we can live in a fairly autonomous way according to our own rules.

This scientific view of nature in the modern period leads to reductionistic materialism, covetousness and acquisitiveness which Heidegger described as standing reserve. On this view, nature is a machine; it harbours no values and expresses no utility. The proposition that there are ends in nature is dismissed as an anthropomorphic smugness. Extrahuman nature is unsympathetic to itself and also to human beings who are cast wandering in it, though human beings may be subjects who posit ends and act in light of purposes, nonhuman organisms are mere objects: matter in motion. And eventually humans, as part of nature, become objects of their own fabrications to be shaped according to the designs of biotechnology. This, also, is symptomatic of Heideggerian *Ge-stell.* If nature presents us with no ethical norms, then no attempt to change our own nature in the name of perfection, convenience, or experimentation could count as a transgression or contravention of crucial limits or a violation of a natural standard of goodness.

With regard to modern technology, Jonas emphasizes the following features: it causes irreversible changes (or damages); it is of cumulative dynamic nature and is a parallel way the natural is absorbed in the sphere of the non-natural.[18] Overall, it can be said that the effects of modern technology have grown further than our scope of insight, so we cannot predict the problems coming from the production and use of a new device and which may come up in the distant future. On account of the rise of the affective potential of human action nature and man himself have

---

16 Hans, J., *The Imperative of Responsibility*, p. 2. We shall note that the word 'wonders' doesn't fully reflect what the original word 'deinos' implies about the dark side of the human nature, with its terrible, awesome potentials and power.

17 *Ibid.*, p. 3.

18 *Ibid.*, p. 32. "Experience has taught us that developments set in motion by technological acts with short-term aims tend to make themselves independent, that is, to gather their own compulsive dynamics, an automotive momentum, by which they become not only...irreversible but also forward-pushing and thus overtake the wishes and plans of the initiators. The motion once begun takes the law of action out of our hands, and the accomplished facts, created by the beginning, become cumulatively the law of its continutation."

become demolishable entities. Consequently, every acting agent has to reflect on the consequences of his deeds from this respect, too.

The mechanically confident, modern thought that has identified technology as the tool of development and growth must be appraised. Likewise, it should be noted that technology is an uncertain implement and its use may in due course set up the ultimate 'good' and the ultimate 'bad'. According to Jonas technology based on natural sciences is not capable of self-restriction. It will rather lead to discomfort, anxiety and, at the limit, despair. In this, rests the deepest source of our cultural debacle: "nihilism."[19] With no grounds for judging nature to be good and strip of any established "reflection of Humanity" to which we are unable to answer the fundamental ethical challenge posed by our technical ingenuity, Heidegger understood this situation not as a supreme practical effect of human inventiveness but as the final consequence of advancements and modernization intrinsic in the inspirational propensities of Western thought. He called this "forgetfulness of Being" from which, he submits, "only a god can save us." Such non-practical and ontological solution seems to eliminate from the field of modern reason our every prospect of regulating further the course of events. This is precisely what Jonas wants to overcome in *The Imperative of Responsibility*, that is, "the retrieval of practical reason as a means of controlling the dynamics of technology and so preserve the integrity of man and the natural world on which his survival depends."[20] Why should we care about the distant future of mankind and the planet? Unable to justify why the presence of humankind on the earth is a categorical imperative, we are ill-equipped for the attitude of guardianship that we must develop if we are not to dissipate the future in the interests of an extravagant present. This is the emergence of ecological ethics today with the aim of showing us the way out of the quagmire. However, Heidegger considers this possibility unachievable without some form of divine intervention.

The first phase of Jonas' way is reached in *The Phenomenon of Life* through an attack on Heidegger, whose existentialism Jonas takes to be the most powerful expression of nihilism in our time. Jonas recognized *Being and Time* as "the most reflective, insightful, thoughtful and important manifesto of existentialism," but criticized Heidegger for restricting existential interpretation of human being.

---

19 Jonas published a text titled "Gnosticism, Existentialism, and Nihilism" as an appendix to his second (1962) edition of *The Gnostic Religion*. Its first version appeared in the journal Social Research ten years before, and then, in expanded form in German, in the theological journal Kerygma und Dogma in 1960. Cf. *The Gnostic Religion,* 320–40; David, J. L., *Hans Jonas: The Integrity of Thinking*, p. 25.

20 David, J. L., *Hans Jonas: The Integrity of Thinking*, p. 26.

He begins his critique of nihilism not by discarding Heidegger's approach overall, but to a certain extent by extending his teacher's categories to capitulate an existential interpretation of the biological facts. From Heidegger's point of view, however it would be inappropriate to interpret nonhuman organisms existentially because only humans "exist." Man is the only being whose Being is an issue: that is, whose conduct and appearance demonstrates a posture towards the sort of being he decide to be within the constraints of his factical existence or "thrownness." Therefore, whatever significance the rest of nature has is through the mirrow of man's care.

Furthermore, according to Jonas, modern technology is a risk to humanity, because technology puts us in a position where ethics do not seem to apply. In other words, ethics is no longer in agreement with its object, human actions. Modern technology requires new ethical thinking, which takes these new conditions into account. Thus, late in his career, Jonas became the philosopher of planetary life in a technological age that seemed bent on its destruction. As Nathalie Frogneux points out, Jonas is interested in the harmful consequences unthought-of in technological development especially based on utopian ideology.[21] Particularly in Germany, the subtitle of the book, *In Search of Ethics for the Technological Age*, gives the impression to answer a question raised philosophically by Heidegger in *The Question Concerning Technology*; "in its invocation of the theme of ethics, it promised a perspective on the matter that was notably absent from Heidegger's own discussion of the issue in terms of the fatality of the history of Being—a discussion that in any terms other than Heidegger's own seemed almost exclusively metaphysical and devoid of ethical content."[22] Lawrence Vogel, in the introductory essay titled "Hans Jonas's Exodus" on the posthumous anthology of theological essays *Mortality and Morality*, describes Jonas's philosophy as presenting "one

21 Frogneux, N., *Hans Jonas ou la vie dans le monde,* Bruxelles: De Boeck Université, 2001, p. 297. "La dimension planétaire et le long terme des effets cumulatifs font penser une menace inédite, et si Jonas a d'abord été sensible a la technique militaire et délibérément destructrice avec la menace de la bombe atomique…, il s'est ensuite intéressé aux consequences néfastes d'une technique de masse purement pacifique et mélioriste portée par une idéologie utopique attribuée a Bacon. Ce qu'il faut craindre, ce n'est pas tant les accidents technologiques, industriels, nucléaires, chimiques ou pétroliers, c'est-a-dire les effets secondaires ou les erreurs de conception et de gestion, que les effets ambivalents des technologies utilisées a bon escient et pour des fins louables. A une certain échelle, la logique du success' invertit et les effets attendus deviennent aux-memes pervers."

22 David, J. L., *Hans Jonas: The Integrity of Thinking*, p. 82.

of the most systematic and challenging rejoinders to the legacy of Heidegger in particular, and to the spirit of the twentieth century as a whole."[23] His writings somehow influenced the birth of the modern environmental movement, providing the philosophical foundations for its most effective political manifestation, the German Green Party.

But, for the most part, the intellectual disenchantment with Heidegger's obscurantism to the question of technology influenced Jonas's philosophy and challenged him into reflecting upon a new philosophy of biology and an ethics for the modern technological civilization. In the words of David Levy, "Jonas provides an alternative, parallel, and yet opposed path of reflection to that of Heidegger."[24] Let us examine his proposition and then see how it helps to answer the question which Heidegger posed.

5.2. Ethical Implication of modern Technology

In *The Imperative of Responsibility*, Jonas states that because of both the temporal reach and the power of modern technologies, traditional ethical theories are inadequate. *Technē* was a measured tribute to necessity, not the road to mankind's chosen goal. "Now, *technē* in the form of modern technology has turned into an infinite forward-thrust of the race, its most significant enterprise, in whose permanent, self-transcending advance to ever greater things, the vocation of man tends to be seen, and whose success of maximal control over things and himself appears as the communication of his destiny."[25]

The old ethical frameworks do not apply anymore because technology has radically changed the scope and effects of human action. Traditional ethics like virtue ethics, utilitarianism, and Kantian Ethics could be considered to be anthropocentric "neighbour" ethics where its scope oversees close human to human interactions. Expert knowledge is not needed to facilitate acting ethically because humans are thought, under traditional ethical frameworks, to be rational animals that seek to act, to make the most of their pleasure and self-interest. Regrettably, human action has changed radically, amplified by technologies that intensify the effects in ways that are not known even today. Hence, "Jonas' ethics is necessarily future-facing and focuses upon the determination of the negative impacts stemming from

23 Hans, J., *Mortality and Morality: A Search for the Good after Auschwitz,* edited with an introduction by Lawrence Vogel, Evanston, Ill.: Northwestern University Press, 1996, p. 4.

24 David, J. L., *Hans Jonas: The Integrity of Thinking,* p. 81.

25 Hans, J., *The Imperative of Responsibility*, p. 9.

rapid technological progress on humanity, animals and nature, in order to prevent global problems such as climate change or the exploitation of non-renewable natural resources."[26]

Drawing on examples that range from nuclear war to human biotechnologies, Jonas remarks that we have the aptitude and propensity to let loose far-reaching changes in our world, the full impacts of which will not be known for generations. Technology not only promises emancipation and seduces us to displace our values; technology can pose a threat to humanity.

> The definitively emancipated Prometheus, to whom science gives forces hitherto unknown, and whom the economy equips with restless incentive, calls for an ethics that voluntarily reins his powers, so they will not be a misfortune for humanity. The thesis of this book is that the golden promises of modern technology have changed into a threat; or that technology is inseparably tied up to this threat."[27]

Technology refers not only to technology in the everyday sense of the term, but also to the creative acting and thinking characteristic of human experience. However, as Jonas explored the meanings of these terms in an effort to disclose an underlying source of value, he moved away from human science, philosophical anthropology toward the study of the organism as such or philosophical biology. In his philosophical biology, the subject was still humanity, however not considered uniquely in its physical characteristics, but from the perspective of the surprising fact of life itself. It is the phenomenon of life, examined under the category of organism that Jonas investigates most analytically in his work, whether the spotlight is on ancient dogma or the predicaments fostered in the contemporary scientific laboratory or technical workshop.

Examining the formal dynamics of technology, Jonas explains that there are several distinctive formal traits of modern technology. Fundamentally, "one major distinction is that modern technology is an enterprise and a process, whereas

---

26 Gordon, J-S., Burckhart, H. & Segler, P., (Introduction), *Global Ethics and Moral Responsibility: Hans Jonas and his Critics*, Gordon, J-S., Burckhart, H. (eds), Surrey, England: Ashgate Publishing Limited, 2014, p. 4.

27 "Der endgültig entfesselte Prometheus, dem die Wissenschaft nie gekannte Kräfte und die Wirtschaft den rastlosen Antrieb gibt, ruft nach einer Ethik, die durch freiwillige Zügel seine Macht davor zurückhält, dem Menschen zum Unheil zu werden. Daß die Verheißung der modernen Technik in Drohung umgeschlagen ist, oder diese sich mit jener unlösbar verbunden hat, bildet die Ausgangsthese des Buches." Jonas, H., *Das Prinzip Verantwortung*, p. 7.

earlier technology was a possession and a state"[28] Conventionally, technologies were advanced and used in connection with given ends of human yearning and want, controlled within "equilibrium of ends and means." However, with modern technology, "every new step in whatever direction of whatever technological field tends not to approach an equilibrium or saturation point in the process of fitting means to ends (nor is it meant to), but, on the contrary, to give rise, if successful, to further steps in all kinds of direction and with a fluidity of the ends themselves."[29] Although this equilibrium was occasionally if not periodically changed, technoscientific invention was usually incremental, comparatively not noticeable, and direct. In contrast to this, modern technological development is totalizing, radical, and circular. Jonas writes that as new technologies "may suggest, create, even impose new ends, never before conceived," they radically transform "the very objectives of human desires"[30] simply by offering their feasibilities. Furthermore, the pace of modern innovation is accelerated, to some extent, on account of the ease and effectiveness of knowledge transmission, itself a product of the technological enterprise, and to some degree as the consequence of the competitive market demands in which technology tends to flourish. Along with these points, Jonas construes modern technological processes as a "'juggernaut', which moves on relentlessly, spawning its always mutated progeny by coping with the challenges and lures of the now."[31] Modernization or novelty is not simply an alternative but is unavoidable and assured by the necessity of modern technology's cumulative-combined formal character. This is of deep moral implication for Jonas, for in view of it he alleges that the sphere of intrahuman "neighbour"ethics "is overshadowed by a growing realm of collective action where doer, deed, and effect are no longer the same as they were in the proximate sphere."[32]

---

28 Hans, J., "Toward a Philosophy of Technology," in *Technology and Values: Essential Readings*, ed. Craig Hanks, West Sussex, UK: Wiley-Blackwell Edition, 2010, (pp. 11–25), p. 13. See also *Readings in the Philosophy of Technology*, ed. David M. Kaplan, (pp. 17–33), p. 19. This article was first published in *The Hasting Centre Report* 9, no. 1 (1979): 40–59.

29 *Ibid.*

30 Here, Jonas made references to the advent of information technology and the giant strides in biotechnology. "Who had ever wished to have in his living room the Philharmonic orchestra, or open heart surgery, or a helicopter defoliating a Vietnam forest? Or to drink his coffee from a disposable plastic cup? Or to have artificial insemination, test-tube babies, and host pregnancies? Or to see clones of himself and others walking about"? Cf. "Toward a Philosophy of Technology," in *Technology and Values*, p. 13.

31 Hans, J., "Toward a Philosophy of Technology," in *Technology and Values*, p. 14.

32 Hans, J., *The Imperative of Responsibility*, p. 6.

According to Jonas, one of the great paradoxes of the contemporary ethos is that, lack of conviction about how morally to be in the world, is in part, a product of modern scientific advances in knowledge about the world. This moral uncertainty is made above all dangerous by its coincidence with the escalating efficiency of technological intervention in the world. The modern situation is one wherein we have enhanced knowledge of, and power over nature, and yet are less convinced about how to morally direct our power than ever before. What now gives us cause for wonder is the question of the extent to which technology has squeezed our moral condition. More specifically, in an age that is on the fast lane, the perplexing question arises: Do these developments have ethical significance to the extent that the moral problems they create cannot sufficiently be addressed by ordinary morality? Does our ability to interact with others, for instance, in information superhighway also pose a new moral condition? Also, does the use of these new technologies that brings about globalization permit humans to act ethically on an individual scale since human action, through modern consumerism, affects others, yet there is no means in which individuals can effectively take responsibility for these actions. This dilemma is perpetuated by intellectual arrogance, free-market capitalism, deregulation, and Western individualism.

Furthermore, Jonas warned that new modern technological progress was thrusting us towards a "utopian" future whether premeditated or unintended. These developments have the latent potential for much good, but also risk changing, harming or even destroying some species, including ourselves. To make the right ethical decisions according to Jonas, "requires supreme wisdom—an impossible situation for man in general, because he does not possess that wisdom, and in particular for contemporary man, because he denies the very existence of its object, objective value and truth. We need wisdom most when we believe in it least."[33] Jonas was referring to the post-modern denial of objective truth that has become so widespread—the idea that all answers are equally valid. On the contrary, ethics searches for superior answers to ethical questions. It acknowledges the boundaries or limitations in current wisdom, and strives to improve our understanding. The way forward is shrouded by our powerlessness to accurately predict the downsides of proposed technological progress. Some argue that we should push ahead and contend with problems as they arise. But given the magnitude of catastrophic devastation that technological mistakes could trigger, Jonas' guiding principle contains much wisdom. He argues that "ignorance of the ultimate

33 *Ibid.*, p. 21.

implications becomes itself a reason for responsible restraint—as the second best to the possession of wisdom itself."[34]

In line with this argument, Jonas maintained that the concept of "responsibility" has chiefly been overlooked in traditional ethical theory—both modern and pre-modern or religious and secular—because of these obsolete interconnected tacit premises:

> All previous ethic—whether in the form of issuing direct enjoinders to do and not to do certain things, or in the form of defining principles for such enjoinders, or in the form of establishing the ground of obligation for obeying such principles—had these interconnected tacit premises in common: that the human condition, determined by the nature of man and the nature of things,was given once for all; that the human good on that basis was readily determinable; and that the range of human action and therefore responsibility was narrowly circumscribed."[35]

All three premises draw attention to one single truth: the triumph of *homo faber* over *homo sapiens*. The abovementioned premises are no longer tenable in our technological age because "the *nature of human action* has changed" with "the realm of making [i.e., technology which] has invaded the space of essential action."[36] Jonas' continual use of the phrase "human condition" in place of "human nature" reveals his understanding of being in terms of "historicity," that human beings can no longer claim to have a definite nature but only an evolving history increasingly twisted by technology. With our escalating use of and dependence on technology, we have "opened up a whole new dimension of ethical relevance for which there is no precedent in the standards and canons of traditional ethics."[37] Traditional ethics cannot help us with its emphasis on virtue because it took for granted that "the entity "man" and his basic condition was considered constant in essence and not itself an object of reshaping *technē*."[38] A monotonously static "human nature" can no longer be spoken of in our age of which "the law of perpetually self-generating change"[39] is the defining attribute, or, as Jonas stated in another place: "the idea of potentially infinite progress permeates the modern ideal of knowledge with the same necessity as it permeates the modern ideal of technical civilization."[40]

---

34 *Ibid.*, p. 22.
35 *Ibid.*, p. 1.
36 *Ibid.*, pp. 1, 9.
37 *Ibid.*, p. 1.
38 *Ibid.*, p. 4.
39 *Ibid.*, p. 124.
40 Hans, J., *The Phenomenon of Life*, p. 207.

Jonas brings to our attention the growing gap between the skyrocketing proliferations of technological power at our disposal and our collective capacity to foresee the potential effects of this power. This indisputable inconsistency in modernity between an increasing magnitude of power and a shortening range of human foresight, Jonas believes, calls for a new ethic of responsibility and foresight: "Our thesis is that the new kinds and dimensions of action require a commensurate ethic of foresight and responsibility which is as novel as the eventualities which it must meet. We have seen that these are the eventualities that arise in the era of technology."[41]

5.3. Toward an Ethics of Responsibility

Hans Jonas is in no doubt that ethics, as experienced in the past, is not capable of coping with the moral problems involved in modern science and technology, because its focal point was only responding to the immediate consequences of human activities. According to Robert Theis, the technosciences developed since the end of the 18th Century are characterized by their dynamic and collective aspects. Dynamic because they are intrinsically associated with the concept of progress, which itself is an ancillary of research and experimentation; collective because of the scientific communication and the technique which that necessitates.[42] Ladriere suggested a reinterpretation of the culture of technoscientific rationality as a culture of harmony with our habitat, a culture of eco-ethics. This eco-ethics aims at a kind of existential appropriation of the living environment in dimensions that are corporeality, coexistence and temporality that seeks to transform the artificial environment created by technology, which has no existential meaning in a habitat in harmony with existence. Technology as a habitat must ensure, as far as possible under the conditions of human life, bodily integrity, participation in the intersubjective life and the possibility of a meaningful history. Thus, proximity could be established between the human and technological environment. "This kind of closeness, familiarity, connaturality that makes a home, an environment, village or even neighbourhood in a big city, a place can be found as existing in its own place […] in which its life can unfold freely."[43]

---

41 HANS, J., *The Imperative of Responsibility*, p. 18.

42 THEIS, R., *Hans Jonas: Habiter le Monde*, Paris: Michalon, 2008, p. 58.

43 LADRIÈRE, J., *L'éthique dans l'univers de la rationalité*, Namur: Artel/Fides, 1997, p. 64: "cette sorte de proximité, de familiarité, de connaturalité qui fait d'une demeure, d'un environnement, d'un village, ou même d'un quartier dans une grande ville, un lieu où

As a result, what we urgently need, in Jonas's conclusion, is an entirely new ethics that does not only recognize the massive increase of human responsibility in connection with the vast development of human supremacy on account of technological development, but also presents an adequate basis for the moral responsibility to guarantee the future existence of human race. In the same vein, Maesschalck discerns that the threat of destruction of nature is emerging as a new constraint which implies, for its taking into account, changing the meaning of responsibility in which the ethics of progress ought to be transformed into ethics of the future.[44] Conserving the conditions for the continuation and protection of human life on our planet has become an ethical question. On the whole, human responsibilities call for reconceptualization to match this drastically new setting, in order that the range of ethics corresponds to the range of impacts of human actions and blunders.

Here, we examine how Jonas advanced responsibility as the principal and perhaps an utterly new category for ethics. On his description, no traditional ethics recognizes the potential for taking the kind of unalterable action made possible by the power of modern technology.

5.3.1. The Question of Responsibility

The ambivalence of the project of modernity, which the proponents of *Critical Theory* in the Frankfurt School described as a "dialectics of enlightenment," has become highly visible due to the 20th century's risk production. Following the devastation of World War I, the German social philosopher and historian, Max Weber, (1864–1920), looking for a sustainable fresh approach to a politic ethic, developed the term "*Verantwortungsethik*" (ethic of responsibility), contrasting it with the term "*Gesinnungsethik*" (ethic of conviction). According to the ethic of responsibility, an action has significance simply as a cause of an effect, namely, purely with respect to its causal connection to the practical world. According to the ethic of conviction, in contrast, a free actor should be capable of choosing

l'existant puisse se retrouver comme en son lieu propre, […] en lequel son existence peut se déployer librement."

44 Maesschalck, M., *Transformations de l'Éthique de la phénoménologie radicale au pragmatisme social*, Bruxelles: P.I.E. Peter Lang, 2010, p. 34. "La menace d'une destruction de la nature apparaît comme une contrainte nouvelle qui implique, pour sa prise en compte, la modification du sens de la responsabilité: l'éthique du progrès doit se transformer en éthique de l'avenir; contre l'idée d'un bien immanent au progrès technoscientifique est avancée l'idée d'un bien transcendant ce progrès et constitué par la marge de décision dont pourront disposer, dans la gestion de leur situation, ceux qui hériteront du monde actuel."

independently both the means and the end; "this concept of personality finds its 'essence' in the constancy of its inner relations to certain ultimate 'values' and 'meanings' of life."[45] Up till today, the fields of politics, economics, and public life are characterized by a mixture of the two terms when it comes to describing individual moral points of view. The discussion among specialists, by contrast, treats the typification implied in the two terms under the rubric of 'deontological' versus 'teleological' ethic.[46] Deontological and teleological approaches to ethics have been there for quite a while, but they no longer fit the vibrant, dynamic and multifaceted phenomenon of modern technology. Technology is no longer only characterized by the relationship between people and their tools. Rather it has become a dynamic system with global ramifications. Modern technology is equally entangled with big business. This amalgamation of scientific, technological, and corporate powers has become a dynamic and reflexive force which calls for an alternative to the older ethical approaches.

Max Weber developed this duet terms (ethics of responsibility and ethics of conviction) when, following the disillusionment of traditional values and striving for a rationalization, he tried to find a new ethical principle that was to bracket religious presuppositions and moral convictions of good and evil. In his renowned lecture of 1919, "*Politics as a Vocation*," he describes the basic difference as follows:

> We must be clear about the fact that all ethically oriented conduct may be guided by one of two fundamentally differing irreconcilably opposed maxims: conduct can be oriented to an "ethic of conviction" (*Gesinnungsethik*) or to an "ethic of responsibility" (*Verantwortungsethik*). This is not to say that an ethic of conviction is identical with irresponsibility, or that an ethic of responsibility is identical with unprincipled opportunism. Naturally nobody says that. However, there is an abysmal contrast between conduct that follows the maxim of an ethic of conviction—that is, in religious terms, "The Christian does rightly and leaves the results with the Lord" *or* conduct that follows the maxim of an ethic of responsibility, in which case one has to give an account of the (foreseeable) results of one's action.[47]

45 Weber, M., *Roscher and Knies: The Logical Problems of Historical Economies*, (1903–06), trans. G. Oakes, New York: Free Press, 1975, p. 192.

46 Spaemann, R., *Grenszen. Zur ethischen Dimension des Handelns,* Stuttgart: Klett-Cotta, 2001, pp. 218–237. Theories about consequences are frequently labeled teleological; a term derived from the Greek word telos, which means "ends" or "purpose." According to consequential theories, the concepts of right, wrong, and duty are subordinated to the concept of the end or purpose of and action.

47 Weber, M., *Wissenschaft als Beruf 1917/1919. Politik LA Beruf 1919*, Wolfgang J. Mommsen & Wolfgang Schluchter, (eds), Tübingen: Mohr, 1994. Published in English as *Politics as a Vocation*, Philadelphia: Fortress Press, 1965. (1992, p. 237).

However, Weber maintained that "the ethics of conviction and the ethic of responsibility are not absolute opposites. They are complementary to one another, and only in combination do they produce the true human being who is capable of having a 'vocation for politics'."[48]

From the above illustration, we can see that the ethics of conviction is that under which one takes actions on the basis of one's pure and good intentions regardless of the immediate consequences—even though these may be the opposite of what is desired. This kind of thinking is identified by Weber as being common among those possessed with religious zeal. The other ethics is that of responsibility, in which one judge's one's actions in terms of the foreseeable results. In other words the first ethic is one of ends predominating over means and the second is of means over ends. With this distinction, Weber neutralizes the utopianism of political fundamentalists prevalent in his era, but also the notion of traditional ethics according to which moral values can determine reality. Since the condition of a "disenchanted world" is no longer valid, Weber postulates, the morality of an action is determined by its consequences.

How did Jonas see the question of responsibility? For Jonas, according to Robert Theis, "Its" right to exist and "ought to be" of the object derived from the "must do" of the subject.[49] In fact, Jonas sees as the purpose of principle of responsibility, to "convince politicians to take strict control measure towards technoscientific activities."[50] Responsibility is seen as responsibility with reference to others (contemporary and future generations). The concept of *the other* is at the center of Jonas' philosophy (and that of Levinas as we have seen in part one of our study). The concern for the future by Jonas is patterned in furtherance of Levinas' ethics. Depending on the point of view that one adopts – in particular the economic one – responsibility takes on a different meaning. Jonas adopts an exceptionally strong responsibility toward future generations, whereas Levinas advances a collective responsibility to present generations.

Jonas adopts a system of meta-norms mechanisms to introduce a hierarchy of priorities. This vision basically implies that we must rethink our relationship with nature, because the responsibility principle takes its essence and its meaning from it. Therefore, Jonas' investigation aims to reverse the relationship between man

48 Weber, M., "The Profession and Vocation of Politics" in *Max Weber: Political Writings,* (1919), ed. and trans. Lassman P and Speirs R., Cambridge: Cambridge University Press, 1994, p. 368.

49 Theis, R., *Jonas: Habiter le Monde*, p. 75.

50 Hottois, G., & Pinsart. M-G., *Hans Jonas: Nature et Responsabilité,* Paris: Vrin, 1993, p. 9.

and nature. In the time of Aristotle, man was nature. In the Christian medieval era, man was left to the fatality of nature as a destiny. With the expansion of technology, he has turn into the master of nature. With this technological supremacy comes responsibility. "Man is the only being that can have responsibility. As he can have it, he has it. The capacity of the responsibility states its imperative: the power itself also carries its duties."[51] Thus, man, in Jona's philosophy, is a responsible being, no longer having nature as a guide, sprouting in a world of radical ambiguity, but also a world of necessary decision. Put in a different way, action is at the heart of responsibility. Hence, according to Jonas, "there can be 'responsibility,' but in two widely differing senses: (a) responsibility as being accountable "for" one's deeds, whatever they are; and (b) responsibility "for" particular objects that commits an agent to particular deeds concerning them."[52] Jonas calls the first *formal* and the second *substantive* responsibility. By the same token, he also made the distinction between *natural* and *contractual* responsibility:

> It is the distinction between natural responsibility, where the immanent "out-to-be" of the object claims its agent *a priori* and quite unilaterally, and contracted or appointed responsibility, which is conditional *a posteriori* upon the fact and the terms of the relationship actually entered into.[53]

Having the ability to act and to make images, man is thrown diving into a moral dimension. He is the only one responsible for responsibility. Jonas' proposals have

---

51 HANS, J., "Zur ontologischen Grundlegung einer Zukunftsethik", in *Philosophischen Untersuchungen und metaphysische Vermutungen*, Frankfurt am Main: Insel Verlag, 1992, p. 130. According to Jonas, the capacity of responsibility is an ethical ability that lays on the ontological aptitude of man to choose among different alternatives with knowledge and will. Responsibility is, therefore, complimentary to freedom. *Ibid.*, p. 131.

52 HANS, J., *On Faith, Reason and Responsibility*, Claremont, California: The Institute for Antiquity and Christianity Studies, 1981, p. 86. Here, Jonas made distinction between different kinds of responsibility. 1. ***Formal responsibility***, in which one is responsible because he did an act. That means that the doer must answer for his deed; he is held responsible for its consequences. So understood "responsibility" does not itself set ends or disallow ends but is the mere formal burden on all causal acting among men (ibid.). 2. ***Substantive responsibility***, which has to do with "the forward determination of what one is to *do*; by whose command, therefore, I feel responsible, not in the first place for my conduct and its consequences, but for the matter that has a claim on my acting" (Ibid., p. 87). 3. ***Natural and Contractual Responsibility***, as exemplified in the responsibility of the parents toward the child or responsibility instituted "artificially" like the appointment to an office. Added to these also is the "self-chosen Responsibility of the politician" (Ibid., pp. 88–89).

53 HANS, J. *The Imperative of Responsibility*, p. 95.

as their goal to procure a new dimension of action, which requires an ethics of foresight and responsibility.

Jonas desires for that reason to relate the ethics of responsibility in the practical aim of persuading politicians to legislate tough procedures that would regulator technical and scientific undertakings. In fact, Jonas pays special attention to the responsibility of the politician, because in the collective acting of the present, statesmen play a vital role in shaping the existence and wellbeing of the future generations. He describes the responsibility of the politician equally to that of the artist as a unique form of *natural responsibility*. The special character of it is that man chooses the burden of responsibility voluntarily.

> The object of responsibility is the *res publica*, the common cause, which in a republic is latently everybody's cause, but actually only in the limits of the general civic duties. These do not comprise the assumption of leadership in public affairs; nobody is formally bound to compete for public office, usually not even to accept an unsought call to it.[54]

The practical demands of the imperative never to compromise the future conditions of human responsibility are radical, and he thinks political willpower to be requisite to its enactment. Resolute, centralized, open-minded, enlightened, political power is necessary to steer the responsible exercise of power upon which the future of humanity and the whole of life depends. Jonas writes, "We are [...] confronted with a dialectic of power which can only be overcome by a further degree of power itself, not by a quietist renunciation of power"[55]

Finally, the word *responsibility*—in the sense of being accountable *for* and accountable *to*—is very appropriate because it also indicates that everyone involved in scientific-technological development must act as a proxy or warden with reference to one another. In other words, every stakeholder must indicate the priorities,

54 Hans, J., *The Imperative of Responsibility*, p. 96. Jonas knows, certainly that power has its own insignificant advantages, too, but as he writes further: "But leaving aside the most blatantly selfish tyranny, for which the "political" is merely a pretext, it will be the rule that the responsibility going, with the power and made possible by it is co-intended in the striving for it, and by the genuine *homo politicus* intended in the first place. The real statesman will see his fame (which he may have quite at heart) precisely in that it can be said of him that he has acted for the good of those over whom he had power, that is, for whom he had it. This—that "over" becomes "for"—sums up the essence of responsibility (*Ibid.*).

55 Hans, J., *The Imperative of Responsibility*, p. 141. The threatening dialectics of power which Jonas made reference to here is the whole lot of the forces of exponential population growth, the demands that growth places on finite natural resources and the always morally double-edged forces of the technological advancements.

values, principles, and norms—the cultural and social portrait—that constitutes the ethos and grounds for one's actions and defines one's input to the scientific-technological event. Consequently, the ethics of responsibility nurtures a positive sense of profession or mission.

5.3.2. Modern Technology and the Imperative of Responsibility

Jonas projected an ethic for modern technological expansion under the anti-utopian "principle of responsibility"—hence the German title of his book (Jonas 1979). He criticized the Marxist utopianism, because according to him, the ideal Marxist society presupposes a rich society where energy and primary resources would be consumed without moderation. Jonas thus comes to the conclusion that humanity will be forced by a vicious circle to "pillage the planet until such time as the planet passes judgment on humanity and saves itself from overexploitation."[56] For Jonas therefore, the utopian ideal connected with the new possibilities of technology is the most dangerous temptation humankind ever faced, since the revolutionary utopia with its link to technology multiplies the hazards to which world and humanity are exposed at least.

The alternative to this utopia is an "ethic of responsibility." Jonas therefore considers it essential "*to unhook the demands of justice, charity, and reason from the bait of utopia*"[57] in order to protect the world and the human race in the future, out of a fear of threatening dangers posed by technology and out of respect and admiration for humans.

If it were true that the world needs to be protected and the immunity of the human fundamental nature or essence to be preserved, then what is required according to Jonas is a "distant ethic" that takes into account the distant consequences of actions. A new, forward-looking ethic would be essential, he reasoned, since the previously existing ethic was a "near ethic," designed solely for the here and now and immediate surroundings of human life, without taking the global conditions, the remote future, or the existence of the species into consideration. The same is true for Kant's "categorical imperative" as formulated in his *Foundations of the Metaphysics of Morals* and *Critique of Practical Reason*. Kant articulates the categorical imperative in this way: "Act so that the maxim of your will can always at the same time apply as a principle of universal legislation."[58]

---

56 Hans, J., *The Imperative of Responsibility*, pp. 193–200.

57 Hans, J., *The Imperative of Responsibility*, p. 201. (Italis in the text)

58 Kant, I., *Kritik der praktischen Vernunft. Die Kritiken*, second edition, (first edition 1974), Frankfurt am Main: Suhrkamp, 1996, p. 140. "Handel so, daß die Maxime deines

The categorical imperative is therefore a moral law that asks for absolute application but, while it prohibits actions that cannot be generalized, it commands actions that could be a principle of collective and universal legislation. What is more, the actor and the acted upon in this ethic are at all times contemporaries, i.e., partakers in a collective present, without taking into consideration the consequences for future generations and times. These quiet assumptions are prevalent to all previous ethics, even though they may vary significantly with reference to experiential facts and judgment in relation to moral actions. It is in any case distinctive for all previous ethics that it concerns recurrent and everyday situations arising between people, and that longer sequences of events are the results of happenstance or perhaps fate. Jonas argues that these assumptions no longer apply, and that it is therefore necessary to rethink and reorganize the ethical situation in light of the new challenges of technology pose by modernity. The assumptions of previous ethics cannot be conserved as an implicit agreement, since the ethical conditions have fundamentally changed with modern technology. The argument is that the essence of human action has changed drastically with recent technological developments, and since ethics has to do with action, the technological developments make it necessary that a new and modern ethics is developed. The changes in human action is fundamental, which means that not only has unfamiliar terrains for human action appeared, which previous ethics for good reasons could not have envisaged, but also that the multiple manifestations of modern technology have changed the nature of moral actions. In other words, all previous ethics rests on assumptions that are obsolete and "out of date" in the light of modern technology, since the previous ethics is no longer in agreement with the conditions and range of moral actions in the age of technology.

Jonas wants to formulate an "up-to-date" categorical imperative, and to justify the need to do so, it must be shown that Kant's categorical imperative is obsolete. Jonas observes that a moral paradox emerges, which displays the weaknesses of the Kantian categorical imperative:

> But there is no self-contradiction in the thought that humanity would once come to an end, therefore also none in the thought that the happiness of present and proximate generations would be bought with the unhappiness or even nonexistence of later ones—as

Willens jederzeit zugleich als Prinzip einer allgemeinen Gesetzgebung gelten könne." "Act in such a way that you treat humanity, whether in your own person or in the person of another, always at the same time as an end and never simply as a means." *Foundations of the Metaphysics of Moral*, 1911, p. 429.

little as, after all, in the inverse thought that the existence or happiness of later generations would be bought with the unhappiness or even partial extinction of present ones.[59]

With this, Jonas presumes to have called attention to a crucial shortfall in the Kantian categorical imperative, since it, as logical principle, does not have a response to the moral paradox mentioned above. Modern ethics faces the task of bringing ethical theory in agreement with the framework for moral actions in the age of technology. Jonas therefore reformulated the Kantian categorical imperative in which he stretched out a dimension of eternity to it:

> "Act so that the effects of your action are compatible with the permanence of genuine human life", or expressed negatively: "Act so that the effects of your action are not destructive of the future possibility of such life"; or simply: "Do not compromise the conditions for an indefinite continuation of humanity on earth"; or, again turned positively; "In your present choices, include the future wholeness of Man among the objects of your will."[60]

With this imperative, Jonas wants to conform to the new dimension of responsibility. He posits the new categorical imperative as a demonstration of a principle of responsibility: The categorical imperative, as formulated by Kant, exclusively has to do with moral action's logical consistency with itself, and in Jonas' reformulation the categorical imperative concerns the constancy of the effects of moral act with the future existence of humanity. Kant's categorical imperative concerned the individual and its decisive factor was the present moment, and Jonas' reformulation of the categorical imperative concerns humanity and its decisive factor is eternity. For that reason Jonas concludes:

> This adds a time horizon to the moral calculus, which is entirely absent from the instantaneous logical operation of the Kantian imperative: whereas the latter extrapolates

---

59 Hans, J., *The Imperative of Responsibility*, p. 11. "Es liegt aber kein Selbstwiderspruch in der Vorstellung, daß die Menschheit einmal aufhöre zu existieren, und somit auch kein Selbstwiderspruch in der Vorstellung, daß das Glück gegenwärtiger und nächstfolgender Generationen mit dem Unglück oder gar der Nichtexistenz späterer Generationen erkauft wird—so wenig, wie schließlich im Umgekehrten, daß Vertilgung gegenwärtiger erkauft wird." (Jonas, 1979, p. 35.)

60 Hans, J., *The Imperative of Responsibility*, p. 11. "'Handle so, daß die Wirkungen deiner Handlung verträglich sind mit der Permanenz echten menschlichen Lebens auf Erden'"; order negativ ausgedrückt: 'Handle so, daß die Wirkungen deiner Handlung nicht zerstörerisch sind für die künftige Möglichkeit solchen Lebens'; oder einfach: 'Gefährde nicht die Bedingungen für den indefiniten Fortbestand der Menschheit auf Erden'; order, wieder positiv gewendet: "'Schließe in deine gegenwärtige Wahl die zukünftige integrität des Menschen als Mit-Gegenstand deines Wollens ein'." (Jonas, 1979, p. 36)

into an ever-present order of abstract compatibility, our imperative extrapolates into a predictable real future as the open-ended dimension of our responsibility.[61]

To Jonas, the will's motivation is the feeling of responsibility as a compulsory counterpart to the unavoidable increase of influence and pressure that modern technology has given humanity. In other words, humanity has extended the range of its potential moral actions, which authenticates the imperative of responsibility.

To the foundation of this human "objective" imperative, Jonas adds a foundation of human "subjective" further imperative of responsibility that is needed to provide the necessary incentive to preserve humankind in future. Therefore, the epitome of responsibility is that felt by humans for humans, which Jonas exemplifies with reference to responsibility of the statesman and parental responsibility for a child. This is a complete and continual responsibility for the child's protection and education, and thus completely designed toward the future. He sees it as the archetype of all responsibility.[62] As he states:

> In this, man has nothing over other living beings—except that he alone can have responsibility also for them, that is, for guarding their self-purpose. But the ends of his fellow sharers in the human condition, whether he shares these ends himself or merely recognizes them in others, and the ulterior self-purpose ("end-in-itself") of their being as such, can in a unique manner be included within his own end: responsibility is first and foremost of men for men, and this is the archetype of all responsibility.[63]

The worth of something is only noticed often when it is threatened with extinction. Thus, this potential annihilation motivates Jonas to employ threats (the heuristics of fear). He believes that the "*heuristics of fear*" can help us to find out *what* it is about humankind that needs to be preserved. Our practical applications

---

61 Hans, T., *The Imperative of Responsibility*, p. 12. "Dies nun fügt dem moralischen Kalkül den Zeithorizot hinzu, der in der logischen Augenblicksoperation des kantischen imperativs gänzlich fehlt: extrapoliert der letztere in eine immer-gegenwärtige Ordnung abstrakter Kompatibilität, so extrapoliert unser imperativ in eine berechenbare wirkliche Zukunft als die unabgeschlossene Dimension unsere Verantwortlichkeit." (Jonas, 1979, p. 37)

62 Hans, T., *The Imperative of Responsibility*, pp. 39–40, 130–135. However, Jacques Taminiaux argued that if this is the paradigm for responsibility, how come it could not concern a great number. Sometimes this archetype of responsibility tends to neglect the cultural diversity which characterizes the parental relationship. "Si (…) la responsabilité parentale a statut de paradigme, on voit mal comment elle pourrait ne pas concerner le plus grand nombre." Jacques Taminiaux, "Sur une éthique pour l'âge technologique" in *Le Messenger Européen*, No. 5, 1991 (pp. 187–202), p. 200.

63 *Ibid.*, p. 98.

of scientific and technological knowledge have to be guided by that "preventive principle," that is founded on the "heuristic of fear".[64] By this Jonas means that we should school ourselves to envisage at all times what may be the most awful consequences of our actions in the pursuit of technological innovation. He considers that a threat is necessary to make man aware of the actions that he must take: "we need *threat* to the image of man."[65] Jonas demonstrates that his *heuristics of fear* is a boundary on expansion and growth, because it ends utopian expansion. And this is necessary since man has the duty to preserve the future of humanity. Thus, if in future there will be mankind, then, according to Depré, we have, by anticipation certain obligations towards the environment.[66]

However, this is not a call back to the Stone Age era of human life. Jonas recognized the intrinsic technological character of man's being in the world. The only nature that man has been able to occupy is nature as changed and modified by culture; and culture even in its most primitive stone age forms, has always made use of and depended on technology. *Homo sapiens* have always been *homo faber*:[67] While others animals must adapt to their environment, man survive by adapting the environment to their requirements. Thus:

> The use of technology, matter rationally fashioned to a given purpose, is what primordially sets humanity apart from other forms of animal life. For while other animals must adapt to their environment or perish, we survive by adapting the environment to our requirements, whether through the first use of furs and fire to protect us from the cold or in our centrally heated and air-conditioned dwellings from which all impact of the natural climate has been excluded.[68]

The overwhelming power of modern science and technology, applied in the framework of global capitalist economy, has given rise to planetary scale environmental

64 Hans, J., *The Imperative of Responsibility*, pp. 26–27, 202–203.

65 *Ibid.*, p. 26.

66 Depré, O., "Ce dont la possibilité contient l'exigence de sa réalité". De l'être au devoir-être chez Hans Jonas in *Étude Phénoménologiques*, Tome XV11, Nos. 33–34, 2001, p. 119. "L' impératif hypothétique commande que *si dans l'avenir il y a des hommes, alors nous avons par anticipation telles et telles obligations* (*comment* exister). Cet impérative hypothétique est au fond le biais privilégié par lequel Jonas est susceptible d' être récuré dans certains secteurs de l'éthique appliqué—qu'on songe aux questions d'éthique environnmentale ou de justice intergénérationnelle: les hommes à venir ont droit à respire un air sain et nous avons donc des devoirs écologiques a leur égard, ou il nous faut veiller à une justice distributive entre les générations. En revanche, l' impératifs catégorique, quant à lui, commande tout simplement *qu' il y ait des hommes* (*que* exister)."

67 Hans, J., *The Imperative of Responsibility*, p. 9.

68 Davis, J. L., *Hans Jonas: The Integrity of Thinking*, p. 86.

problems. These threaten the reliability of ecosystems upon which human society depends, and scientific experts are now documenting several aspects of nature which human society is permanently and irreversibly changing. Some scientists and philosophers now argue that humanity is in a new ethical moment, that the scientific, technological and economic revolutions of the past two centuries have raised new challenges for human society and its ability to come to grips with the consequences of these revolutions. Jonas proposed that humans currently suffer from an ethical fracture, and that traditional understandings of ethics do not provide satisfactory guidance. The ethics of our time must also look to the future. For instance, nuclear energy poses a problem, as humankind will be leaving behind concealed poisonous, toxic, deadly waste for its children and grandchildren. And as Depré summarized it, the contingency situation of the new technology, that is, the possibility of self-destruction calls for a new ethics of the future which ought to be a fundamental ethics that answers the metaphysical questions in which this new ethics is found in the last instance.[69]

Heidegger has spoken of the two-faceted science-technology as a destiny; Jonas began to call it a catastrophe, which only a higher degree of self-awareness and moral commitment by the actual persons involved could counter. In order words, ends has to be rethought both individually and socially by the supposedly 'value-nonaligned' or unbiased scientists themselves that search for fresh, rapid, uninterrupted growth which often aim at not only self-exaltation but also a risky one. Technology is no longer simply a matter of objects ready-to-hand as Heidegger might insinuate; it has become a very complex system in which our everyday lives are embedded. The question is not about the object but rather the question of subjectivity. The whole application of modern technology challenges us with practically new questions and dimensions of human responsibility. And this responsibility must not be merely private, but must be for the other person. This is where Levinas and Jonas echoes one voice against Heidegger. Levinas places the Other in the present, such as when he talks of *the face of the other*, while Jonas uses the future as his point of emphasis, for instance *future generations*. Jonas, essentially, put the last touches on Levinas' philosophy, seeing the Other from a general outlook.

---

69 Depré, O., *Philosophy Moral*, Louvain-la-Neuve, Beligum: Academia Bruylant, 1999, pp. 115–116. « À ce niveau, on peut donc resumer l'argumenttion dans les termes suivant : — la contingence d'une situation technologique nouvelle (la possibilité technique de l'auto-destruction), — appelle une éthique nouvelle (une éthique de futur), — qui doit etre une éthique fondamentale (pae souci de légitimité conceptuelle), — et qui appelle dès lors une question métaphysique en llaquelle cette nouvelle éthique se fonde en dernière instance. »

Heidegger's reflections on enframing do not concretely address the problem of the imminent technological dangers devastating our universe. Although by adopting the lines of Hölderlin, Heidegger, however critical, can be described as a technological optimist. Levinas and Jonas call attention to the question of ethical responsibility and precaution towards technological advancement. But since today technology is tied to economic advancement and virtually every political decision making is remotely controlled by economic wellbeing, and since the money making faculty in man is very active, the question of precaution and responsibility is consciously or unconsciously thrown overboard. We therefore look to new sociological sciences, especially Ulrich Beck and Anthony Giddens' theory of reflexive modernization which calls for the question of reflexivity towards technology.

5.4. Technology and Reflexivity in Second Modernity

New technology has always been applied in ways that transcend the intentions and the purpose of its creators, and new technologies have revealed consequences that were not anticipated. Experience has shown that the potential consequences of the power created by modern science and technology are obviously dangerous to the modern world and to the future of man. The result has been to seriously question the implications of scientific progress on human welfare, technical change, and economic growth.

While some say we are already postmodern, proponents of second modernity like Ulrich Beck and Anthony Giddens says modernity has not vanished, we are not beyond it. In reality, Ulrich Beck assert that we now live in a qualitatively new kind of society, one he calls 'second modernity' or 'risk society', in which technological risks have become the key organising factor. Central to this theory is the concept that the world is still modern rather than postmodern, but this modernity is radicalizing itself as a conscious response to the innumerable risks and unanticipated side effects generated by the processes of modernization.The scale and spread of risks have increased markedly in the contemporary era due principally to the global spread of technologies of increased complexity and connectivity. Industrialization as a whole is putting the world's climate at risk of calamitous change of incalculable cost to human society. This has indeed redefined the question of risk, for according to Beck, "risk is not the same as catastrophe, but the anticipation of the catastrophe."[70]

The consequences of such risks, Beck echoes, stretch far beyond physical level, into human consciousness as well as the social institutions which seem today more

70 Beck, U., *World at Risk*, Cambridge, UK: Polity Press, 2009, p. 188.

and more incapacitated to contend with risk and safeguard her citizens. In the long run there is no institution that would be prepared for the worst conceivable mishap. Unlike the earlier phase, second modernity tackles rather than presuppose the foundations of modern growth that are now destabilized by events defined as unclear, fluid and lacking in certainty; where old beliefs, ideals and traditions are being challenged through processes of de-institutionalization, individualization and reflexivity. According to these authors and those swayed by them, these changes make conventional party-political ideologies unfashionable, and Giddens specifically is renowned for his promotion of a 'third way' in contemporary politics, away from the long-standing skirmish between Left and Right. According to them, the 'third way' is found in the vista of 'sub-politics' of the new social movements in new modernity. This view of modernity attempts to demonstrate the significance of reflexivity or *gain in reflexivity*[71] in the processes of social change.

Although criticisms have been levelled against the meaning of reflexivity in radicalizing modernity, ranging from the lack of sensitivity to cultural issues (Alexander, 1996; Argyrou, 2003) to the problem of boundaries (Latour, 2003; Shields, 2006), our concern here is not to dispute whether reflexivity or second modernity is happening but to see reflexive technology as offering a pragmatic key to the existential gap created by Heidegger's ontologico-metaphysical interpretation of modern technology. As Barbara Adam, Beck and Joost van Loon explains, "reflexivity requires us to be meditative, that is, looking back upon that which allows us to reflect in the first place. In the context of technologically-induced hazards, this means that we appropriate technology in Heideggerian sense as 'that which reveals'."[72] This is further expressed in the following way:

> In the risk society, what stands-in-reserve is revealing itself to us as our destiny, which we can no longer 'enframe' (organize) ourselves [...]. Indeed, only in the risk society, we are perhaps enabled to 'unconceal the essence of technology'. It is this meditative insight that might be the saving power emerging with the growing danger.[73]

Here we want to argue for a more open approach which involves greater public involvement through the entire cycle of a technological product as well as call for environmental ethics and responsibility. But first, we need to get a clearer understanding of what reflexivity means as used without this framework.

---

71 Maesschalck, M., *Normes et Contextes. Les Fondements d'une Pragmatique Contextuelle*, Zürich/New York: Olms, Hildescheim, 2001, p. 283–85.

72 Adam, B., Beck, U. & Joost van Loon, *The Risk Society and Beyond: Critical Issues for Social Theory*, London: Sage Publications, 2000, p. 7. See also QCT., pp. 28–29.

73 Ibid., p. 8.

5.4.1. *The Question of Reflexivity*

Within the context of our study, the concept of reflexivity refers to the process through which the rationality that epitomizes Modernity critically evaluates the suppositions, hypotheses and logic that lies beneath scientific and technological invention. This critique articulates itself both in the latest social movements and in a questioning of scientific and technological progress. According to Lenoble and Maesschalck, within the framework of reflexive governance, "Reflexivity is an operation specific to every act of judgement which consists in grasping this act itself in its activity, in its semantic productivity with regard to a determined context, that is, a power of 'potentiation'."[74]

The notion of 'reflexivity' is interpreted in a different way by Giddens, Beck and Lash.

a. Anthony Giddens' view on the notion of 'reflexivity' is an important ancillary interpretation to the 'factor' approach used by Beck. According to him, high modernity features the growing importance of reflexivity. This has crucial consequences for the individual, both for the "construction" of the individual, for the relationship of the self to itself and for the relationships to other people or the state. As Giddens states:

> The reflexivity of modernity extends into the core of the self. Put in another way, in the context of a post-traditional order, the self becomes a *reflexive project*. [...] In the settings of modernity, by contrast, the altered self has to be explored and constructed as part of a reflexive process of connecting personal and social change.[75]

---

74 Lenoble, J. & Maesschalck, M., *Towards a Theory of Governance: The Action of Norms*, trans. John Paterson, The Hague: Kluwer Law International, 2003, p. 5. This definition is based within the context of "reflexive governance" and the "pragmatique contextuelle" (pragmatic and contextualized approach), a theoretical framework developed by Belgian philosophers Jacques Lenoble, Marc Maesschalck and the team of researchers at the *Centre de philosophie du droit* (UCL). In the context of this reflexivity of the concept of governance, the notions of coordination, complex interdependence of actors and shared project are central. This new governance model requires both groups (government and the governed) to engage in a social learning process. Such a joined participation in collaborative problem-solving can lead to critical scrutinizing of governing variables: goals, values, plans and rules. In this perspective, "reflexive governance" reviews its own mechanisms to insure institutional learning. Hence, it results in the co-design of institutions and the elaboration of common social representations.

75 Giddens, A., *Modernity and Self-Identity*, pp. 32–33. Giddens sees modernity's reflexivity as referring to the susceptibility of most aspects of social, and material relations with nature, to chronic revisions in the light of new information or knowledge. Such knowledge is not incidental to modern institutions, but constitutive of them. See p. 20.

Accordingly, Giddens proposes that the individual in high modernity becomes involved in the construction of his/her own identity, something that was formerly founded in the local and concrete social ambiance of everyday life. However, Giddens reasons that a different concept of reflexivity could be ascribed to post-traditional social order. He defines this reflexivity thus:

> The reflexivity of modern social life consists in the fact that social practices are constantly examined and reformed in the light of incoming information about those very practices, thus constitutively altering their character [...] only in the era of modernity is the revision of convention radicalized to apply (in principle) to all aspects of human life [...].[76]

As a result of this, the modern epoch is characterized by what Giddens describes as a high degree of 'social reflexivity'. "Social reflexivity refers to a society where the conditions in which we live are increasingly a product of our own actions and, conversely, our actions are increasingly oriented towards managing or challenging the risks and opportunities that we have created."[77] Obviously this is a process extending back from the establishment of modern society itself, where the destiny and way of life of individuals was pulled out of policy and regulated structures.

Further, Giddens compares late modernity to a juggernaut, a run away device of massive power threatening to obliterate itself, and with it collective humanity, in quest to achieve control over it. As he puts it:

> The ride is by no means wholly unpleasant or unrewarding; it can often be exhilarating and charged with hopeful anticipation. But, so long as the institutions of modernity endure, we shall never be able to control completely either the path or the pace of the journey. In turn, we shall never be able to feel entirely secure, because the terrain across which it runs is fraught with risks of high consequences. Feelings of *ontological security* and existential anxiety will coexist in ambivalence.[78]

In a rough way, this refers to emotional security and is based on the notion that one's self identity is linked to one's biography. Consequently an ontologically *insecure* individual is, according to this supposition, a person who displays certain characteristics: firstly, they "lack a consistent feeling of biographical continuity;" secondly, "in an environment full of changes the person is obsessively preoccupied with apprehension of possible risks to his or her existence, and paralyzed in terms

76 Giddens, A., *The Consequences of Modernity*, pp. 38–39.

77 Giddens, A. & Pierson, C., *Coversations with Anthony Giddens: Making Sense of Modernity*, Stanford, California: Stanford Univeristy Press, 1998, p. 16.

78 Giddens, A., *The Consequences of Modernity*, p. 139.

of practical action"; and thirdly, "the person fails to develop or sustain trust in his own self-integrity" and may subject him or herself to constant self-scrutiny.[79]

"Reflexive" may, in other words, have various connotations. Primarily, reflexive means that we draw awareness to the intended, as well as the unintended consequences of our action. Reflexivity involves the recognition that "the subject itself forms a large part of the object"—as a matter of "self-awareness."[80] In that way we become aware of the dangers we face, not as natural disasters, but as products created by our technological, socio-economic and political systems. This means that a reflexive approach sensitizes our consciousness not only to the question of how risks are produced but also to the question how risks so produced can be sensibly and realistically managed. Production of risks is one thing and its management is another. "Reflexivity thus requires attention not just to the 'representation' of the object to the subject, but also to the way in which the attributes of the subject help condition the representations of the object and how these representations themselves can help recondition the subject."[81]

b. Ulrich Beck, on his part, makes the distinction between "reflexivity" and "reflection."[82] As he explained:

> ... the 'reflexivity' of modernity and modernization in my sense does not mean reflection on modernity, self-relatedness, the self-referentiality of modernity, nor does it mean the self-justification or self-criticism of modernity in the sense of classical sociology; rather (first of all), modernization undercuts modernization, unintended and unseen, and therefore also reflection-free, with the force of autonomized modernization. ... [R]eflexivity of

---

79 Giddens, A, *Modernity and Self-Identity*, pp. 53–54.

80 Giddens, A., *The New Rules of Sociological Method*, London: Hutchinson, 1976, p. 17.

81 Stirling, A., "Precaution, Foresight and Sustainability: Reflection and Reflexivity in the Governance of Science and Technology," in Von Jan-Peter Voss, Dierk Bauknecht, René Kemp, (eds.), *Reflexive Governance for Sustainable Development*, Cheltenham, UK: Edward Elgar Publishing Ltd., 2006, p. 227.

82 Beck, U. et al., "The Reinvention of Politics: Towards a Theory of Reflexive Modernization" in *Reflexive Modernization: Politics, Tradition and Aesthetics in the Modern Social Order*, pp. 5–6. Reflexivity tends to undermine the foundations of modern self-descriptions. It is a problem for reflection. In the phase of global risk society, the incalculable nature of the perils unleashed by technical-industrial development comples to reflection upon the foundations of the social whole and to re-examination of accepted conventions and shared criteria of rationality. Society becomes, where it begins to understand itself as a risk society, reflective (....). That is to say, society itself becomes, for itself, a theme and a problem. Cf. Beck U., "Reflexive Governance: Politics in the Global Risk Society," in *Reflexive Governance for Sustainable Development*, pp. 34–38.

> modernity can lead to reflection on the self-dissolution and self-endangerment of industrial society, but it need not do so.[83]

The 'factors' of progress leads instinctively and unobserved to self-destruction. Beck traces the main role played by reflexivity in today's world down to the insight that science and technology produce more risks than "progress" and to the demythification of science; fallout from processes of differentiation in science; unanticipated interactions between scientific and technological innovations; and the dissemination and transmission of non-desired effects of scientific discoveries. In his words, "We live in an age of unintended consequences, and it is this state of affairs that must be decoded and shaped methodologically and theoretically, in everyday life and politically."[84] If modernity was an age of goal-aspiration, the age of reflexive modernity is an age of crisis, managing the side-effect of the older goals.[85] By '*disembedding*' the socially accepted patterns and '*reembedding*' a new order, new social movements or organs emerge that are challenging, skeptical but coherent. According to Beck, citizens' initiative groups or grassroots activists will more and more play an important role in risk society especially in redefining political agendas.[86] In particular, they will have a strong influence on the definition of what is 'safe'.

So, citizens' initiative groups will operate outside the corporate and state institutions that have accepted 'legitimacy' and are in fact the 'laity'. These citizens and groups will be eloquent and able to communicate at complicated echelons. They will take an issue that will require them to become experts in varied and intricate fields, like disposing of toxic waste or forecasting wind patterns in the diffusion of smoke emissions. They will confront the rights and standpoints of conventional establishments and create unconventional or avant-garde knowledge. Citizen initiative groups will critique and evaluate society and its affiliation to science and technology in an act of dynamic reflexivity. Beck describes an era where the industrial social forms of nowadays are at the moment substituted by the intrinsically dynamic reflexive modernity, not in an act of revolution but of

83 Beck, U., *Ecological Enlightenment: Essays in the Politics of the Risk Society*, Atlantic Highlands, NJ: Humanities Press, 1994, pp. 176–177.

84 Beck, U., *World Risk Society*, Cambridge, Uk & Massachusetts, USA: Polity Press in association with Blackwell Publishers, p. 119.

85 Delanty, von, G., *Social Theory in a Changing World: Conceptions of Modernity*, Cambridge, UK: Polity Press in association with Blackwell Publishers Ltd., 1999, p. 151.

86 Beck, U., *The Reinvention of Politics: Rethinking Modernity in the Global Social Order*, trans., Mark Ritter, Cambridge: Polity Press, 1997, p. 100; First published in German as *Die Erfindung des Politischen*, Frankfurt am Main: Suhrkamp Verlag, 1993.

evolution. So "reflexivity within Beck's discourse does not produce fundamental change; it produces more risks and more attempts to control/manage/limit risks by the same means."[87] Risk induces reflexivity because there are no absolute or self-evident answers, only discursive outcomes. In the shift from class politics to ecological politics, the political field opens up to an ever widening public.[88] In Beck's works, an essential part in this relationship is performed by "subpolitics" that shapes society from below as we shall see later.

c. Scott Lash's view of reflexivity, which puts in the picture his media and technology theory, marks a different approach from both Giddens and Beck. Unlike their somewhat quick dismissal of postmodernism, Lash takes the postmodern challenge to social theory very seriously. Yet, Lash compliments Beck for influening his own more technologically informed, notion of "reflexivity as non-linearity." He argues that Beck's distinction of "first modernity" and "second modernity" can be structured through socio-technical changes in institutions and subjectivity. In first modernity, the social order was preserved through social structures, whose influence and power were mostly maintained through their determinacy. In the second modernity, the social order is dependent not on the rigid nature of structure, but on currents that are only partially determinate and unquestionably immanent. While first modernity is characterized by linear systems, second modernity is distinguished by non-linear or reflexive systems. Lash esplains:

> Linear systems have a single point of equilibrium, and only external forces can disturb this equilibrium and lead to system change. The reflexivity of the second modernity presumes the existence of non-linear systems. Here system dis-equilibrium and change are produced internally to the system through feedback loops [...]. Reflexivity now is at the same time system *destabilization*.[89]

Scott Lash reiterates the view that reflexivity not reflection is the defining moment of Second Modernity. He explores the contrast between reflection and reflexivity:

---

87 MCKECHNIE, B. R., & WELSH, I., "When the Global Meets the Local: Critical Reflections on Reflexive Modernization" in VON RILEY E. DUNLAP, BUTTEL, H. FREDERICK, DICKENS, P. & GIJSWIJT, A., (eds.), *Sociological Theory andthe Enviromental: Classical Foundations, Contemporary Insights,* Maryland, USA: Rowman & Littlefield Publishers, Inc., 2002, p. 288.

88 DELANTY, VON G., *Social Theory in a Changing World: Conceptions of Modernity,* p. 154.

89 LASH, S., "Reflexivity as non-linearity," in *Theory, Culture, and Society*, Vol. 20, No. 2, 2003, (49–57), p. 50. (Empasis in the original)

[T]he individual of the first modernity is reflective while that of the second modernity is reflexive. The idea of reflective belongs to the philosophy of consciousness of the first modernity [...]. To reflect is to somehow subsume the object under the subject of knowledge. Reflection presumes apodictic knowledge and certainty. It presumes a dualism, a scientific attitude in which the subject is in one realm, the object of knowledge in another.[90]

In first modernity, the subject is peceived to have been *reflective,* reminiscent of the subject found in the emergence of consciousness in Descartes, while the subject in second modernity is *reflexive,* comparable to the subject found in phenomenology. Descartes' philosophy maintained a strict dichotomy of mind and body, choosing the former as in control of the later. According to Lash, the reflexive consciousness of the second modernity is devoid of the dualism between subject and object: "Reflexive ... has more to do with reflex than reflection. Reflexes are indeterminate. They are immediate. They do not in any sense subsume."[91] Lash sums up the perspective introducing a distinction between *self* and *structural reflexivity*:

First there is *structural reflexivity* in which agency, set free from the constraints of social structure, then reflects on the 'rules' and 'resources' of such structure; reflects on agency's social conditions of existence. Second there is *self-reflexivity* in which the actor reflects on itself. In self-reflexivity previous hetronomous monitoring of agents is displaced by self-monitoring.[92]

Structural reflexivity basically means that the consistency and power of ancillary institutions and expert-systems has become ineffective or no longer appear obvious, infallible or sacrosanct.[93] Van Loon puts it briefly:

---

90 *Ibid.*, p. 51.

91 *Ibid.*

92 LASH, S., "Reflexivity and its Doubles: Structure, Aesthetics, Community" in *Reflexive Modernization: Politics, Tradition and Aesthetics in the Modern Social Order,* pp. 115–116.

93 BECK, U., *The Brave New World of Work*, Cambridge: Polity Press, 2001, pp. 23–24. Here, Beck explains that while the collective plannings and extremely controlled or ordered practices of organized modernity still meant a relatively firm closure of biographical emergency and uncertainty, these industrial modern institutional barriers have been replaced by secondary institutions that are considerably less dependable and transmit the final decisions and judgments to the individuals. In the reflexivity of modernity, therefore, "the apparent outside of the institutions becomes the inside of the individual biography." (Beck, U., *Risk Society: Towards a New Modernity,* p. 130). This means that the shortcomings and biographical risks must now be acknowledged, identified, evaluated and procecced by individuals themselves. Cf. Beck, U. and Beck-Gernsheim, E., "Individualization and 'Precarious Freedoms': Pespectives

> All 'expert systems' of modern society have been forced to surrender more of their previously unchallenged claims to authority. This is not to say that their authority was never challenged before, but simply that such challenges have become normal rather than exceptional. The effect of these challenges is an increased disembedding of individuals from positions assigned to them by these authoritative institutions within the structures of modern life.[94]

In this respect, Van Loon explains that there is nothing free about this individualization as it "is *a default outcome of a failure of expert systems to manage risks*; neither science, governance, media, commerce, law nor even the military are able to provide sufficient closures of risks to enable people to place their thrust in these institutions. As a consequence, people are thrown back onto themselves, they are alienated from traditional communal systems but have nothing else instead."[95]

Self-reflexivity, on the other hand, is the process by which agency reflects upon itself, as in the autonomous monitoring of individual life narratives or the individual reflection of these changing institutional configurations and conditions. The newly formed social relations and social networks, individual lifestyles and self-images are becoming 'reflexive' so that they have to be established, maintained, and constantly renewed by individuals.[96] In an epoch of reflexive modernity, "the institutional conditions that determine individuals are no longer just events and conditions that happen to them, but *also consequences of the decisions they themselves have made*, which they must view and treat as such."[97] Self-reflexivity, as a result, is not one and the same with an individual agency released from social formations, but refers to an institutional environment that is much more vulnerable to individual differences and alterations in social practices and life tracks.

Finally, reflexivity is related to a policy of development which is aware of its cultural foundation. Crucial to this is not just an instrumental but a reflexive approach to policy programs by asking what development or modernization serves for in the end. The theorists of reflexivity are in effect declaring that we have reached

---

and Controversies of a Subject-oriented Sociology" in HELAAS, P., et al., (eds.), *Detradionalization*, p. 27.

94 VAN LOON, J., *Risk and Technological Culture, Towards a Sociology of Virulence*, London: Routledge, 2002, p. 32.

95 *Ibid.*

96 BECK, U., *Risk Society: Towards a New Modernity*, p. 97. trans. MARK RITTER, Los Angeles & London: Sage Publications, 1992. First published in German as *Risikogesellschaft: Auf dem Weg in eine andere Modern,* Frankfurt am Main: Surbkamp Verlag, 1986. See also, BECK, U. & BECK-GERNSHEIM, E., *Individualization*, London: SAGE publications, 2002, p. 35.

97 BECK, U., *Risk Society: Towards a New Modernity*, p. 136. (emphasis in the original)

that point: "Modernity has not vanished, we are not past it. Radical social change has always been part of modernity.What is new is that modernity has begun to modernize its own foundations. It has become directed at itself."[98] Different theorists highlight different aspects of reflexive modernity. While Beck is concerned with politics and Giddens with culture and social structures, Lash is perhaps more abstract and theoretical.

The main questions to be asked are then as follows: Given that technology has become part of our daily lives and given the complexities of risks we face in this technological environment, how can we better cope with the imminent problems or the ones at hand to guarantee the protection, safety, and welfare of the citizens? Should we focus on 'overcoming' risks in the sense of getting rid of them? Or should we be unassuming to explore an enhanced way of coping with risks at hand? What is the function of the government as a major actor of interference? What about scientists, experts, as well as journalism? While modernity sought answers to questions, reflexive modernity seeks to reframe the questions. Reflexive (modernization) is meaningful when it raises these questions as plainly as possible and can provide good answers or guidelines.

When science and technology has become suspect; when institutions are no longer trusted; when social systems are unresponsive and insensitive; how are individuals in risk socicties to comprehend the amplified risks that define those societies? For Beck the answer is "reflexive modernization"—a set of processes developing within the contemporary age of uncertainty where the continuous threat of a mounting range of local risks and global mega-hazards necessitates the constant reinvention of scientific and political institutions.

5.4.2. Transition from Modernity to Risk Society (Reflexive Modernity)

Ulrich Beck wrote his highly influential work, *Risk Society: Towards a New Modernity*, whose original German text, *Risikogesellschaft* appeared in 1986, in the aftermath of the Chernobyl catastrophe. It became "one of the most influential European works of social analysis in the late twentieth century."[99] Beck's

---

98 Beck, U., Wolfgang, B. & Lau, C., "The Theory of Reflexive Modernization: Problematic, Hypotheses and Research Programme," in *Theory, Culture and Society,* Abstracts, Vol. 20, No. 2. April, 2003.

99 Lash, S. & Wynne, B., *Introduction Risk Society: Towards a New Modernity*, p. 1. Other contributions to risk dialogue and theory have been made by Douglas, M and Wildavsky, A., Risk and Culture: An Essay in the Selection and Interpretation of Technological and Environmental Dangers, Berkeley, CA: University of California Press, 1982; Giddens, A., *The Consequences of Modernity*; GIDDENS, A, *Modernity*

conceptualization has stimulated research that focuses on the implications of science and technology for the social and natural environment and on the increasing use of risk analysis in deliberations of public policies linked to science and technology, and which involve ethical questions. His key concern is to comprehend and to sociologically hypothesize modern Western civilization and novel uncertainties in the last part of the 20th century. According to Beck, in late modernity our society is changing from an industrial to a risk society. Thus, the concept of risk is central to his thesis, and he defines risk as follows:

> Risk is the modern approach to foresee and control the future consequences of human action, the various unintended consequences of radicalized modernization. It is an (institutionalized) attempt, a cognitive map, to colonize the future.[100]

In radicalized modernity the new risks are manufactured or made-up uncertainties and dangers, because the array of impending disasters and uncertainties grows with technological and scientific progress and more industrialization, more automobiles and more prosperity also cause more environmental problems. Again, it is not post-modernity but more modernity radicalized, which produces world risk society.

Beck distinguishes three stages of modernity: simple modernity; second, or late modernity (the transition stage between industrial and risk society); and reflexive modernity, or risk society. According to Lenoble and Maesschalck, Beck's 'reflexive modernity' "seeks explicitly to differentiate itself from the existing approaches by emphasizing the reflexivity of the risk situations confronting contemporary societies and of the collective strategies that must be deployed to deal with them."[101]

In précis, Beck's argument is predicated upon differences between the types of hazards which affect 'pre-industrial', 'industrial' and 'risk' societies.[102] We have made a transition from 'simple modernity' to 'reflexive modernity' or from 'industrial society' to 'risk society'. This transformation began at the late 1960s. Contrary to postmodern theories that present late 20th Century social transformations

---

and Self-Identity, Cambridge: Polity Press, 1991; Mary Douglas, *Risk and Blame: Essays in Cultural Theory*, London: Routledge 1992; Niklas Luhmann, *Risk: A Sociological Theory*, New York: Aldine De Gruyter 1993; Barbara Adam,*Timescapes of Modernity: The Environment and Invisible Hazards*, London: Routledge, 1998, etc.

100 Beck, U., *World Risk Society*, p. 3.

101 Lenoble, J. & Maesschalck, M., *Towards a Theory of Governance: The Action of Norms*, p. 253.

102 Beck, U., *Ecological Politics in an Age of Risk*, Cambridge: Polity Press, 1995, p. 78.

as going beyond modernism, Beck argues that modernity is passing through an inadvertent stage that is compelling it to challenge the principles and bounds of its own ideal. What points out the risk society are *manufactured uncertainties*, specifically a series of new risks, for instance, environmental concerns, which are inadvertent side effects of technological and economic progress. These manufactured uncertainties stem from scientific and technological advancement, which hypothetically should solve, not cause problems.

While the traditional industrial society was worried with the distribution of social inequalities, reflexive modernity is concerned with the distribution of risks. Beck seems to be saying that we are moving from the fight against poverty and against income inequality, to all-embracing fear across social classes and countries because of the risk or uncertainties of new technologies. We are moving from conflicts on the distribution of wealth to conflicts on the distribution of environmental and technological dangers. Often such dangers have an effect on all and sundry. Modernization exalts science and technology but it has launched new environmental hazards that it does not know how to manage. Environmental movement is a response to these new hazards. Modernization has turn out to be, in Beck's words, "reflexive." This reflexive modernity or risk society progresses outside its conventional or classical industrial society. Beck states this as:

> *Just as modernization dissolved the structure of feudal society in the nineteenth century and produced the industrial society, modernization today is dissolving industrial society and another modernity is coming into being [...].* We are witnessing not the end but the *beginning* of modernity—that is, of a modernity *beyond* its classical industrial design.[103]

Beck's argument is that "while in 'classical industrial society, the 'logic' of wealth production dominates the 'logic of risk production; in the risk society this relationship is reversed."[104] To distinguish risk society from industrial society, the harsh ways of existence in both epochs are juxtaposed: "The driving force in the class society (industrial) can be summarized in the phrase: *I am hungry!* The movement set in motion by the risk society, on the other hand, is expressed in the statement: *I am afraid!* The *commonality of anxiety* takes the place of the *commonality of need*."[105] The specific result of this new modernity is the emergence of risk society. This dynamics of second modernity, namely a more reflexive agency, has brought about a more risk inhabited society. This risk, Beck states, is ambivalent.

---

103 Beck, U., *Risk Society: Towards a New Modernity*, p. 10. (Italics in the text)

104 *Ibid.*, p. 12.

105 *Ibid.*, p. 49.

Technology comprises three major apparent features in late modernity: (1) a sharp rate of transmission due to the globalization of production processes and to the increased flow of communication; (2) an increase in scale and intricacy; and (3) an escalation in the magnitude and distribution of risks. According to Beck and his colleagues, therefore, a new form of modernity is now emerging. It is distinguished by increasing speed, more and more extreme transnational interdependence, and the emergence of economic, cultural, political and societal "globalisation" in which the separating lines among states and even among Western and non-Western societies are ever less applicable. As the old bipolar order has recoiled, new challenges arising out of technological change and global integration are convincing governments and societies to deal not so much with adversaries as with risks. Giddens is right when he remarkes:

> Globalization is changing the way the world looks, and the way we look at the world. By adopting a global outlook, we become more aware of our connections to people in other societies. We also become more conscious of the many problems the world faces at the start of the 21st century. The global perspective opens our eyes to the fact that our increasing ties with the rest of the world means that our actions have consequences for others and that the world's problems have consequences for us.[106]

In *The politics of Risk Society*, Beck insists that society "has become a laboratory where there is absolutely nobody incharge". The risks being created in the laboratory, he insists, are different from the risks that earlier generations encountered.

> There was once a time when risk was something that you indulged in for a bit of excitement. A bet on the Grand National, a spin of the wheel—it was all meant to add a bit of spice to an otherwise orderly and predictable life. Now manufactured uncertainty means that risk has become an inescapable part of our lives and everybody is facing unknown and barely calculable risks. Risk becomes another word for 'nobody knows'. We no longer choose to take risks, we have them thrust upon us. We are living on a ledge—in a random risk society, from which nobody can escape. Our society has become riddled with random risks. Calculating and managing risks which nobody really knows has become one of our main preoccupations. That used to be a specialist job for actuaries, insurers and scientists. Now we all engage in it, with whatever rusty tools we can lay our hands on—sometimes the calculator, sometimes the astrology colume.The basic question here is: how can we make decisions about a risk we know nothing about?[107]

Beck contends that it is necessary to separate the notion of risk from hazard or danger. Hazard refers to those naturally occurring events that are not the invention

---

106 GIDDENS, A., *Sociology*, 4th edition, Cambridge: Polity Press, 2001, p. 51.

107 BECK, U., "The Politics of Risk Society", in *The Politics of Risk Society*, ed. Jane Franklin, London: Polity Press, 1998, p. 12.

of human activities (for instance, seismic activity, tropical storm, famine caused by drought, tsunamis, hurricane, infernos, etc). Thus, hazards "differ essentially from 'risks'...since they are not based on decisions that focus on techno-economic advantages and opportunities." Risks, on the other hand, "presume industrial that is, techno-economic decisions and considerations of utility."[108] Contrary to natural hazards, manufactured risks are 'side effects' of the goal of progress within business, science, technology and medicine. For instance, in quest for good life, the capitalist system of mass production and mass consumption has caused environmental dis-equilibrium through the exhaustion of limited resources, industrial pollution and consumer waste. More blatantly, techno-scientific expansion has yielded nuclear, chemical and biological technologies that are capable of exterminating human life.

Today the question of risk has become a political issue. Beck writes that "Risks always depend on decisions – that is, they presuppose decisions".[109] Societies are at the moment beginning to use risk awareness to update decisions at every level, be it social, political or environmental because, the prevailing mechanisms are not equipped to contend with the new levels of risk. As Beck states: "This concept describes a phase of development of modern society in which the social, political, ecological and individual risks created by the momentum of innovation increasingly elude the control and protective institutions of industrial society."[110]

---

108 Beck, U., "From Industrial Society to the Risk Society: Questions of Survival, Social Structure and Ecological Enlightenment" in *Theory, Culture and Society*, Vol. 9, 1992, p. 98.

109 Beck, U., *Democracy Without Enemies*, Cambridge: Polity Press, 1997, p. 30. "They arise from the transformation of uncertainty and hazards into decisions (and compel the making of decisions, which in turn produce risks). The incalculable threats of pre-industrial society (plague, famine, natural catastrophes, wars, but also magin, gods, demons) are transformed into calculable risks in the course of the development of instrumental rational control, which the process of modernization promotes in all spheres of life. This characterizes the situation and the conflicts in early, classical industrial and bourgeois society. In the course of its expansion it is true not only for the 'feasibility' of production capacities, tax revenues, the calculation of export risk and the consequences of war, but also for the vicissitudes of individual lives: accidents, illnesses, death, social insecurity and poverty. It leads...to the emergence of diverse systems of insurance, to the extent that society as a whole comes to be understood as a risk group in insurers' terms—as a *provident state* and *providing state*." Cf. Beck U., *World Risk Society*, pp. 75–76.

110 Beck, U., "Risk Society and the Provident State", trans. Martin Chalmer, in Lash, S., Szerszynski, B & Wynne, B. (eds.), *Risk, Environment and Modernity: Towards a New Ecology*, London: SAGE Publications, 1996, p. 27.

The world at the moment discovers herself agitated because industrial and post-industrial risks and dangers are becoming more and more part of any decision-making, in so doing appropriating a reasonable expanse of the authority intrinsic to industrialisation and modernization:

> The concepts of 'industrial' or 'class society', in the broadest sense of Marx or Weber, revolved around the issue of how socially produced wealth could be distributed in a socially unequal and *also* 'legitimate' way. This overlaps with the new *paradigm of risk society* which is based on the solution of a similar and yet different problem. How can the risks and hazards systematically produced as part of modernization be prevented, minimised, dramatised, or channelled? Where do they finally see the light of day in the shape of 'latent side effects', how can they be limited and distributed away so that they neither hamper the modernization process nor exceed the limits of that which is 'tolerable' – ecologically, medically, psychologically and socially?.[111]

Technological developments have thus increased "our capacity to act upon the world in such an intensive and extensive way that the consequences of our actions have escaped our capacity to foresee them."[112] This again, calls for the question of ethical responsibility.

5.4.3. Risk and Responsibility

In the above state of affairs, human responsibility for technological expansion is an ethical issue that is both pertinent and complex. The processes and techniques of risk management block out responsibility. Modern society functions as a "laboratory" in which no one specifically must answer for the negative effects of technological experimentation. The responsibility for global environmental degradation on an accelerating degree is however characteristically sheeted home to the transnational beneficiaries of economic liberalisation who are now seen as beyond the 'control' of the nation state. Without a doubt the external economic interaction and the heightened rivalry between nation states for foreign investment in the current deregulatory atmosphere is invariably blamed for the failure of environmental standards and the threat to non renewable natural resources. The rise of new totalitarianisms, the disintegration of the world's ecosystems, and a fortress society of the affluent in permanent struggle with the improverished majority, is simply a disaster scenario generated by contemporary globalization.[113] This has led to the

111 Beck, U., *Risk Society: Towards a New Modernity*, p. 19. (Emphasis in the original)

112 Shlomo, G., "Living in a World Risk Society: A reply to Mikkel V. Rasmussen," in *Millennium: Journal of International Studies*, Vol. 31, No. 1, 2002, p. 149.

113 Giddens, A., *Beyond Left and Right: The Future of Radical Politics*, p. 253.

institutionalization of risk environments, for instance, the chemical pollution of the oceans, nuclear radiation from accidents, destruction of rain forests, and global climate change.

Early phases of globalization were controlled primarily by the "advanced" societies of the West. But present-day globalization is no longer a process of one-directional imperialism. In fact, globalization has no clear direction; it is disorderly and chaotic.[114] So, typically, nobody can be held personally responsible for producing risks. They are a part of the normal functioning of all social institutions together and so appear unstoppable. Although nobody is responsible, everybody is responsible! Whenever something takes place, inevitably an agent did it. Beck postulates the response as the "*organized irresponsibility*" of industrial modernity, that is to say, disregarding and being indifferent to the probable risks that emerge due to industrialization.[115] Everyone sees himself as a victim, no one as a perpetrator. Environmental pollution, together with its increasingly global impact in the form of climate change, graphically demonstrates this paradox. The greater the environmental degradation is, the more laws and environmental regulations there are, but at the same time no institution seems to be exclusively responsible. Thus, the escalating profile of bads has not been met with fitting counter measures at the level of public policy. Political parties have failed to develop articulate strategies or policies to contend with the pervasiveness and mobility of social bads. In fact such social bads are usually being met by a combination of rebuff, obfuscation and contrived pledges.

Modern "manufactured" technological risks are sometimes not tangible and cannot be smelled, heard, tasted or touched.

> Many of the new risks (nuclear or chemical contaminations, pollutants in foodstuffs, diseases of civilization) completely escape human powers of direct perception. The focus is more and more on hazards which are neither visible nor perceptible to the victims [...] imperceptibility abandon the victims completely to the judgments, mistakes and controversies of experts, while subjecting them to terrible psychological stresses.[116]

This situation is inevitable because modern risk judgments are made on the basis of "expert knowledge". Essentially, formal power holders discover themselves in

---

114 GIDDENS, A., *Runaway World: How Globalization is Reshaping our Lives*, 2nd Edition, London: Profile Books, 2002.

115 BECK, U., *World at Risk*, pp. 27–29; See also, ELLIOT, A., "Beck's Sociology of Risk: A Critical Assessment," in *Sociology*, 35(2), 2002, pp. 293–315; BECK, U., *World Risk Society*, p. 32.

116 BECK, U., *Risk Society: Towards a New Modernity*, p. 27.

charge of making 'choices' from multiple options in a state of defective insufficient knowledge:

> The ultimate deadlock of risk society … resides in the gap between knowledge and decision: there is no one who really knows the global outcome—at the level of positive knowledge, the situation is radically 'undecidable'—but we nonetheless have to decide […] risk society is provoking an obscene gamble, a kind of ironic reversal of predestination: I am held accountable for decisions which I was forced to make without proper knowledge of the situation.[117]

Modern risks set the stage for societal self-annihilation and Beck's writings signify the awful consequences of contemporary technological hazards for the future and undoubtedly illuminate the destructive side of progress.

The recognition that risks of unparalleled magnitude abound in advanced modernity is also acknowledgment that the social framework itself is at risk and is perhaps a pointer that ours is an era where all societies are in a state of unprecedented fragility and vulnerability. This ubiquity and magnitude of the risks of advanced modernity pose a critical challenge for all societies, captured in the unsettling questions: (i) Are societies creating risks more rapidly than we can understand and manage them? (ii) What is our knowledge about these risks? And (iii) how can societies develop the institutional and political means for managing them effectively? Reflexive modernization will help us come to grips with with these questions.

5.4.4. Towards a Reflexive Modernization

In Beck's theory of reflexive modernization, the term reflexivity does not in the first place imply reflection, but self-confrontation: "Reflexive modernization means self-confrontation with the effects of risk society that cannot be dealt with and assimilated in the systems of industrial society—as measured by the latter's institutionalized standards."[118] It is the sovereign, compelling dynamic of second or reflexive modernization that, according to Beck, stimulates modern men and women into 'self-confrontation' with the consequences of risk that cannot sufficiently be addressed, measured, controlled or overcome, at least according to the standards of industrial society. Reflexive modernization begins from being aware of the fact that the very sense and processes of modernization have created

---

117 Beck, U., *World Risk Society*, p. 78.

118 Beck, U., "The Reinvention of Politics: Towards a Theory of Reflexive Modernization," in *Reflexive Modernization: Politics, Tradition and Aesthetics in the Modern Social Order*, p. 6.

an enormous diversity of risks menacing the stability and safety of average citizens. We now face, according to Beck:

> The possibility of creative (self-) destruction for an entire epoch: that of industrial society. The 'subject' of this creative destruction is not the revolution, not the crisis, but the victory of Western modernization [...]. This new stage, in which progress can turn into self-destruction, in which one kind of modernization undercuts and changes another, is what I call the stage of reflexive modernization.[119]

Accordingly, reflexive modernization is not about drastic changes taking place on account of certain critical dysfunction of modernity. It includes an accidental change of industrial society which synchronizes with existing political and economic orders. Beck argues that we are between industrial society and advanced modernity, between simple modernization and reflexive modernization. He develops these distinctions thus:

> In view of these two stages and their sequences, the concept of 'reflexive modernization' maybe introduced. This precisely does not mean reflection (as the adjective 'reflexive' seems to suggest), but above all self-confrontation. The transition from the industrial to the risk epoch of modernity occurs unintentionally, unseen, compulsively, in the course of a dynamic of modernization which has made itself autonomous, on the pattern of latent side-effects. One can almost say that the constellations of risk society are created because the self-evident truths of industrial society (the consensus on progress, the abstraction from ecological consequences and hazards) dominate the thinking and behaviour of human beings and institutions. Risk society is not an option which could be chosen or rejected in the course of political debate. It arises through the automatic operation of autonomous modernization processes which are blind and deaf to consequences and dangers. In total, and latently, these produce hazards which call into question - indeed abolish – the basis of industrial society.[120]

The purported change in the direction of reflexive modernization and risk society offers a framework wherein a different political and economic environment is shaped, now questioning the erstwhile prevailing attitude, accepted views, beliefs and guarantees of first or simple modernity. As Beck explains further:

> ...(The first) modernity based on nation-state societies, where social relations, networks and communities are essentially understood in a territorial sense. The collective patterns of life, progress and controllability, full employment and exploitation of nature that were

---

119 Beck, U., "The Reinvention of Politics: Towards a Theory of Reflexive Modernization," in *Reflexive Modernization: Politics, Tradition and Aesthetics in the Modern Social Order*, p. 2.

120 Beck, U., "Risk Society and the Provident State" in *Risk, Environment and Modernity: Towards a New Ecology*, p. 28.

typical of this first modernity have now been undermined by five interlinked processes: globalization, individualization, gender revolution, underemployment and global risks [...]. If the five processes are considered more closely, it becomes clear what they have in common: namely, they are all unforeseen consequences of the victory of the first, simple, linear, industrial modernization based on the national state. This is what I mean by talking of 'reflexive modernization.'[121]

Or as Giddens puts it: "Simple modernization is old-type unilinear modernization; reflexive modernization, by contrast, implies coming to terms with the limits and contradictions of the modern order."[122]

Whilst acknowledging Beck's 'second' modernity and Giddens 'late modernity' Baumann uses the term 'liquid' to emphasis the changes and continuities. His overall examination of modernity steered him to see modern-day society as a "liquid" modernity. Bauman defines a liquid modern society as:

> A society in which the conditions under which its members act change faster than it takes the ways of acting to consolidate into habits and routines. Liquidity of life, and that of society, feed and reinvigorate each other. Liquid life, just like liquid modern society, cannot keep its shape or stay on course for long.[123]

This definition is even made explicite when Mark Deuze states that "a liquid modern society is one where uncertainty, flux, change, conflict, and revolution are the permanent conditions of everyday life."[124]

Within this context, therefore, Beck argues for a new model of risk assessment, one that identifies the benefits of technology advancement, but at the same time recognizes the many diverse and likewise genuine ways that technology can be rationally evaluated. In other words, reflexive modernization also connotes self-reflection. More accurately, reflexive modernization is not a plain negation. It is a reconstruction as well as a deconstruction. It is reconstructive in the sense that it opens up its own new potentiality from inside by listening carefully to the success

---

121 BECK, U., *World Risk Society*, pp. 1–2.

122 GIDDENS, A., "Risk Society, the Context of British Politics," in FRANKLIN, J. (ed.) *The Politics of Risk Society*, Oxford: Polity Press, 1998, p. 31.

123 ZYGMUNT, B., *Liquid Life*, Cambridge: Polity Press, 2005, p. 1. In an earlier work, Bauman examines how we have moved away from a 'heavy' and 'solid,' hardware-focused modernity to a 'light' and 'liquid', software-based modernity. What distinguishes liquid modernity from early modernity is the lack of stable institutions. There is no condition; everything is process. It implies a sense of rootlessness to all forms of social construction. See ZYGMUNT, B., *Liquid Modernity*, Cambridge: Polity Press in association with Blackwell Pub. Ltd., 2000.

124 DEUZE, M., *Media Work*, Cambridge: Polity Press, 2007, p. 43.

story of modernization, so to speak, and by reconstituting it within a new structure of advancement or development which is normatively supported and practically feasible. Simultaneously, it is deconstructive in the sense that it endeavours to cut loose from the very suppositions or hypotheses under which the past model of development was made possible. 'Reflexive modernization means here that modernity of our societies must again be inquired into, determined, fought for and obtained by interpreting and reinterpreting past, present and future'. As Beck states further:

> The sociology of reflexive modernization discovers industrial society as a contradictory historical symbiosis between modernity, pre-modernity and counter modernity, a semi-modern society that is being disembedded because of continued modernization and radicalization of modernity and re-embedded as different types of 'modern' or 'Counter-modern' societal forms.[125]

The methodical and regular occurrence of risks which cannot be envisaged nor controlled, results in a fundamental institutional crisis, where "the recognition of the unpredictability of the threats provoked by techno-industrial development necessitates self-reflection on the foundations of social cohesion and the examination of prevailing conventions and foundations of 'rationality.'"[126] In constantly changing the structure of society and the environment, modern science and technology keep undermining the very system they set out to control. Science and technology are generating and discerning fresh problems at a parallel pace as they are unravelling old ones. We have to point out that "Reflexive modernization", says Beck "means not less but more modernity, a modernity radicalized against the paths and categories of the classical industrial setting."[127]

Using the term manufactured uncertainty; Anthony Giddens established his idea of reflexive modernization around our shifting environment. We are, according to him, in the middle of stirring from a traditional society in which risk is predictable to the world beyond our control, which he describes as a "runaway world," a world that is becoming de-traditionalized. Giddens stresses that the ultimate menacing risks in society does not come "from the fixities of tradition or

---

125 Beck, U. *The Reinvention of Politics: Rethinking Modernity in the Global Social Order*, p. 38.

126 Beck, U., "The Reinvention of Politics" in *Reflexive Modernization: Politics, Tradition and Aesthetic in the Modern Social Order*, p. 8.

127 Beck, U., *Risk Society: Towards a New Modernity*, p. 14.

nature" any more, but instead from "the very impact of our developing knowledge upon the world."[128]

Though it still remains explorative in the main, the idea of reflexive modernization would have to become more existential or actual in terms of institutional blueprints, policy options, citizens' involvement, and developmental roadmap. The issue of citizens' participation is of greatest importance for reflexive modernization since at the end of the day, it is neither the government nor experts who are particularly affected by the environmental and ecological crisis caused by modern technology, who define what risks are and how to respond to these, but citizens themselves can and should do it democratically. Ted Benton thus remarks that 'the theory of reflexive modernization is somewhat ambiguously placed between the notions of sustainable development and environmental/ecological modernization, on the one hand, and the more politically radical advocacies of "emancipator eco-politics," on the other'.[129] Later in the next chapter, we shall see how such citizen's political participation can help to confront the deficiencies of modern technological advancement. At the moment, let us examine how Beck moved from reflexive self-confrontation to self-monitoring which also forms the nucleus of reflexive modernization.

5.4.5. Reflexive Individualization

The advent of second modernity is not entirely about risk; it is in addition about an intensification of choice. There is an emerging context in which individual's 'self-monitoring' capability becomes increasingly permeated. Thus, in the course of reflexive modernization, Beck argues, more and more areas of life are on the loose or disembedded from the hold of tradition. This he sees as a 'second modernity' transforming what had become the 'traditional' aspects of industrialism:

> ... reflexive modernization dissolves the traditional parameters of industrial society: class culture and consciousness, gender and family roles. It dissolves these forms of the conscience collective, on which depend and to which refer the social and political organizations and institutions in industrial society. These detraditionalizations happen in a *social surge of individualization.* At the same time the *relations* of inequality remain stable [...]. Against the background of a comparatively high material standard of living and advanced

128 Giddens, A., *The Runaway World,* Cambridge: Polity Press, 2002, p. 26.

129 Benton, T., "Social Theory and Ecological Politics: Reflexive Modernization or Green Socialism," in Von Riley E. Dunlap, et. al., *Sociological Theory and the Enviromental: Classical Foundations, Contemporary Insights*, p. 257.

social security systems, the people have been removed from class commitments and have to refer to themselves in planning their individual labour market biographies.[130]

Specifically, "people living in the modernized societies of today develop an increasing engagement with both the intimate and more public aspects of their lives, aspects that were previously governed by tradition or taken-forgranted norms."[131] This set of developments is what Beck calls 'individualization.' According to him, "individualization must be clearly distinguished from individualism. Whereas individualism is commonly understood as a personal attitude or preference, individualization refers to a macro-sociological phenomenon, which possibly—but then again perhaps not—results in changes in attitude in individuals."[132] Very different from such concept as individuation or individualism, therefore, individualization is a concept which describes a structural, sociological transformation of social institutions and the relationship of the individual to society.[133]

According to Beck's work, *The Normal Chaos of Love*, "the term individualization covers a complex, manifold, ambiguous phenomenon, or more precisely a social transformation."[134] Individualization means the fragmentation or the growing instability of hitherto prevailing social forms. It is "the disintegration of the certainties of industrial society as well as the compulsion to find and invent new certainties for oneself."[135] Or as Baumann states: "'Individualization' consists in transforming human 'identity' from a 'given' into a 'task'- and charging the actors with the responsibility for performing that task and for the consequences (also the side-effects) of their performance."[136] Individualization emancipates or liberates

---

130 BECK, U., *Risk Society: Towards a New Modernity*, p. 87.

131 ELLIOT, A., Contemporary Social Theory: An Introduction, New York; Routledge, 2009, p. 290.

132 BECK, U., World at Risk: The New Task of Critical Theory, in *Development and Society*, *37*(1), 2008, 1–21, p. 3.

133 See BECK, U., & WILLIAMS, J., *Conversation with Ulrich Beck*, Cambridge: Polity press, 2004, pp. 62–108; BECK, U. & BECK-GERNSHEIM, E., *Individualization: Institutionalized Individualism and its Social and Political Consequences*, London: SAGA Publications Ltd., 2002.

134 BECK, U. & BECK-GERNSHEIM, E., *The Normal Chaos of Love*, Cambridge: Polity Press, 1995, p. 7.

135 BECK, U., "Reinvention of Politics: Towards a Theory of Reflexive Modernization" in *Reflexive Modernization. Politics*, Tradition and Aesthetics in the Modern Social Order, p. 14.

136 ZYGMUNT, B., *The Individualized Society*, Cambridge: Polity Press, 2002, p. xv. This is a collection of papers and essays that cover a diverse range of themes divided into three subheadings: "The Ways We Are", "The Way We Think", and "The Way We Act". The

people in different ways from traditional roles and limitations in such institutional domains as class, social rank, family, gender roles, nation, and so on. As Beck says:

> As this liberation or 'disembedding' occurs, new forms of reintegration and control are created ('re-embedding'). With the decline of class and status groups the individual must become the agent of his or her own identity making and livelihood. The individual, not his or her class, becomes the unit for the reproduction of the social in his or her own lifeworld. Individuals have to develop their own biography and organize it in relation to others.[137]

To give a simple definition, Beck states that individualization means *"disembedding"* of the way of life of industrial society without "*re-embedding*."[138] The '*disembedded*' self is progressively liberated from traditional and communal contexts as it attempts to construct his or her own narratives. People are freer to reflect on their own selves: "The newly formed social relationships and social networks now have to be individually chosen; social ties, too, are becoming *reflexive*, so that they have to be established, maintained, and constantly renewed by individuals."[139] As the old institutions of industrial society—family, community, social class—are undermined by the process of global modernization, individuals must learn to navigate society for themselves and become designer, stage director of his or her own "biography, identity, social network, commitments and convictions."

> Opportunities, threats, ambivalences of the biography, which was previously possible to overcome in a family, in the village community or by recourse to a social class or group, must increasingly be perceived, interpreted and handled by individuals themselves. To be sure, families are still to be found, but the nuclear family has become an even more rare institution.[140]

Beck sees structural change in the modern world as the process by which individuals are "set free from the certainties and modes of living of the industrial epoch – just as they were 'freed' from the arms of the church into society in the age of the

---

major concern of the collection is what Baumann refers to as the "process of relentless individualization" (p. 6). Tracing the shift in emphasis from social individual to the individualized individual is "of the utmost sociological relevance" (p. 6). According to Baumann, in the 'society of individuals' supra individual factors shape the belief that the individual is responsible for their choices and any situation they happen to find themselves in.

137 Beck, U., Beck-Gernsheim, E., *Individualization*, p. 203.

138 *Ibid.*,Preface, p. xxii.

139 Beck, U., *Risk Society: Towards a New Modernity*, p. 97.

140 Beck, U., "Reinvention of Politics: Towards a Theory of Reflexive Modernization," in *Reflexive Modernization: Politics, Tradition and Aesthetics in the Modern Social Order*, p. 8.

reformation."[141] The traditionally homogeneous life course has been supplanted by a "do-it-yourself-biography" which individuals have to create themselves. Scott Lash once summarized what Beck has done in the field of sociology into two theses: an environmental thesis or the 'risk thesis,' and an 'individualization thesis.'[142] From this outlook, the individual agency, although *disembedded* with or disconnected from the collective grounds, is regarded as a hub and measure of everything, and in addition, as the basis of the most reliable value. The agency is accordingly considered as a normative subject of institutions and so perhaps negating his or her own instincts in addition to the authority given from the outside.

> The decisive feature of these modern guidelines is that, far more than earlier, individuals must, in part, supply them for themselves, import them into their biographies through their own actions. This has much to do with the fact that traditional guidelines often contained severe restrictions or even prohibitions on action [...]. By contrast, the institutional pressures in modern Western society tend rather to be offers of services or incentives to action – take, for example, the welfare state, with its unemployment benefit, student grants or mortgage relief.To simplify: one was born into traditional society and its preconditions (such as social estate and religion). For modern social advantages one has to *do* something, to make an active effort. One has to win, know how to assert oneself in the competition for limited resources – and not only once, but day after day.[143]

The significance of this approach is that it offers a broader concept of reflexivity and also addresses the key question of institutionalization. The vanishing of tradition and the collapse of previously existing social institutionalized forms leads people into making resolutions about their own lives and prospective lines of action in new 'life worlds' such as new social movements that are chosen rather than given. In this view, individuals are continuously confronted with a plurality of uncertain life-course options and are forced to develop a "calculative attitude" to potential actions. As traditional ways of doing things become problematic, people must choose routes for a more gratifying life – all of which entails planning and rationalization, contemplation and dedication. The desires and choices

141 Beck, U., *Risk Society: Towards a New Modernity*, p. 14.

142 Lash, S., "Forward: Individualization in a Non-Linear Mode" in Ulrich B. & Beck-Gernsheim, E., *Individualization*, Sage Publication, 2003, p. vii. See also, Beck, U & E Beck-Gernsheim, "Individualization and 'Precious Freedom': Perspectives and Controversies of a Subject-Orientated Sociology", pp. 23–48 in Heelas P., Lash, S., & Morris, P., eds. *Detraditionalization*, Oxford. Blackwell Publishers, 1996.

143 Beck, U. & Beck-Gernsheim, E., "Individualization and 'Precarious Freedoms': Perspectives and Controversies of a Subject-oriented Sociology," in Helaas, P., et.al. (eds.) *Detraditionalization*, p. 25.

of individuals increasingly produce distinct institutions where "individuals must produce, stage, and cobble together their biographies themselves."[144] A dynamic commitment with the self, with the body, with relationships and marriage, with gender norms, and with work: this is the subjective scenery of the risk society. "The risk society is potentially also a *self*-critical society [...]. Precisely where traditions and hence values *have deteriorated*, risk come into being."[145] The idea of individualization is the foundation whereupon Beck constructs his vision of a 'new modernity', of fresh personal experimentation and cultural novelty against a social ambiance of risks, dangers, hazards, reflexivity, and globalization. As he states it: "Individualization of life situations and processes thus means that biographies become *self-reflexive*; socially prescribed biography is transformed into biography that is self-produced and continues to be produced."[146]

That means that the unleashing of experimentation and choice which individualization brings is undoubtedly not without its complexities, progressive and regressive elements. "....it is, simply, that in modern society's new demands, controls and constraints are being imposed on individuals. Through the job market, the welfare state and institutions, people are tied into a network of regulations, conditions, provisos."[147] Individualization, therefore, is both releasing individuals from traditional structures and *re-embedding* them in new social relations in such a way that the gains of today's individualization might be tomorrow's limitation, as advantage and progress turn into their opposite.

> Seen from one angle it means freedom to choose and from another pressure to conform to internalized demands, on the one hand being responsible for yourself and on the other being dependent on conditions which completely elude your grasp.[148]

Giddens examines the question of individualization by stressing the importance of this question of *de- traditionalization,* or the freedom from traditional ties and affiliations and the dissolution of collective structures as the root of individualization. According to Giddens:

> For most of its history, modernity has rebuilt tradition as it has dissolved it. Within Western societies, the persistence and recreation of tradition was central to the legitimation

144 Beck, U., "The Reinvention of Politics: Towards a Theory of Reflexive Modernization" in *Reflexive Modernization: Politics, Tradition and Aesthetics in the Modern Social Order*, p. 13.

145 Beck, U., Risk *Society: Towards a New Modernity*, p. 176.

146 *Ibid.*, p. 135.

147 Beck, U. & Beck-Gernsheim, E. *Individualization,* p. 2.

148 Beck, U., Beck-Gernsheim, E., *The Normal Chaos of Love*, p. 7.

> of power, to the sense in which the state was able to impose itself upon relatively passive 'subjects.' For tradition placed in stasis some core aspects of social life- not least the family and sexual identity- which were left largely untouched so far as 'radicalizing' Enlightenment was concerned.[149]

So, if individuals are to be emancipated from rigid, legalistic social positions to be able to construct and determine their own biographies reflexively there is need for *de-traditionalization,* in which, the individual, due to the inadequacies of traditional institutions, (family, state, work etc), questions their legitimacy. Ordinarily, such traditional institutions, according to Bauman, "determined identity by birth and hence provided few if any occasions for the questions of 'who am I' to arise."[150] Giddens labels this process of questioning traditional institutional power *institutional reflexivity* which means that "there is a continuous filter-back of expert theories, concepts and findings to the lay population."[151] Succinctly put:

> Institutional reflexivity means that with the advent of modern institutions knowledge about social life is no longer knowledge about an external, given environment of action, but increasingly comes to constitute what that environment of action is. Institutional reflexivity means that constant use of information to organize and reorganize the environments of action which that information describes.[152]

Thus, unbound from the institutional and social monitoring and normative bondage, individuals are mandated to organize and reorganize their activities taking into consideration their new understanding of the environment. Individual personal identity is not engraved inside traditional or institutional rules anymore; instead, it becomes a dynamic venture of reflexivity, transformation and progress. Giddens argues: "The self is not a passive entity, determined by external influences; in forging their self-identities [...] individuals contribute to and directly promote social influences that are global in their consequences and implications".[153]

Life in the contemporary welfare is one where every decision becomes a personal risk. As a result of the institutional failure to deal with this risk, there has been a new alignment of interests and the emergence of a new form of politics—a

---

149 GIDDENS A., "Living in a Post-Traditional Society", in *Reflexive Modernization*, p. 56.

150 ZYGMUNT, B., *Identity*, Cambridge: Polity Press, 2004, p. 49.

151 GIDDENS, A., "Living in a Post-Traditional Society" in *Reflexvie Modernization: Politics, Tradition and Aesthetics in the Modern Social Order,* p. 91.

152 GIDDENS, A., Industrialization, Ecology, and the Development of Life Politics, in W. V. D'ANTONIO, M. SASAKI & Y. YONEBAYASHI, (eds.), *Ecology, Society & The Quality of Social Life*, New Brunswick, New Jersey: Transaction Publishers, 1994, (1–10), pp. 2–3

153 GIDDENS, A., *Modernity and Self-Identity*, p. 2.

sub-politics that leads to a sharing of power between established and informal politics, and the government and society. Society, in effect, seeks to regain 'the political' from its modernist demotion to the institutional sphere, and this, according to Beck, is achieved primarily through sub-political means – that is, locating the politics of risk at the heart of forms of social and cultural life.

Concluding Remarks

Hans Jonas provides a useful basis for the formulation of a new ethics. The most interesting uniqueness of his thought lies in his aptitude to expound the new challenges facing the modern world. His summon to acknowledge that new challenges necessitate new strategies and attitude is one that ought to resonate before the Heideggerian as well as modern ethicists.

In *The Imperative of Responsibility*, Jonas attempts to meet this challenge and the first step is to formulate the specific demands that ethical theory must meet. Since the conditions for ethical theory has changed radically, and collective activities that covers extensive time periods need to be incorporated in the ethical thinking. Ethics in the age of modern technology seems to be subjugated to at least three new conditions, which according to Jonas requires the unfolding of three "dimensions of responsibility."[154] Firstly, the range of moral actions is extended by modern technology, and their costs have, in the long run, the possibility of being apocalyptic. Moreover, the planet must intrinsically be incorporated in every human responsibility. Secondly, information, knowledge and awareness become an appealing obligation since moral actions could have wide implications. Ethics is therefore not a matter for "even the most common mind", as famously claimed by Kant, but moral responsibility must be a commitment to attain as much knowledge and insight into the consequences of moral actions. Thirdly, the anthropocentric restraint must be excluded, as the non-human is dominated and placed beneath human responsibility in an outright sense. Humanity can no longer be sheathed in the delusion that nature is motionless and boundless environs for man; it is essential to assume the responsibility of the array of moral actions and therefore imagine nature as an entity very priceless and inimitable. In the age of technology, traditional ethics cannot be able to accomplish these conditions, and as such, ethics must enlarge its range from ethics of nearness to *ethics of distance*.

Jonas' examination of the modern situation is reasonably perceptive but his "unique" theory of responsibility replicates the unavoidable spherical thinking

154 Hans, J., *The Imperative of Responsibility*, pp. 6–8.

that entire protagonists of new ethics have got to depend on. However, Jonas' "ethics of responsibility" is merely a historicized version of the Kantian categorical imperative and thus fails to meet the challenge of the collective character of technological action. Also, one can object that the responsibility presented by Jonas does not allow us to distinguish the degree of responsibility in a universal way, because this responsibility acts, in general, paternalistical, in which authority and control is in the hands of the elite (politicians, scientific experts).

According to Hottois, Jonas posited three phases in practical decision-making, namely: the awareness of the danger, the heuristics of fear and finally, the role to be played by experts. Hottois agrees relatively with the first, but criticizes Jonas on the other two. For him, heuristics of fear is unacceptable because he perceives it as a threat to technoscience wheras Jonas considers the absolute as the goal to justify absolute power.[155] Along the lines of Hottois also, it is dangerous to give power to the experts because they are not the absolute custodians of virtue and history has shown that humanity have often suffered while placing absolute trust on experts.[156]

Finally, the new studies in social science have shown that modern "risk society is inherently reflexive in the sense that its consequences contradict its premises. As it becomes conscious of the threat it poses for its own survival, reflexivity becomes self-reflection, leading to new kinds of political intervention aimed at transforming industrialism."[157] Today, the theoretical construct of modernity and its reliance in technology has emerged as a reflexive political practice of modernization. In the following chapter; we therefore look up to this new politicization as a way forward in the present technological deficiency.

---

155 "Mais les absolutisations, en tant que produit de la volonté, ne sont pas de veritable absolus; ells sont l' expression d' intentions et d'attitudes humaines qui, par diverses techniques (de persuasion, notamment), tendent á mettre au-dessus du débat et de la diversité des conceptions et valeurs humaines." Hottois, Gilbert, *Aux Fondements d' une éthique contemporaine: H. Jonas et H.T. Engelhart en Perspective*, p. 23.

156 *Ibid*. "L' humanité, au cour de son histoire, ná que trope u á souffrir des experts en absolu".

157 Feenberg A., "Modernity Theory and Technology Studies," in *Modernity and Technology*, p. 84.

Chapter VI: Towards a Reflexive Political Modernization

Introduction

In our previous sections, we have presented Heidegger's ontologico-poetic interpretation of technology in which a certain destining is at work in the modern quest to be technological. We have equally seen that Hans Jonas concretized Heidegger's ontological solution when he offered a new dimension of responsibility in accessing the question of technology. While Jonas' solution is more of *forward-looking*, that is, preservation for the future generation, the new social sciences, especially as articulated in Giddens view of technology is more of *backward-looking*, namely, examining inwardly, the areas of tension within technology itself where modern man has overstepped the bounds of negligence.

Today, the new questions identified by Giddens and Beck as 'risk society' are becoming the crucial fields wherein political arrangement and battle take shape. The long-standing left/right mutual politics that were supposedly created from inside the social affairs that are engraved in modernity are not effective anymore, assuming they were before. This, obviously, likewise entails that the conventional platforms of politics, namely, state, parliament, political parties, etc., are not the distinctive ground of the political any more. Chantal Mouffe summarizes Beck's prophetic vision of a new democracy as follows:

> In a risk society, which has become aware of the possibility of an ecological crisis, a series of issues which were previously considered of a private character, such as those concerning the lifestyle and diet, have left the realm of the intimate and the private and have become politicized.The relation of the individual to nature is typical of this transformation since it is now inescapably interconnected with a multiplicity of global forces from which it is impossible to escape. Moreover, technological progress and scientific development in the field of medicine and genetic engineering are now forcing people to make decisions in the field of 'body politics' hitherto unimaginable [...]. What is needed is the creation of forums where a consensus could be built between the experts, the politicians, the industrialists and citizens on ways of establishing possible forms of co-operation among them. This would require the transformation of expert systems into democratic public spheres.[1]

1 Mouffe, C., *On The Political*, London: Routledge, 2005, pp. 40–41.

It is precisely the side-effects (the risks) of modernising technology that needs management that require new politicization, namely, new democratic public sphere, as we shall present in this chapter.

Such political intervention, otherwise known as political modernization refers to processes of transformation within the political realm of society. "…Is about changing relations between state, market and civil society, and about the way governance and policy making are affected by these changing relations."[2] Contemporary society as we have seen is in transition as a consequence of processes of globalization and individualization, and demonstrate a new or amplified scale of disintegration, pluralism and individualism, while political, economic and cultural life is sturdily influenced by developments at the global stratum.

> The irreducible pluralism and diversity of contemporary society is not denied […] that pluralism is not ordered and integrated according to any discernible principle. There is not, or at least no longer, any controlling and directing force to give it shape and meaning […]. There is simply a more or less random, directionless flux across all sectors of society. The boundaries between them are dissolved, leading however not to a neo-primitivist wholeness but to a post-modern condition of fragmentation.[3]

The formation of new coalitions is perceived both within and outside the nation state model, and influenced by local-global interconnectedness.

From Heidegger's assessment, the entire available political structures in the developed Western world are in the clutch of modern technology. Liberalism, socialism or communism as well as modern democracy are merely same coin of different sides. Modern democracy operates on the supposition that technology is prone to human management and control. In Heidegger's view, this postulation is fundamentally flawed in the fact that modern technology is the result of the destiny of Being. Modern political systems like democracy do not really present the whole truth as regards technology. As Heidegger explains, "I would characterize them as half- truths because I do not see in them a genuine confrontation with the technological world, because behind them there is in my view a notion that technology is in its essence something over which man has control. In my opinion, that is not possible. Technology is in its essence that which man cannot control by himself."[4] Consequently, modern democracy is deficient because

---

2 Leroy, P., Jan van Tatenhove, "Political Modernization Theory and Environmental Politics," in A. Von Gert Spaargaren. P. J. Mol & F. H. Buttel, (eds.), *Environment and Global Modernity*, London: Sage Publications, 2000, p. 188.

3 Kumar, K., *From Post-industrial to Post-modern Society. New Theories of the Contemporary World*, Oxford/Cambridge: Blackwell, 1995, pp. 102–103.

4 Heidegger, M., "Only a God Can Save Us," p. 105.

it sticks to the delusion that technology is a ready-to-hand tool that is subject to human manipulation and mastery. This delusion, ultimately, hides man from the essence of being. As Heidegger states:

> Today it seems that the withdrawal of the essence of being is complete. We say "today" and mean the atomic age which is beginning, an age through which modernity supposedly comes to completion insofar as the initial, basic trait of this epoch unrestrictedly unfolds to its furthest extremity.[5]

In this respect, the human social scientists, especially Beck, emphasizes the incapacity of conventional political institutions to come to meaningful centralized decision making, in particular where the basic issues implicated in judgments over environmental problems and highly technological processes are concerned. In fact, coping with environmental risks in an age of reflexive modernity is akin to what might be labelled the end of traditional politics. The crucial decisions pertaining to the route of societal modernization wriggle out of centralized management or decision-making structures and take place in such officially *apolitical* realm as science, commerce or, in effect, the family circle. Experimental and investigational work in scientific laboratories, the invention of new technologies, or the semi-autonomous changes in cultural values and consumer behaviour, make the traditional political realm into a sporadic sub-system that seeks to control social developments but enters at a phase at which such developments have already created substantial and significant force.

In this chapter, as a way of alternative approach to the problem posed by modern technology, we examine some of the emerging concepts or political theories that go a long way to address the problems more existentially and more contemporaneous to the present technological altitude than Heidegger's ontological interpretation. Most distinctive of these new practices is that they blur the conventionally strict boundaries between the public and the private. They stand for practices of "responsibility taking for common well-being through the creation of concrete, everyday arenas on the part of citizens alone, or together with others to deal with problems that they believe are affecting what they identify as the good life."[6] At the end, we shall call for new environmental ethics/policies.

---

5 Heidegger, M., The Principle of Reason, trans. Iilly Reginald, Indianapolis: Indiana University Press, 1996, p. 56.

6 Micheletti, M., *Political Virtue and Shopping: Individuals, Consumerism and Collective Action*, New York: Palgrave MacMillan, 2003, pp. 25–26.

6. Towards a Political Reflexivity

6.1. Reflexivity and Subpolitics

Reflexive modernization necessitates the reinvention of politics. At the vanguard of Beck's articulation on enhanced democratization is his emphasis on sub-politics. Sub-politics is an evolving structural component where political processes are executed outside of conventional representative institutions at social sites formerly considered non-political. Beck's *raison d'être* for introducing the term *sub-politics* in contrast to *politics* is that this facet lacks, whatever the case may be, one essential distinguishing trait of politics: the need to *justify* or *legalize* itself via democratic procedures. The conventional view of politics is oriented toward the rule-directed struggle between parties for privileges and grips of power. The environmental issues highlight the need for a more distinguished analysis of politics. The risk society is understood to be a phase of development of modern society in which the amalgam of individual, social, economic and ecological risks, created by the momentum of innovation, increasingly eludes the control and protective institutions of industrial society.[7] In the technological world of risk society, new types of clashes come into view and fresh coalitions turn out to be thinkable. Subpolitics therefore questions the status of existing systems, calls for a rethinking of the different systems of categorization according to which people are used to perceive their organizational environment, and asks for the discovery of new institutional ways to deal with technological risk.

This is politics outside the established state-related institutions. More generally, sub-politics can be conceptualized as a form of politics 'outside and beyond the representative institutions of the political system of nation-states'.[8] Or put in another way, subpolitics is a revitalization of politics that symbolizes an option to the classical breakdown of politics in political science on three different features, namely; polity, policy and politics: i.e., (i) the institutional aspect of the political community (polity), (ii) the content of political programs (policy) and (iii) the political process about distribution of power and values (politics). This tells us that politics has through the 1980s stirred to new arenas outside Parliaments, that there are new political actors, in particular dynamic citizen groups, and that there

---

7 TATENHOVE, VAN J., BAS A. & LEROY, P. (eds), *Political Modernization and the Environment: The Renewal of Environmental Policy Arrangement,* Dordrecht, The Netherlands: Kluwer Academic Publishers, 2000, p. 45.

8 BECK, U., "World Risk Society as Cosmopolitan Society? Ecological Questions in a Framework of Manufactured Uncertainties", in *Theory, Culture & Society,* Vol. 13, No. 4, 1996, (pp. 1–32), p. 18.

is a tussle to create a new dimension in politics, for instance, in connection with ecological and moral problems. As Becks explains:

> *Sub-politics* is distinguished from 'politics' first, in that, agents *outside* the political corporate system are allowed to appear on the stage of social design (this group includes professional and occupational groups, the technical intelligentsia in plants, research institutions and management, skilled workers, citizens' initiatives, the publicsphere, and so on), and second, in that not only social and collective agents but individuals as well compete with the latter and each other for the emerging shaping power of the political.[9]

What this entails is that sub-politics is an idea about a society shaped 'from below' with new political proxies at the heart of democratic politics. This includes interactions between individual organizations ranging from NGOs like grass root movements, citizen groups, expert and professionals, corporate organizations, diverse governmental and semi-government bodies as well as private citizens. It is an expansion of democracy in the sense that it gives social groups that have not been involved with elite politics 'a voice and a share'. Such outside voice in the mainstream political arena is described by Marlen Buizer as energizing the society.

> The concept of 'sub-politics' refers to politics outside and beyond the representative institutions of the political system of nation-states. It focuses attention on signs of an (ultimately global) self-organization of politics, which tends to set all areas of society in motion. Sub-politics means 'direct' politics—that is, *ad hoc* individual participation in political decisions, bypassing the institutions of representative opinion-formation (political parties, parliaments) and often even lacking the protection of the law.[10]

While the agents of traditional politics are almost always collective agents, the agents of subpolitics are largely unconnected individuals. Circumventing existing political establishment, or institutions, subpolitics shapes political action from below through spontaneous, *ad hoc* individual participation in political decisions. The concept includes a wide range of political activities outside of institutionalized channels. Frequently it comprises spontaneous, evolving activities that attract the amalgamation of unrelated factions over particular issues.

According to Beck, the individualization of politics ensuing from the retreat of traditional institutions will lead to more comprehensive, extraordinarily

---

9 Beck, U., "The Reinvention of Politics" in *Reflexive Modernization*, p. 22. (Emphasis in the original)

10 Buizer, M., "Local Initiatives Challenging Mainstream Policies: Signs of Sub-Politics?" in Henk de Haan, René van der Duim, (eds.), *Landscapes, Leisure and Tourism*, Delft, The Netherlands: Uitgeverij Eburon, 2008, p. 218.

heterogeneous decision-making in a new 'sub-political', extra-parliamentary arena, aggravated by varied aims and value sets. In this arena, it is argued that temporary and numerous stakeholder networks functioning as decentralized, self-determining, pliable public squares for decision-making will enable new and more democratic ways of decision-making around areas of technological risk. This does not mean political insurrection or mutiny against those in power but the expression of fundamental human rights of the citizen for freedom of thought, expression and dialogue.

Although sub-politics does not demand an articulate, logical or shared ideology, a major factor in this challenge to conventional political systems is increasing acknowledgment that long-established systems of authority, be they political, executive, legal or scientific have facilitated and legitimated modernization processes associated with an institutionalized underestimation of risk. The conceptualization of this new conditions and challenges for democracy indicates that politics tend to go beyond "political left-right metaphor" as well as the public/private crack. There is an inclination toward a 'reinvention of politics' i.e. 'to create new rules', instead of just 'following rules'. According to Beck, the three critical questions of great concern here are: "What is your attitude towards, first, uncertainties, second, toward strangers and third, towards the possibility of shaping society."[11]

These individuals and groups outside the official political structure, in contributing to the welfare of the society, engage in action on moral issues relating, for example to ecology, the family, gender issues and ethnicity in addition to act in the interests of economic growth, technological development and ambitious scientific progress. Beck argues that citizen groups have taken power politically, placing such issues on the agenda regardless of chronic stiff opposition from the established official parties. Whereas people have turn out to be detached from conventional forms of politics, they are more and more occupied in subpolitics and these groups use the same machineries—freedom of the press and expression and the legal structure—to advance their preferred aspirations and intentions. As Beck states, it may even seem that "the political constellation of industrial society is becoming unpolitical, while what was unpolitical in industrialism is becoming political."[12] Scores of these groups have also been extremely triumphant in lobbying political institutions, operating in joint effort with these institutions supplying

---

11 Beck, U., "The Reinvention of Politics" in *Reflexive Modernization*, p. 42.

12 Beck, U., *The Reinvention of Politics: Rethinking Modernity in the Global Social Order*, p. 99.

statistics and support for harsh and societally harmful activities or creating scientific imprecision or ambiguity to counter or defuse other sub-political groups.

The politicization of science and technology is swiftly establishing reflexive culture whereby politics and moral ethics are disrupting the knowledge base of scientific experts. It is more and more evident that a 'status quo' approach is generating the conditions for extensive environmental and social degradation. It has for the most part deflated the concept of the "the most appropriate solution" to problems in the accelerating sophistication of late modernity. Subpolitics have produced the notion that solutions should be a process of collective actions that observe citizens democratic rights. The movements have, therefore, fashioned systems of extra-parliamentary watch-dog and observation of potentially everything and everyone. Global climate change and legacies of toxic-waste are examples of issues that threaten the survival of species including our own. Through rallies and negotiations, governments, NGOs and corporations are pursuing agreements aimed to counter the effects of environmental and social risk at national and international levels.

> The response to threats against strategic oil reserves in the Middle East, the efforts to expand the GATT framework, the coercive implementation of the nuclear nonproliferation regime, the containment of South-North migration and refugee flows [...]. The legal implications of globalization-from-above would tend to supplant interstate law with a spacies of global law, but one at odds in most respect with 'the law of humanity'.[13]

Those affected by globalization from above have begun to converge, brought together by common interests, goals and a number of specific campaigns. This emerging network is the iceberg of which the 'globalization from below'—or as Beck would have it, subpolitic—form the most visible angle.

Often structured by trans-national NGOs, business lobby groups, 'epistemic communities' assisted by activist individuals and spread on the Web, the 'globalisation from below' movement reflects a re-emphasis on ideals supporting sustainability and championing the protection of the destruction of local environments around the world. They do this with emphasis on behaviour adjustment, and most especially through their use of diversified forms of civil action such as

13 Falk, R., "The Making of Global Citizenship", in B. van Steenbergen (ed.), *The Conditions of Citizenship*, London: Sage, 1994, p. 137. Cited in Beck, U., *World Risk Society*, p. 38. (GATT stands for *The General Agreement on Tariffs and Trade.* It was negotiated during the UN Conference on Trade and Employment and was the outcome of the failure of negotiating governments to create the International Trade Organization (ITO). Formed in 1949 and lasted until 1993, when it was replaced by the World Trade Organization in 1995).

protest march, rallies, publications, petition writings etc. The major goal here is to rally mass opinion against the dangers of or the effects of a particular technological development on the environment, especially in time of disaster, through mass communication. "Increasingly, such groups became involved in campaigning about global environmental risks such as those associated with nuclear technology, global warming and acid rain. Also, in contrast to many of the earlier traditional organizations, they adopted a proactive approach towards generating public sympathy and political support."[14] Beck believes that mass media—particularly newspapers, television and radio—help to raise the risk consciousness of individuals, and thus empowers social movements, through dramatic coverage of catastrophes.

> Even the right to *freedom of the press*, with all its opportunities and problems of interpretation, offers numerous occasions for the differentiation of large and partial public sphere (from the global television network to the school newspaper) with individually very particularized, but overall considerable opportunities to influence the definition of social problems.[15]

Thus in developed democracies of the West, a media-saturated publicity can perform an actual or potential monitoring function with regard to political decision. In this manner, centers of sub-politics or social movements are created and because of their resilience, organizational acumen, and focus, thcy play a central role in this democratizing process. Braithwaite and Drahos argue that NGOs do not necessarily need large budgets to be effective but "they do have to be able to convince regulatory policy-makers that they might, if push comes to shove, be able to mobilize mass public around their concerns."[16] Certainly, much domestic and foreign policy-making over the past several decades has been shaped from below by grass-roots pressure from the environmental, women's, and peace movements as we can see below.

6.1.1. Consumer Protest as Subpolitics – the question of Political, Ethical, and Green Consumerism

Political consumerism has been defined as "consumer choice of producers and products with the goal of changing objectionable institutional or market

---

14 Alison, A., "Environmental Pressure Politics and the 'Risk Society'" in Stuart Allen, Barbara Adam and Cynthia Carter (eds.), *Environmental Risks and the Media*, London: Routledge, 2000, p. 94.

15 Beck U., *Risk Society: Towards a New Modernity*, p. 197.

16 Braithwaite, J. & Drahos, P., *Global Business Regulation*, Cambridge: Cambridge University Press, 2000, p. 500.

practices."[17] In late modernity, the market place has appeared as an important terrain for the exercise of politics. Anxieties about personal and family welfare in addition to ethical or political appraisal of favourable and unfavourable commercial and government practices become an essential part of the political marketplace. Once the public take part in boycotts, they use the market as an apparatus for politics and engage in political consumerism, which can be described as consumer preference of companies, manufacturers and products based on a mixture of ethical and political considerations. "It represents actions by people who make choices among producers and products with the goal of changing objectionable institutional or market practices."[18] This bears a close similarity to what is refered to, in UK, as *ethical consumption* which has recently been described as "rather than rejecting the persona of 'consumer', ethical consumption campaigning represents a distinctive strategy for connecting the politics of consumption with the practices of being a discerning, choosy, responsible consumer."[19] Outside the electoral podium, the intensifying uses of consumer campaigns are transforming public interest in politics by applying direct public pressure on government officials as well as national or multinational corporations to adopt higher environmental standards. "Ethical consumption, understood as an organized field of strategic interventions, seeks to use everyday consumption as a surface of mobilization for wider, explicitly political aims and agendas."[20] Beck and Gernsheim argue that today, "citizens discover the act of shopping as one in which they can always cast their ballot—on a world scale, no less".[21] And further than the nation state, the globalization of economic and communication systems has aggravated and enabled citizen consumers to make political claims in international stages. Thus sub-politics is related to processes of globalization and to citizen's perceptions about government's inability to control and take responsibility for the new uncertainties created by civic and corporate policy. Beck has eminently argued that if modernity is a democracy oriented to producers, late modernity is a democracy oriented to consumers: a

---

17 Micheletti, M., Follesdal, A. & Stolle, D., (eds.), *Politics, Products, and Market: Exploring Political Consumerism Past and Present,* New Brunswick, New Jersey: Transaction Publishers, 2004, p. xiv.

18 Micheletti, M., *Political Virtue and Shopping: Individuals, Consumerism, and Collective Action,* New York: Palgrave Macmillan, 2003, p. 2.

19 Clive Barnett, Paul Cloke, Nick Clarke & Alice Malpass, *Globalizing Responsibility: The Political Rationalities of Ethical Consumption*, West Sussex, UK: Blackwell Publishing, 2011, p. 13.

20 Ibid.

21 Beck, U., and Gernsheim, E., *Individualization*, p. 44.

pragmatic and cosmopolitan democracy where the quiet giant of the "independent citizen-consumer movements" is becoming a counterweight to big transnational corporations in which "a democratization of divided democracy could offer one future perspective of the political organization of individualized society."[22]

Beck's famous examples are the world-wide movement against nuclear testing at Mururoa Island, the mass consumer protests against the disposal in the North Sea of an obsolete oil rig (Brent Spar) and the BSE crisis, the public perception of which, according to Beck, gave rise to the fastest passage of laws in the history of the German Republic.[23]

Through grassroot mobilization, petition signing, protest marching, boycotting and local campaigning, local actions are generating international effects and "citizens are discovering that the act of purchase can be a direct ballot which they can always use in a political way."[24] The mass consumer boycott of Shell Oil in Western Europe in the mid-1990s over the *Brent Spar* case illustrates the present efficacy and potential strength of subpolitics when divergent actors mobilize. Shell had secured authorization from the British government to deposit an old and no longer used oil-storage buoy, the *Brent Spar*, at the floor of the Atlantic Ocean. Shell, the British government and a great number of scientists in the field were in no doubt that a deep-sea dumping in a particular part of the Atlantic was the best way, economically as well as environmentally, to dispose of the *Brent Spar*. Nevertheless, Greenpeace strongly disagreed with this view. They consequently started a campaign against Shell in order to stop the dumping of the *Brent Spar*. The battle of Greenpeace against Shell (and the British government) was stimulated by politicians in other countries in the region of the North Sea and by consumers who began a mass boycott of Shell service stations, for the most part in Germany but also in other northern European countries.[25] Eventually Shell was

---

22 Beck, U., *The Reinvention of Politics: Rethinking Modernity in the Global Social Order*, p. 50.

23 Beck U., Bonss W., & Lau C., "The Theory of Reflexive Modernization: Problematic, Hypotheses and Reasearch Programme" in *Theory, Culutre and Society*, Vol. 20, No. 2, 2003, p. 14 (pp. 1–33).

24 Beck U., *World Risk Society*, p. 42.

25 Anni Olsen from Shell Denmark remarked that the quantity of petrol sold at German Shell petrol stations reduced by almost 30 percent in the last days before the decision was taken not to dump the Brent Spar. Cf. Sorensen, M.P. "Den politiske forbruger som subpolitiker", in Fenger-Gron and Kristensen J. E., (eds.) *Kritik af den okonomiske fornuft*, Copenhagen: Has Reitzels Forlag, 2001, p. 127. Also Wätzold reached a parallel conclusion when he states that 'according to one opinion poll, 8.5% of the population were in favour of the boycott. Peter Duncan, Chairman of German Shell, admitted that

forced to discontinue the disposal of the *Brent Spar* in the Atlantic. Instead, she commenced a string of dialogue conferences with NGOs and experts concerning the future of the oil buoy. The objective of these meetings was to find a superior and more acceptable alternative for the disposal of the *Brent Spar*. Finally, all the parties involved agreed on a more labour-intensive but environmentally better solution: instead of being discarded in the Atlantic Ocean the *Brent Spar* came to be used in a harbour extension project in the Norwegian town Mekjavik.[26]

One reason behind the outcome was that Greenpeace managed to push the controversy to the media public sphere as a normative question rather than a solely scientific one and in the course of action sign up a great number of actors to their network. This successful conscription process might somewhat be explained by Beck when he states that: "The Greenpeace people are multinational media professionals."[27] However, Beck stresses further that it was primarily the consumer boycott that forced Shell to change its mind regarding its discarding plans. Hence, for him, the *Brent Spar* case is an illustration of how the new political cultures of the second modernity, in this case consumers and an environmental protection organization, openly and directly, without using conventional political participation channels, tries to induce the techno-economic sphere to legitimize its acts. Greenpeace's campaign against Shell accentuated the existing vacuum of legitimization

---

the filling stations had a drop in sales of between 20 to 30%, sometimes even more than 50%'. Cf. WÄTZOLD, F., "When Environmentalists Have Power: The Case of the Brent Spar", in MADSEN H., & ULHOI J.P., (eds.) *Industry and the Environment: Practical Applications of Environmental Management Approaches in Business*, Gylling: Naryana Press, 1996, pp. 327–38.

26 *Ibid.*

27 BECK U., *World Risk Society*, p. 45. In an article on the controversy between Shell and Greenpeace over a disposal of an offshore oil platform in the North Sea, Tsoukas Haridimos observed that the fundamental rationale behind the decision to reject the dumping was the different rationalities underlying the public and Greenpeace response versus approach by Shell. Shell symbolized a more technocratic view of the clash and the public, a more common-sense rationality. "Shell made sense of the conflict in terms of "reason" against "emotion", and "head" versus the "heart"...." Thus both the scene and the underlying suppositions were crucial. Cf. TSOUKAS HARIDIMOS, "David and Goliath in the Risk Society: Making Sense of the Conflict between Shell and Greenpeace in the North Sea", in *Organization*, vol. 6, no. 3, 1999, p. 523 (pp. 499–528). See also. JORDAN, G. *Shell, Greenpeace and the Brent Spar*, Basingstoke, England: Palgrave, 2001; HOLZER, B., *Transnational Subpolitics and Corporate Discourse: a Study of Environmental Protest and the Royal Dutch/Shell group*, Department of Sociology, London: London School of Economics, 2002, Ch. 6.

and power in the political scheme. Political institutions are said to be more and more illusory in their programs and foundations, and blank of political content. This explains why politics—when seen as the need for collective action to provide for our common well-being—is stirring out of the political sphere and into the everyday life sphere that transverses the public and private. Individual citizens act increasingly political in their every day lives. As Beck remarked, "What appeared to be a 'loss of consensus', an 'unpolitical retreat to private life', 'new inwardness' or 'caring for emotional wounds' in the old understanding of politics can, when seen from the other side, represent the struggle for a new dimension of politics."[28] The repercussion is that daily acts by citizens have the influence to potentially reorganize society. Citizens are seen as the major players in shaping new democratic structures.

Also it should be pointed out that the reason for the triumph of this consumer boycott of Shell lies in the fact that the consumers did not incur additional cost for taking part in the boycott. They were able to merely buy 'morally good petrol' at other petrol stations. It was therefore comparatively simple for the consumers to think normatively and act morally in this situation.

To sum up, sub-politics is much more likely to be characteristically shaped through the domains of consumption, small screen media and the repoliticization of science. As the public deal with the daily involvement, risk products are boycotted and positions swiftly adopted and discarded in what Beck calls "the world fairground of symbolic politics".[29] A shared environment of global risk facilitates the formation of an ecological politics that seeks to recover democratic exchange. Here the ecological movement has sought to develop a "cultural Red Cross consciousness."[30] Organizations like *Greenpeace* and *Friends of the Earth* have fostered a sense of public trust in their own assertions, taking a moral posture that is apparently above the daily oddments of political parties. The emergence of sub-politics where members of the public adopt oppositional political values as a basis for active debate and protest, will continue to call into question those traditional political institutions that are increasingly unable to effectively manage the omnipresent risks of second modernity. This drastic democratization eventually means not only the reinvention of politics but also the prospective obliteration of those modern political institutions that are ineffective. In the democratic supervision of the risks, all parties whose welfare or interests will be considerably affected need

28 Beck, U., *The Reinvention of Politics: Rethinking Modernity in the Global Social Order*, p. 101.

29 Beck, U., *World Risk Society*, p. 46.

30 *Ibid.*, p. 44.

to have the chance to contribute and no one actor should have control over others. That means that the continued appearance of subpolitical movements which extend beyond left and right raises forth the possibility of a more deliberative and inclusive form of democracy in management of technological shortcomings.

6.1.2. Dialogic/Deliberative Democracy

Similar to Beck, Giddens uses the term "dialogic democracy" to explain his vision of dialogue in a public space.[31] The concept of a public sphere wherein citizens freely and bluntly engage in deliberation on policy matters and problems is for the most part strongly linked with the work of Habermas. A great deal of the text on new varieties of public participation glances back to *The Structural Transformation of the Public Sphere* (1989), in which Habermas expound a concept of the public sphere as a realm of dialogue based on reason, a public mode of deliberation and decision-making and a spot of communicative action by participatory publics.

Giddens provides a straight prescriptive device that is analogous to the one offered by Beck. He argues that we must move from emancipatory politics to *life politics*. The life politics schema brings to the vanguard moral and existential questions founded on the more all-embracing questions of "How shall we live?" that surfaced mostly at the outset of the Enlightenment. These sorts of profound and incisive questions in late modernity call for what Giddens dubs *dialogic democracy*: "recognition of the authenticity of the other, whose views and ideas one is prepared to listen to and debate, as a mutual process."[32] It is a concept involving open communicative exchanges, independent from formal political institutions, and spreading social reflexivity in ways that condition both everyday life and collective action.[33] Fundamentally, this entails "creating a public arena in which controversial issues—in principle—can be resolved, or at least handled, through dialogue rather than through pre-established forms of power."[34] As Giddens states it further, his dream of dialogic democracy is based in part on the expectation that "dialogue in a public space provides the means of living along with others in a relation of mutual tolerance....Dialogue would be understood as the capability to create active trust...a means of ordering social relations across time and space."[35]

---

31 GIDDENS, A., *Beyond Left and Right:The Future of Radical Politics*, p. 112.

32 GIDDENS, A., "Living in a Post-Traditional Society," in *Reflexive Modernization: Politics, Tradition and Aesthetics in the Modern Social Order*, p. 106.

33 GIDDENS, A., *Beyond Left and Right: The Future of Radical Politics*, p. 115.

34 *Ibid.*, p. 19.

35 *Ibid.*, pp. 115–116.

Deliberative democracy is one more effort to resolve decision-making in questions of huge complexity where various diverse outlooks on the sustainability of modern form of consumption and production are involved. Also sometimes called discursive democracy, it is a term used by political theorists, such as, Jon Elster or Jürgen Habermas, to refer to any system of political decisions based on some swapping of consensus decision-making and representative democracy. Representative democracy, it is argued, is associated with a modern, techno-scientific rationality, whereas deliberation is more adjusted to 'reflexive modernity'. The appearance of what Taylor Gooby et al. calls "querulous citizens", who are prepared to make choices rather than accept tradition or habit, "leads to demands for a more dialogic democracy, in which government, to retain its legitimacy, must win an active trust rather than simply presuming a habitual loyalty."[36] Representative democracy is seen as a desirable but deficient means of connecting citizens with governing institutions and processes:

> [C]ontrary to the classic form of 'government', contemporary governance is not imprisoned in closed institutions and is not the providence of professional politicians. Though rarely defined with precision, it refers to patterns of decision-making taking place in a larger set of institutions, with a broader range of actors and processes. One of the ambitions of those who defend this new concept is indeed to enlarge the accepted notion of civic participation beyond the well established and constantly declining procedures of representative democracy.[37]

Deliberative democracy, therefore, aims to surmount problems linked with traditional systems of democracy by moving from an attention on votes to a concentration on the quality of the process of public deliberation. Rather than consensus, the objective is to attain a neutral stance and complete knowledge of critical issues, concepts similar to Habermas' ideal of full communication. As two commentators jointly explained; "broadly defined, *deliberative democracy* refers to the idea that legitimate lawmaking issues from the public deliberation of citizens. As a normative account of legitimacy, deliberative democracy evokes ideals of rational legislation, participatory politics, and civic self-governance. In fact, it presents an ideal of political autonomy based on the practical reasoning of citizens."[38] Deliberating

---

36 Taylor-Gooby, P., Hastie, C. & Bromley, C. "Querulous Citizens: Welfare Knowledge and the Limits of Welfare Reform", in *Social Policy and Administration*, vol. 37, no. 1, 2003, p. 2 (pp. 1–20).

37 Magnette, P., "European Governance and Civic Participation: Beyond Elitist Citizenship", in *Political Studies*, vol. 51, no. 1, 2003, p. 144 pp. (144–160).

38 Bohman J. & Rehg, W., editors, *Deliberative Democracy: Essays on Reason and Politics*, Cambridge, MA: MIT Press, 1997, p. ix.

citizens seek relevant information; reflect on the issues, and exchange views with others. According to theorists of deliberative democracy, one of the main values of public participation in institutional politics is that it renders issues publicly visible, thereby forcing them onto the political agenda.[39]

The dialogic processes which shape the backbone of deliberative democracy are derived from rational and unprejudiced discussion. As Seyla Benhabib clarifies:

> According to the deliberative model of democracy, it is a necessary condition for attaining legitimacy and rationality with regard to collective decision-making processes in a polity, that the institions in this polity are so arranged that what is considered in the common interest of all results from processes of collective deliberation conducted rationally and fairly among free and equal individuals.[40]

These processes are democratic because the participants in the discussions of deliberative processes are those who will, after the ruling has been made, be subject to the cost or consequences of that decision.

Science and technology present enormous challenges to today's reality and these can only be met by rethinking the way deliberation on these subjects is carried out. Processes which are more comprehensive, all-inclusive need to be experimented which can help us restructure ourselves, and our institutions. Even though science continues to be a decisive basis for decision making, many theorists of science and technology today appear to agree that scientific experts need to substantiate their knowledge claims to much wider communities to regain public trust and legitimacy.[41] Indeed, ever since its commencement, Science and Technology Studies (STS) has been concerned with the issue of the 'democratization of science', namely, of how practices of techno-scientific research and development

---

39 Habermas, J., *The Postnational Constellation: Political Essays*, trans.by Max Pensky, Cambridge & Oxford: Polity Press and Blackwell Publishers, 2001 (1998), p. 111; See also Dryzek, John, *Deliberative Democracy and Beyond: Liberals, Critics, Contestations*, Oxford: Oxford University Press, 2000, p. 131.

40 Benhabib, S., "Towards a Deliberative Model of Democratic Legitimacy", in Benhabib, S., (ed), *Democracy and Differences: Contesting the Boundaries of the Political,* Princeton, New Jersey: Princeton University Press, 1996, p. 69.

41 See Fischer, F., *Citizen, Experts, and the Environment*, Durham: Duke University Press, 2005. On the issue of the nature of expertise, the framing of knowledge, processes of public engagement and the issues of right, justice and democracy, see, Leach, Melissa, Ian Scoones & Brian Wynne (eds.) *Science and Citizens: Globalization and the Challenge of Engagement,* London: Zed Books, 2005.

can be brought in agreement with democratic ideals of inclusive opinion-making and accountable decision-making.[42]

The Directorate-General (DG) of Research in the European Commission (EC) in 2007, published an outstanding report titled "Taking European Knowledge Society Seriously."[43] The report was the product of an expert group commissioned by DG Research to assess how to increase the legitimacy of science and technology governance in an age of public mistrust. The mandate was among other concerns, "how to respond to the widely-recognised problem of European public unease with science, especially in relation to new science-based technologies." Although the expert group consisted of a limited number of academics, their conclusions reverberate with a much wider debate in studies of science and technology.

Three Mile Island and Chernobyl accidents in the late 1970s and mid-1980s, and more recent food scares such as the Bovine Spongiform Encephalopathy (BSE) crisis in Uk; genetically modified organism (GMO) debate in Europe as well as the controversy over the use of Asbestos in European countries; the 2010 oil spill in the Gulf of Mexico, USA, as well as the 2011 Fukushima Daiichi nuclear accidents,[44] are all examples of decisions and measures taken in a time when experts were trusted to have knowledge and politicians were trusted to decide. In the end, very hazardous, precarious and controversial results came to light. These events can be characterized as having huge publicity coverage from the media, a high level of risk to the wellbeing of the public and, usually with international implications. Consequently, today scientific know-how or expertise is facing a crisis of authenticity and legitimacy and calls for democratization of scientific expertise. Active participation in the process of deliberation makes citizens better informed

---

42 De Vries, Gerard de, "What is Political in Sub-politics?: How Aristotle Might Help STS", in *Social Studies of Sciences,* 37 (5), 2007, pp. 781–809.

43 Ulrike F., Wynne, B., *Taking European Knowledge Society Seriously: Report of the Expert Group on Science and Governance to the Science, Economy and Society Directorate, Directorate-General for Research, European Commission,* 2007. The working group was composed of scholars from the academic field of science and technology studies (STS) and related areas of philosophy, sociology, policy analysis and law, as well as participants from public interest and labour organizations. Accssed from http://ec.europa.eu/research/science-society/document_library/pdf_06.

44 Fukushima Daiichi nuclear accidents are the results of release of radioactive isotopes from the Fukushima Daiichi Nuclear Power Plant after the 11 March 2011 9.0 magnitude Tōhoku earthquake and tsunami. It was estimated that the total amount of iodine-131 and caesium-137 released into the atmosphere exceeded 10% of the emissions from the Chernobyl accident. Cf. "Fukushima radioactive fallout nears Chernobyl levels," *New Scientist* 24 March 2011. Retrieved 26 May 2011.

and better able to contribute to the critical assessment of policy issues. In essence, it increases their political efficacy. As Graham Smith argues,

> Deliberative democratic theory has evolved in response to the perceived weaknesses of liberal democratic theory and practices and offers a challenge to and a critical perspective from which to judge, the institutions of contemporary liberal democratic states.[45]

Contrary to aggregative forms of democracy that have recognized voting and representation as the measures for reaching collective decisions, deliberative democrats emphasize the need to justify collective decisions through an open argumentation and reasoned dialogue among free and equal citizens.[46] As Cohen explains:

> The notion of a deliberative democracy is rooted in the intuitive ideals of a democractic association in which the justification of the terms and conditions of association proceeds through public argument and reasoning among equal citizens. Citizens in such an order share a commitment to the resolution of problems of collective choice through public reasoning, and regard their basic institutions as legitimate in so far as they establish the framework for free public deliberation.[47]

Therefore, if deliberation is to be democratic; it must depend upon substantive *political equality*: all concerned interests should have access to the process, and each participant should enjoy an equal opportunity to present viewpoints, hear contributions, and take part in debates and decisions. Public interest cannot come into sight simply by summing pre-existing inclinations like what good technology has done in the past, but only can make headway through a deliberative process, which generates new insights and transforms initial perspectives. In other words, deliberative interaction allows the democratic citizenry to create a collective path forward. As Graham argues further:

---

45 Graham S., *Deliberative Democracy and the Environment*, London: Routledge, 2003, p. 56.

46 Cf. Bohman, J., *Public Deliberation: Pluralism, Complexity, and Democracy*, Cambridge, MA: MIT Press, 2000; Gutmann, A, & Thompson, D., *Why Deliberative Democracy?* Princeton: Princeton University Press, 2004. Argumentation plays a vital role in deliberation. Cooke defines it as "an unconstrained exchange of arguments that involves practical reasoning and always potentially leads to a transformation of preferences." (Cf. Cooke Maeve, "Five Arguments for Deliberative Democracy," in *Political Studies*, vol. 48, no. 5, p. 948 (pp. 947–969).

47 Cohen, J., "Deliberation and Democratic Legitimacy," in Bohman J. & Rehg W., *Deliberative Democracy: Essays on Reason and Politics*, p. 72 (pp. 67–91). See also Benhabib, S., "Towards a Deliberative Model of Democratic Legitimacy", p. 69.

> Deliberative democracy promises much: more trustworthy and legitimate forms of political authority based on inclusive and unconstrained dialogue, more informed political judgements and decisions, and a more active account of citizenship. It promises a political environment within which the plurality of environmental values can be effectively and sensitively assessed and considered in decision-making processes.[48]

Accordingly, each time deliberation is juxtaposed with negotiating, it inclines to be regarded as a considerably decent practice of social interaction. Deliberation is therefore "a particular way of thinking: quiet, reflective, open to a wide range of evidence, respectful of different views. It is a rational process of weighing the available data, considering alternative possibilities, arguing about relevance and worthiness, and then choosing about the best policy or person."[49]

6.1.3. Importance of Epistemic Communities

Amid growing interest in participatory/deliberative forms of governance and environmental concern and decision making, as well as public engagement in science, Peter Haas has emphasized the importance of epistemic consensus emergent among communities of scientists in the development of a broader political agreement. These are "knowledge-based networks of specialists who share beliefs in cause-effect relations, validity tests, and underlying principled values, and pursue common goals."[50] Haas text, focusing on learning in international organizations, has identified the "epistemic community" that exists both within and outside of formal organizational structures as an important actor in policy-making processes. Citing several empirical studies, Dedeurwaerdere stressed that "the independent character of the epistemic community enhances the influence of the ideas and so their transmission to the policy process."[51] Community members may be scientists or academics, interest-group analysts, decision-makers, governmental or intergovernmental personnel and other professional from different disciplines who share a set of normative and principled beliefs, as well as causal beliefs and cause-and-effect understandings. Here, the importance of Intergovernmental

---

48 GRAHAM S., *Deliberative Democracy and the Environment*, p. 61.

49 WALZER, M., "Deliberation and What Else"? In STEPHEN MACEDO (ed.), *Deliberative Politics: Essays on Democracy and Disagreement*, Oxford-New York: Oxford University Press, 1999, p. 58, (pp. 58–68).

50 HAAS, M. P., "Banning Chlorofluorocarbon: Epistemic Community Efforts to Protect Stratospheric Ozone," in *International Organization*, 46 (1), Winter, 1992, p. 187.

51 DEDEURWAERDERE, T., *The Contribution of Network Governance to sustainable Development*. Institute du Developpment et des Relations Durable Internationales (Iddris), Paris: Les seminaires de I'Iddri, no 13, 2005, p. 9 (1–15).

Panel on Climatic Change (IPCC) cannot be over-stressed in today's complex technological environment.

Referring to the influence of the "ecological epistemic community" in policy-making associated with the United Nations Environmental Programme (UNEP) Plan for the Mediterranean, Peter Haas writes;

> These new actors led governments to recognize and follow new interests in environmental protection, so that they were willing to resist systemic forces that would push them to pursue more constrained and transitory arrangements. International environmental cooperation is generated by the influence wielded by specialists with common beliefs, contrary to conventional approaches which stress the role of interstate powers.[52]

Epistemic communities like the IPCC play an increasingly important role in policy formulation about the climate and our environment on national and international levels because decision-makers need technical information to handle complex issues and the uncertainty they generate. In addition to information, community members provide an interpretative framework of cause-and-effect relationships that may alter decision maker's views of their national interests.[53] Today, a growing number of science and technology scholars are asking scientific experts to justify their data claims to much broader groups than merely their

---

52 HAAS, M. P., *Saving the Mediterranean: The Politics of International Environmental Cooperation*, New York, Oxford: Columbia University Press, 1990, p. xxii. According to Hass, epistemic communities possese a special blend of traits. In an article, in lays out a four-step definition, which states: "...Although an epistemic community may consist of professionals from a variety of disciplines and backgrounds, they have (1) a shared set of normative and principled beliefs, which provide a value-based rationale for the social action of community members; (2) shared causal beliefs, which are derived from their analysis of practices leading or contributing to a central set of problems in their domain ans which then serve as the basis for elucidating the multiple linkages between possible policy actions and desired outcomes; (3) shared notions of validity—that is, intersubjective, internally-defined criteria for weighing and validating knowledge in the domain of their expertise; and (4) a common policy enterprise—that is, a set of common practices associated with a set of problems to which their professional competence is directed, presumably out of the conviction that human welfare will be enhanced as a consequence." HAAS, M. PETER, "Introduction: Epistemic Communities and International Policy Co-ordination," in International Organization, 46 (1), 1992, p. 3 (pp. 1–36).

53 BROWN M. LEANN, "Policymaking in International Organizations: The European Union and Ozone Protection" in ROBERT V. BARTLETT, PRIYA A, KURIAN, MADHU MALIK (eds.) *International Organizations and Environmental Policy*, Westport, CT: Greenwood Press, 1995, p. 54.

scientific associates.[54] Jasanoff, for instance suggests that it would be better "to recast the role of experts in terms that better lend themselves to political critique," that is, to subject expert decision making to "notion of delegation and representation." In such a structure, experts will act "not only in furtherance of technical rationality, but also on behalf of their public constituencies, under cognitive and normative assumptions that are continually open to wider review."[55] But the problem according to Jasanoff is:

> How to institutionalize polycentric, interactive, and multipartite processes of knowledge-making within institutions that have worked for decades at keeping expert knowledge away from the vagaries of populism and politics. The question confronting the governance of science is how to bring knowledgeable publics into the front-end of scientific and technological production—a place from which they have historically been strictly excluded.[56]

Nevertheless, answers to this crucial question are not implausible. Jasanoff herself pointed out as a way forwad, the initiative of the U.S. Congress to "concede unchecked autonomy to the scientific community in the collection and interpretation of data," as well as the European Union's pledge to "involving public in technically grounded decision," or the various "experiments...such as citizen juries, consensus conferences and referenda."[57] Be that as it may, there is also the danger that public participation might become "an instrument to challenge scientific points on political grounds." If citizens are invited to question how experts frame an issue

---

54 Cf. Weingart, P., "How Robust is 'Socially Robust Knowledge'"? in *The Challenge of the Social and the Pressure of Practice: Science and Values Revisited*, eds. M. Carrier, D. Howard, & J. Kourany, Pittsburgh: University of Pittsburgh Press, 2008, p. 139 (131–145).

55 Jasanoff, S., "Judgment Under Siege: The Three-Body Problem of Expert Legitimacy" *in* Sabine Maasen and Peter Weingart, (eds.), *Democratization of Expertise? Exploring Novel Forms of Scientific Advice in Political Decision-Making*, Dordrecht, The Netherlands: Springer, 2005, p. 222. Howerver, Jasanoff maintained that "equally, citizens need to recognize that governmental experts are there to make judgments on behalf of the common good rather than as spokespersons for the impersonal and unquestionable authority of science." The end result is "that a fullfledged political accountability—looking not only inward to specialist peers but also outward to engaged public—must become integral to the practices of expert deliberation" Ibid. For an interdisciplinary as well as a global approach to public communication of science and technology, see: Massimiano Bucchi & Brian Trench (eds.), *Handbook of Public Communication of Science and Technology*, London & New York: Routledge, 2008, especially, chaps. 10, 12 & 14.

56 Jasanoff, S., "Technologies of Humility: Citizen Participation in Governing Science" in *Minerva*, vol. 41, no. 3, 2003, p. 235. See also David, M. Kaplan, ed., *Readings in the Philosophy of Technology*, 2nd Edition, Rowman & Little field Pub. Inc., 2009, pp. 570–583.

57 Jasanoff, S. *Ibid.*, pp. 236–37.

by identifying opaque hypotheses and tactic value choices, they may perhaps, it is argued, dispute science-based claims made by social elites and thus repoliticize a technocratic policy dialogue. But Jasanoff was quick to remark that "participation alone…does not answer the problem of how to democratize technological societies," but the question is "how to promote more meaningful interaction among policy-makers, scientific experts, corporate producers, and the public."[58] She suggested that there should be a reciprocal complementarity between the existing "technologies of hubris" with new "technologies of humility"[59] These will give birth to "social technologies" that "would give combined attention to substance and process, and stress deliberation as well as analysis," or, in other words, would "seek to integrate the 'can do' orientation of science and engineering with the 'should do' question of ethical and political analysis."[60]

6.1.4. *Hybrid Forums and Collectives*

Michel Callon, Pierre Lascoumes, and Yannick Barthe go so far as to treat the articulation of matters of public concern as itself a principle merit of public involvement in science and technology. They elaborate the concept of "the hybrid forums" to characterize the type of democratic process performed during public controversies over techno-scientific issues.

> …forum because they are open spaces where groups can come together to discuss technical options involving the collective, hybrid because the group involved and the spokespersons claiming to represent them are heterogeneous, including experts, politicians, technicians, and laypersons who consider themselves involved. They are also hybrid because the questions and problems taken up are addressed at different levels in a variety of domains, from ethics to economics and including physiology, nuclear physics, and electromagnetism.[61]

As Callon et al. explains here, "Hybrid forums host deliberative processes in which heterogeneous actors—those belonging to affected groups, experts, politicians and officials—collectively define the problems in which they are all implicated."[62]

---

58 JASANOFF, S. *Ibid.*, pp. 237–38.

59 *Ibid.*, p. 238, 240.

60 *Ibid.*, p. 243.

61 CALLON, M., LASCOUMES, P., & YANNICK B., *Acting in an Uncertain World: An Essay on Technical Democracy*, translated by GRAHAM BURCHELL, Cambridge, MA: The MIT Press, 2009, p. 18. First published as *Agir dans un monde incertain: Essai sur la democratie technique* by Editions du Seuil, 2001.

62 *Ibid.*, p. 36.

Thus, democratic legitimacy has increasingly come to be seen as the result of "free and unconstrained deliberation of all about matters of common concern."[63]

Bruno Latour, in his *Politics of Nature,* apparently in the same way calls for such new really democratic cosmo-political constitution through which both human and non-human *actants* enter in a new public sphere, where matters of fact are turned into matters of concern, articulated and brought together through wide-ranging and plane networks of connected and relationally constituted human/non-human assemblages.[64] Latour explores the taken for granted nature of technology by tracing the social and technological relations involved in the development and implementation of new technologies through application of Actor-Network Theory (ANT).[65] Actor-Network Theory conceptualizes technologies in terms of the different political, social, and material factors which unite to make entities meaningful and helpful to us. In Latour's terminology, these factors should be understood as the various "actants" which are pulled together into "networks" that enable us to take an entity for granted.

His book begins as a reaction against the split between nature and society, between human and thing, the split between facts and values. What Latour wants to do then is to construct a 'new collective' – a 'political ecology of collectives consisting of humans and non-humans.'[66] He does this by 'exchanging properties' between humans and nonhumans, with a view to exposing characteristics that they have in common.

---

63 BENHABIB, S., "Toward a Deliberative Model of Democratic Legitimacy," in *Democracy and Difference: Contesting the Boundaries of the Political*, p. 67.

64 LATOUR, B., *Politics of Nature: How to Bring the Sciences into Democracy*, trans., CATHERINE PORTER, Cambridge, Mass: Harvard University Press, 2004, pp. 80–81; originally published as *Politique de la Nature*, Paris: Editions La Découverte, 1999. See also; LATOUR BRUNO, *Reassembling the Social: An Introduction to Actor-Network-Theory,* Oxford: Oxford University Press, 2005.

65 Actor-network theory (ANT) is associated with the analysis of scientific and technological artefacts by Bruno Latour and Michel Callon in the late 1980s. ANT declares that the world is full of hybrid entities containing both human and non-human elements, and was developed to analyze situations where separation of these elements is difficult. Cf. Callon, M., "Techno-economic networks and irreversibility", in LAW, J. (ed.), *A Sociology of Monsters: Essays on Power, Technology and Domination*, London: Routledge, 1991, pp. 132–161; Latour Bruno, *Science in Action*, Cambridge: Harvard University Press, 1987; LATOUR, B., *Reassembling the Social: An Introduction to Actor-Network-Theory,* etc.

66 LATOUR, B., *Politics of Nature,* p. 61.

Thus, beside the sanitizing practice that defines modernity, there exists another apparently conflicting one: the construction of systems that mix politics, science, technology, and nature. The ozone question is such a hybrid, in Latour's investigation, in addition to global warming, deforestation, etc. As these hybrids multiply, the prospect of keeping nature and culture in their separate rational compartments becomes overwhelming—and rather than make the effort, Latour suggests, we should rethink our distinctions, rethink the definition and constitution of modernity itself. In an earlier work '*We Have Never Been Modern*,' Latour examines how modernity has accelerated the production of hybrids by successfully denying their very existence:

> The moderns think they have succeeded [...] only because they have carefully separated Nature and Society (and bracketed God), whereas they have succeeded only because they have mixed together much greater masses of humans and nonhumans, without bracketing anything and ruling out any combination![67]

The more skillful we have developed at imagining a superior Nature and an excellent Society, the tougher it has turn out to face up a progressively weird hybrid. So, according to Latour, we cannot afford to continue to be modern in the old-fashioned way. We must change our ways of changing and open our eyes to the reality of hybrid lives. Our political imaginations must be stretched far enough to see that "half of our politics is constructed in science and technology. The other half of Nature is constructed in societies."[68] Latour therefore calls for the need to govern technology through politicization. Politicization is to make things enter the parliament of politics, but the dialogic-democratic condition does so in a consensual conversation.

Nothing is permanent, certain, or specified; everything is constantly in doubt, negotiated, and brought into the political arena. Political space is not a contingent space where what has no name is brought into the discussion, given a name, and is counted; but rather, things and people are 'summoned' to become part of the compromized dialogue, of the dialogic community. Latour argues that: "Contrary to the claims of those who want to hold either the state of technology or that of society constant, it is possible to consider a path of an innovation in which all the actors co-evolve."[69] To do this successfully, Latour proposes some requirements:

---

67 LATOUR, B., *We Have Never Been Modern*, London: Harvester Wheatsheaf, 1993, p. 41.

68 *Ibid.*, p. 144.

69 LATOUR, B., "Technology is Society made Durable", in LAW JOHN. (ed.), A *Sociology of Monsters: Essays on Power, Technology and Domination*, London: Routledge, 1991, p. 117 (103–131).

Perplexity: This says that "you shall not simplify the number of propositions to be taken into account in the discussion."[70] It means that:

> First, the number of candidate entities must not be arbitrarily reduced in the interests of facility or convenience. In order words, nothing must stifle too quickly the perplexity into which the agents find themselves plunged, owing to the emergence of new beings.[71]

Consultation: also known as *the requirement of relevance* states that; "You shall make sure that the number of voices that participate in the articulation of propositions is not arbitrarily short-circuited."[72] That is to say:

> The number of those which participate in this process of perplexing must not itself be limited too quickly or to arbitrarily. The discussion would of course be accelerated, but its outcome would become too easy. It would lack broader consultation, the only form capable of verifying the importance and the qualification of the new entities. On the contrary, it is necessary to make sure that reliable witnesses, assured opinions, credible spokepersons have been summoned up, thanks to a long effort of investigation and provocation (in the etymological sense of "production of voices."[73]

Here, Latour emphasized the role of spokesperson that he defines thus: "The Spokesperson is precisely the one who does not permit an assured answer to the question, 'Who is speaking?'"[74]

Latour wants to restructure how the roles of politician, scientists, moralists and economists are complimentary and only become chaotic and problematic when they aspire to do each others jobs, much like the parallel between science and religion. Emphasis should not be devoted to a hierarchy of importance but on a

70 Latour, B., *Politics of Nature*, p. 109.
71 *Ibid.*, p. 110.
72 *Ibid.*, p. 109.
73 *Ibid.*, p. 110.
74 *Ibid.*, p. 250.We are very used to the idea of spokespersons in the human realm – particularly in the context of our representative democracies. Less so, perhaps, in the realm of science, but Latour claims that, 'we have to acknowledge that the notion of the spokesperson lends itself admirably to the definition of the work done by scientists in lab coats … the lab coats are the spokespersons of the nonhumans, and, as is the case with all spokespersons, *we have to entertain serious but not definitive doubts* about their capacity to speak in the name of those they represent' (pp. 64–5). Therefore, it is wrong to think that, 'Humans and humans alone are the ones who speak, discuss, and argue' (p. 66).On the contrary, he says, 'lab coats have invented *speech prostheses that allow nonhumans to participate in the discussions of humans, when humans become perplexed about the participation of new entities in political life*' (p. 67; emphasis in the original).

mutual relationship which the collective must utilize to the utmost capability. This will allow them to articulate and represent matters of common concern.

Finally, from what we have seen in this sub-section, it is clear that sub-politics, be it through dialogic democracy, epistemic community or hybid forums and collectives, is a way of engaging a greater number of the public in decision making concerning technological development, especially as it affect the society at large. These bodies outside the main-line politics provide the alternative avenue through which different actors and the general public can be involved in finding how to live in a technologically induced modern society with its attendant depravities.

6.2. Towards a New Environmental Ethics

From what we have seen so far from the above analyses, it is obvious that through the aid of modern technology, predominantly in the past 100 years, our faculty for action has stretched out, significantly altering our planet, and raising quite a lot of ethical questions about the expansion and use of modern technology. The dislocation of the environment which Heidegger decried still revibrates today in a larger scale with more advanced technology. The questions ocassioned by this advancement in technology and the boost in market efficiency and output set in motion questions about the environmental issue. There is a wide variety of views, however, on the necessity for an ethics of technology. Nowadays, the ecology question has penetrated the political and economic spheres. In addition, it raises vital ethical questions at this time that humanity is itself at risk by the modernization it has generated. However ethical appraisal will be of maximum use if it is contextualized to particular realm of activity, relevance for technological innovation; for instance, bio-medical ethics, computer, information technology ethics, engineering ethics, and environmental ethics. In spite of being stretch across a variety of fields, these approaches to technology and ethics tend to have an established series of ethical themes; they reflect on the probable problems and profits posed by a new technology, the impact the technology may possibly have on individual sovereignty, and its sound effects on justice and fairness. Sufficiently addressing technology and ethics needs a much broader view of the function of technology in society and how the interaction between technology and environment changes our world. Such descriptions are not susceptible to calling for distinctive approaches to ethics and technology, but somewhat imply that we can employ a constant set of ethical concepts or ideas and apply them to the particular technologies in sight. This presents a means of making decisions as regards whether a technology is good or bad, based above all on our assessment of the exploits to which human put the technology.

Through an extensive study of the status of "nature" and allied concepts such as "earth" in the thought of Heidegger, Bruce Foltz attempts to show how Heidegger's understanding of the natural environment and our relation to it offers a more potential starting point for environmental philosophy than others that have so far been put forward. Heidegger suggests that refocusing on the prime questions that make it significant to be human—the question of Being—could create the means for alternative dialogues that both challenge and circumvent the attempt for total surveillance and total control. He advocates recognizing the problematic relationship humanity has with the environment and reinventing new trajectories of understanding ourselves and our planet. Foltz argue that the relation between human being and the natural environment can be more drastically understood by critically examining the character of the relation itself with regard to how it has been fashioned and expressed by the tradition of Western metaphysics, especially taking into consideration the manner in which this tradition contains the central assumptions of both modern natural science as well as contemporary technology.[75]

MortonWinston, in *Children of Invention*, writes impressively about technology's unprecedented sway and impact on ethics, and how we have to build up a different outline and approach as we press forward in our modern technological mind-set. Such ethical challenges are expressed eloquently by Morton is this following long passage:

> Our previous ethics has not prepared us to cope with such global threats. Traditional ethics has focused primarily on the moral requirements concerning individual action, on the direct dealings between persons, rather than on the remote effects of our collective action. This problem is particularly important with respect to widely distributed technologies, such as the internal combustion engine, whereby the cumulative effects of individual decisions can have a major impact on air quality even though no single individual is responsible for the smog.
>
> By and large, traditional moral norms deal with the present and near-future effects of actions of individual human beings and do not prepare us to deal with cumulative effects and statistical deaths. Traditional ethics, above all, has been anthropocentric—the entire non human world has been viewed as a thing devoid of moral standing and significance except insofar as it could be bent to satisfy human purposes. We have assumed the natural world

75 Foltz, V. B., *Inhabiting the Earth: Heidegger, Environmental Ethics, and the Metaphysics of Nature*, see especially chapter 8, pp. 154ff.

was our enemy and that it did not require our care (for what could we possibly do to harm it really?) and nature was not regarded as an object of human responsibility.

In the past, we have attempted to fashion out ethical theories in terms of these assumptions. The traditional maxims of ethics—for example, "Love thy neighbour as thyself," "Do unto others as you would have them do unto you," and "Never treat your fellow man as a means only but always also as an end in himself"— are in keeping with the individualistic present- oriented, and anthropocentric assumptions of our ethical traditions. Even the Christian ethics of universal love does not transcend the barriers of time, community, and species. Even more modern ethical theories such as utilitarianism and Kantian ethics do not provide particularly good guidance when it comes to the sort of ethical concerns raised by technology. In part this is because they were designed to be used to evaluate individual actions of particular moral agents.

But the socio-technological practices that comprise our collective action are not only made up of many individual choices—such as the choice to have a child, to eat a hamburger, or to invest in a mining stock—but also the aggregation of these individual choices, plus those of organized collectivities such as corporations and governments. In most cases, the individuals, business executive, or politicians who are making the choices that add up to our collective insecurity do not intend these threats to result, and neither they nor we consequently feel any sense of responsibility of them.

Although individuals view themselves as moral agents and consider themselves bearers of responsibility in all the roles in which they participate, the collectivities of which we belong do not.

All the threats we face are in part the result of this diffusion of responsibility. How then should we, the citizens of Earth, be responding to these environmental questions? Do people in richer countries have any responsibility to those in poorer ones? Do we, in general, have any responsibilities to future generations concerning the long-term social and environmental effects of our present economic lifestyle, and political choices?

The notion of responsibility that we need to cultivate is not the backward-looking notion of responsibility as liability, which seeks to allocate blame for past harms, but the forward-looking sense of responsibility in which each of us and every organization and institution "take responsibility" for future generations of humans and then on human species with whom we share this planet. This notion of social responsibility, although it is voluntary and discretionary, places real demands on us as individuals and members of communities and requires that we think carefully about the decisions and choices that we make.[76]

---

76 WINSTON, E. M., EDELBACK, D. R., Society, *Ethics, and Technology*, 4th Edition, Introd. "Children of Invention Revisited" by MORTON WINSTON, Boston, MA: Wadsworth Publishing, 2011, pp. 22–23. The contributors to this text stress the latest technological innovations and how these modern advancements represent new ethical challenges and dilemmas for society as a whole. Drawing from a variety of contemporary social

Here Morton tells us what is going on currently in our society within the areas of ethics, technology, and science; where we were, where we are now, and why we need to look at this for our future. Primarily, he stresses that the impact of technology is different from other ethical questions since it is not individual decisions but the collective of those resolutions that have an ethical impact. Winston maintains that the damage attributable to technology is not direct, as such, but extensive and elusive, frequently carried out without any awareness from the group executing the harmful actions. Certainly, industrialization and technological efficiency has estranged us from nature – we live in era of fast-food chains, prefabricated houses, already-made clothes, etc. This is not intrinsically appalling in itself; obviously, it has given us incredible freedoms and time for other leisure leading to subjective temporary state of happiness. However, it encourages alienation, chemically induced delirium, frenzy deliberation from personal responsibility and aggressive anticipation of freedom of action, or what Winston calls a "diffusion of responsibility", wherein we don't associate our actions with their side effects, because we are so detached from them. Winston also raises the concern that we have to consider that in the technological age, the effects of our actions are not all the time felt immediately, but in reality influences many generations yet to come. The question of environmental ethics is on the phase not of an individual human being, living now, but somewhat on human race and the survival of the planet all together.

Hans Jonas, as we saw earlier, has contentions about traditional ethics' ineptness to handle environmental plights, but his point of emphasis was on technology. "Modern technology has introduced actions of such novel scale, objects, and consequences that the framework of former ethics can no longer contain them."[77] Jonas is aware that the compass of technology has advanced in a manner that we can no longer have discussions about it and the environment simultaneously with valuable outcomes under the ethics of antiquity. He calls for a new kind of ethics, an ethics of the future, which seems to anticipate the voice of today's environmental and Green movements.

> And what if the new kind of human action would means that more than the interest of man alone is to be considered—that our duty extends farther, and the anthropocentric confinement of former ethics no longer holds? It is at least not senseless anymore to ask

---

issues, the contributors examined technological change and its social consequences from a variety of historical, societal and philosophical perspectives. See also, Ethical Reflection on Modern Technology: *Do We Need Ethics To Handle Modern Technology?* in http://thetechnologicalcitizen.com/?p=1975.

77 Jonas, H., *The Imperative of Responsibility*, p. 6.

> whether the condition of extrahuman nature, the biosphere as a whole and in its parts, now subject to our power, has become a human trust and has something of a moral claim on us not only for our ulterior sake but for its own and in its own right. If this were the case it would require quite some rethinking in basic principles of ethics. It would mean to seek not only the human good but also the good of things extrahuman, that is, to extend the recognition of "ends in themselves" beyond the sphere of man and make the human good include the care for them.[78]

Here, Jonas argues that we must recognize the wellbeing and fundamental value of nature, and we have to make the human good include taking into account what is good for the biotic community. This new ethic is rooted in humanity, but with the environment as a concern. Consequently we could liken his imperative, "Act so that the effects of your action are compatible with the permanence of genuine human life" to ecological imperative of technology: approach technology with an aura that is beneficial to the stability of authentic human life, which demands ecological sustainability. At this point Heidegger's categorization of technology as *revealing* and *enframing* comes to mind. Revealing technology involves humanity discovering the concealed potential that nature has, such as the windmill. One distinguishing quality of the windmill is that it joins forces with nature by demonstrating how it and nature produce power together. It respects the ecological integrity of nature since the windmill does not damage the environment to a considerable extent. Enframing technology, on the contrary, is a sort of revealing technology that challenges nature devoid of respect for its ecological integrity "attuned only to the practical use one might put things to or the energy one might extract from them"[79] with the hydroelectric dam and modern 'mechanized food industry'as examples. "Such industrial institutions violet billions of non-human animals yearly by ignoring those animal's subjectivity and the ethical demands it generates, and they debase humans by shutting out other ways that we might appreciate non-human animals—artistically, poetically, and so on."[80] As a result of humanity's use of enframing technologies, we can argue that natur's ecological integrity has not been considered. This is because in such managerial, technological mode of thinking, "no plant, no animal, no ecosystem has life of its own, has

---

78 Jonas, H., *The Imperative of Responsibility*, p. 8.

79 Turner, D., "Humanity as Shepherd of Being: Heidegger's Philosophy and the Animal Other," in Ladelle McWhorter and Gail Stenstad, (eds.), *Heidegger & the Earth: Essays in Environmental Philosophy*, 2nd Edition, Toronto: University of Toronto Press Inc., 2009, p. 159.

80 *Ibid.*, p. 160.

any significance, outside the bounds of human desire and need. Nothing, we say, other than human beings, has any intrinsic value."[81]

Winston, Hans Jonas as well as others, insists on a new ethical structure that takes into cognizance the future of the planet and the generations to come, a structure that did not lay emphasis solely on the individual Dasein, or the pressing moment, but on human race generally, and its continued existence.

Concluding Remarks

To sum up this chapter, we want to state that while the participation of both individual citizens and organized groups in the policy subsystem is important, the expansion of social movements or group-based deliberative exchanges—which assemble actors from government, business, and civil society to address specific problems—has the maximum potential. This is particularly correct given the new and more challenging and complicated forms in which technological and environmental problems are presenting themselves in the first half of this century. Social movements bear a considerable responsibility in influencing and shaping the visions of institutions responsible for managing risks. They raise awareness to unwanted risks while providing an image of the social transformations required to eradicating or addressing them. For instance, environmental movements will bear majority of the responsibility for bringing about the humanizing or improving of technology since they have created an intense consciousness of the soaring effect or risks associated with globalization. Social movements are in addition the means for bringing public awareness to *life politics* in all societies.

We have to point out also that collective choices are to be made through *articulate and rational discussion*, rather than by blind acceptance of the views of established conventional authorities, by contracts or deals concluded among vested interests, or even recourse to threats or coercions. As they deliberate, participants press forward arguments and listen to counter arguments; they make use of critical reason to evaluate options and make decisions. Definitely, every individual cannot be involved personally in each deliberative context; representative devices are indispensable for the working of modern democracy. But the overall formation of decision-making processes must be fair for all. Specifically, disparities of wealth and status should not distort deliberative interaction, and systemic prejudice on account of class, race, creed, gender, and the like, must be proscribed. Whereas some theorists interpret the dominance of reason as not

---

81 McWhorter, L., "Guilt as Management Technology: A Call to Heideggerian Reflection," in *Heidegger & the Earth*, p. 12.

including appeal to the emotions, others recognize a genuine place for an extensive variety of communicative approaches provided they are aimed towards the reasoned resolution of collective problems. Such a model is more pertinent to decision-making for governance in conditions of environmental sustainability where decision-making is made difficult because of intrinsic social and scientific uncertainty. All have the same opinion that coercion, threats, and manipulation are to be excluded. And contrary to traditional risk governance that has been inclined to turning citizenship into a submissive inactive affair, citizen's involvement in science, the "hybrid- forum" or technical democracy are thought to encourage a lively public sphere and reflective citizens enthusiastic to participate in a critical discussion on the future direction of scientific and technological progress. So, instead of approaching disagreement as sign of failure or a division that reason should transcend, the authors of the 2007 EC report describe it as a resource that will keep public engagement with science alive.[82] Rather than aiming for a final accord, the very rationale of opening up expert judgments to public analysis is to make explicit the divergences of perspectives.[83] Rather than structuring the legitimacy of science and technology governance exclusively on the supremacy of recommendation from a closed cadre of experts, academics of science and technology today ask scientific experts to substantiate their knowledge claims in view of alternative ways of thinking and knowing. Also the likelihood of rejecting certain techno-economic decisions and actions has aggravated a persuasive on-going debate about the desirability of the "precautionary principle" at a time of rapid technological change.

---

82 Ulrike, F. & Wynne, B., *Taking European Knowledge Society Seriously: Report of the Expert Group on Science and Governance to the Science, Economy and Society Directorate, Directorate-General for Research, European Commission,* p. 61.

83 Stirling, A., 'Opening up' and 'Closing down': Power, Participation, and Pluralism in the Social Appraisal of Technology, in *Science, Technology and Human Values,* vol. 33, no. 2, 2008, pp. 262–294.

Chapter VII: General Evaluation and Conclusions

7. General Evaluation and Conclusions

The primary aim of this work has been to show how Heidegger's critique of modern technology is derived from his fundamental ontology, namely, the question of Being, and to go beyond this metaphysical understanding into a more pragmatic approach that is contemporaneous with the present technological predicaments facing modern humanity, namely, the question of technology and ethical responsibility and the call for reflexivity towards technology.

Every civilization has its own idiosyncratic view of man, reality and the relation between the two. Our civilization is technological. That technological understanding has, in the course of time become globalized, making any alternatives virtually impossible. Heidegger puts forward an alternative. A civilization grounded in the way Heidegger proposes will understand how human existence stretches into the enigma. His history of being culminates in technological enframing. Heidegger presents a critique of modernity we cannot simply pay no attention to or sidestep. He opens a path to a new way of thinking, but perhaps he goes too far. That notwithstanding, the first thing to be observed about Heidegger is his originality of thought and method. He has to his credit the re-introduction of the question of being which was relegated into oblivion by modern metaphysics. In the context of post-Cartesian and post-Kantian philosophies, this was an enduring achievement.

Heidegger maintained that his reflection is marked by a linear connection and continuity, and that the attempt to differentiate between its various phases is obligated to take proper account of this fact. As Heidegger explains, such a distinction is "justified only on the condition that this is kept constantly in mind: only by way of what [Heidegger] I has thought does one gain access to what is to-be-thought by [Heidegger] II."[1]

7.1. Heidegger I

In the first chapter, we examined the nature of Dasein as Being-in-the-world. The history of being is structurally grounded in the ontological conformation of

1 Heidegger M., Preface to William Richardson, *Heidegger: Through Phenomenology to Thought*, p. xxii.

Dasein. His being-in-the-world is characterized by care. At this stage, he is caught up in a concernful association with entities and other Daseins which he encounters in his everydayness as being-in-the-world. Heidegger's phenomenology of "Being-in-the-world" (*In-der-Welt-Sein*) and "Being-with" (*Mitsein*) represents his early attempt to combat the influence of Cartesian subjectivism. We showed also how Heidegger successfully illustrated that the disclosure of beings are most fundamentally encouraged through the phenomenon of the ready-to-hand in which he displayed his early critique of modernity and technology.

Heideger claimed to be dealing with ontology in *Being and Time*. But according to Husserl, he only dealt with ontology in the first few pages of the book. Husserl noticed that, Heidegger, not having anything extra to add to ontology independent of human existence, changed the subject to *Dasein*. Although Heidegger argued that the question of human existence is fundamental to the quest for the question of being, Husserl criticized this as reducing phenomenology to "philosophical anthropology" and presenting a conceptual and inaccurate depiction of the human being.[2]

According to Levinas, Heidegger's Being epitomizes another maneuver employed by Modern Western metaphysics to disregard and inadvertently eradicate the alterity of the *Other* in the guise to uphold the autonomy of the individual. In fact his conclusion is that Heideggerian ontology, his fundamental ontology as first philosophy unavoidably leads to philosophy of power and tyranny: "Even though it opposes the technological passion issued forth from the forgetting of Being hidden by the existent, Heideggerian ontology, which subordinates the relationship with the Other to the relationship with Being in general, remains under obedience to the anonymous and leads inevitably to another power, to imperialist domination, to tyranny. Tyranny is not the pure and simple extension of technology to reified men."[3] The end result of Levinas's argument as we saw is essentially quite simple: Heidegger prizes ontology over ethics.

---

2 Cf. Husserl, E., *Psychological and Transcendental Phenomenology and the Confrontation with Heidegger* (1927–1931) Dordrecht: Kluwer, 1997.

3 Levinas, E., *Totality and Infinity*, pp. 46–47. It has to be said that not every person would concur with this appraisal of Heidegger. For instance, Julian Young states that; "Emmanuel Levinas complains, in 'Is Ontology Fundamental' that Heidegger fails to treat 'the Other' as person in an authentically neighbourly fashion. What is lacking, he claims, is a 'religious' stance to the Other person. Such a stance, he implies, is confined, by Heidegger, to being. Man is, then he concludes, 'neighbour' to the Other as being but not as person." Young concludes that "this is a false dichotomy, an elevation of

Furthermore, Hans Jonas sees as immoral "seeming, false humility of Heidegger's shifting initiative to Being." He interprets Heidegger's Being as shrouded in paganism which he described as "the most enormous hubris in the whole history of thought." Jonas states:

> Man: the shepherd of being - not, mind you, of beings! Apart from the blasphemous ring which this use of the hallowed title must have to Jewish and Christian ears: it is hard to hear man hailed as the shepherd of being when he has so dismally failed to be his brother's keeper. The latter he is meant to be in the Bible. But the terrible anonymity of Heidegger's "being," illicitly decked out with personal characters, blocks out the personal call. Not by the being of another person am I grasped, but just by "being"! And my responsive thought is being's own event. But called as person by person—fellow beings or God—my response will not primarily be thinking but action (though this involves thinking), and the action may be one of love, responsibility, pity; also of wrath, indignation, hate, even the fight to the death. Such calls are drowned in the voice of being to which one cannot say No.[4]

The bottom line of looking at this interpretation of Heidegger's question of being is that Heideger's ontology is to say the least unscrupulous, amoral and pagan.

Likewise, Heidegger exhibited a philosophy of obscurity by being more ambiguous in his concept of Being. For him, Being seems to have something of a holy or divine character. Certainly, Heidegger does not identify Being with God. Yet, we think it would be true to say that in his thought, Being has taken the place of God. Being undoubtedly is furnished with most of the attributes that has traditionally been assigned to God and Being seem to perform most of the functions that belongs to God. This is because, in his quest for Being, Heidegger supplements his phenomenological analysis of man's being with a religious or even mystical experience of being itself. Heidegger thinks that the Biblical God is an entity (*Seindes*) rather than being (*sein*) itself. Therefore he was unfair to Christian theology by labeling God, the highest of the existents as a man-made entity in contrast to the absoluteness of His being as such. Christian theology deals with the Supreme Being who is the foundation of Himself and of all existence. Heidegger was not able to grasp this clearly or he simply feined ignorant of it.

7.2. Heidegger II

In part II of our study, we showed how Heidegger's philosophy of technology stems directly from his fundamental ontology and the question of being. In the third

---

modern individualism into a form of idolatory..." See *Heidegger, Philosophy, Nazism*, Cambridge: Cambridge University Press, 1997, p. 211.

4 Hans, J., "Heidegger and Theology" in *Phenomenon of Life*, pp. 257–258.

chapter, we examined Heidegger's critique of Western metaphysics, his critique of Descartes whose primacy of consciousness, according to Heidegger, led to the forgetfulness of Being in modern philosophy. In his quest to provide the contemporary society with the real meaning of being, Heidegger regarded being as the presencing of the presence which gathers all things into the presence. He situated the question of being in the context of the ontological difference between being and what-is. Being is the presencing of the presence, the *legein* of the *logos* which gathers what-is onto presence and gives it warrant to be. Since metaphysics constitutes the forgetfulness of Being, Heidegger calls for the *destruktion* and overcoming of metaphysics. In his view, technology is the logical consequence of modern Western metaphysics. Philosophy stimulates modern science, and the pragmatic result of modern science is modern technology: "Philosophy turns into the empirical science of man, of all that can become for man the experiential object of his technology, the technology by which he establishes himself in the world by working on it in the manifold modes of making and shaping."[5] This is what we saw in Chapter four in which Heidegger lunched his critique of modern technology after we highlighted the nature of modernity and some trajectories on the question of technology from different schools of thought.

Heidegger's *Being and Time* and *The Question Concerning Technology* contain a commanding critique of modernity and technology. Ours is a world abysmally wrought by the sway and dominance of modern technology in a network described by Max Weber as "iron cage" of bureaucracy and rationality.[6] In his works on modern technology, Heidegger recognized various dangers that are associated with the growth of modern technology especially the 'thingification' and uprooting of man as well as dislocation of nature, among other things. When it was pointed out to him in *Der Spiegel* interview that people in highly technological part of the world are well provided for, and we live in prosperity, Heidegger objects:

> Everything is functioning. This is exactly what is uncanny, that everything is functioning and that the functioning drives us more and more to even further functioning, and technology tears men loose from the earth and uproots them. I do not know whether you were frightened, but I at any rate was frightened when I saw pictures coming from the moon to the earth. We don't need any atom bomb. The uprooting of man has already taken place. The only thing we have left is purely technological relationships. This is no longer the earth on which man lives. As you know, I recently had a long conversation

5 Heidegger, M., The End of Philosophy and the Task of Thinking, in *Basic Writings*, p. 434.

6 Weber, M., *The Protestant Ethic and the Spirit of Capitalism,* New York: Charles Scribner's Sons, 1958, p. 181.

> with René Char in Provence, the poet and Resistance fighter. Rocket bases are being built in the Provence and the country is being devastated in an incredible way. This poet, who certainly cannot be suspected of sentimentality or a glorification of the idyllic, tells me that the uprooting of man which is taking place there will be the end, if poetry and thought do not once more succeed to a position of might without force.[7]

The functionalization of man and society is thus a destiny from which there is no escape.[8] Is there anything that can be done about this? In Heidegger's observation, little can be done because the technological uprooting of man is the result of the history of Being. Therefore, human thought and action cannot change what has already been destined by Being itself. He calls for resignation and passivity rather than an active program of reform which would simply constitute a further extension of modern technology. As Heidegger asserts, "philosophy will not be able to effect an immediate transformation of the present condition of the world. This is not only true of philosophy, but of all merely human thought and endeavour. Only a god can save us."[9] This is because "human activity can never directly counter this danger. Human achievement alone can never banish it."[10] To be saved, he says, is to be brought back to an awareness of who we are: as those to whom has been entrusted the safekeeping or guardianship of the world.

Heidegger's immense contribution to metaphysical questioning is indisputable. His profound analysis of Dasein is lusciously illuminating. However, we have to point out that Heidegger's philosophy started from phenomenology; he discarded phenomenology and turned to metaphysics. Later, he left metaphysics for poetry. Perhaps he realized the failure of philosophy to name Being. This indefinite misunderstanding led to his failure to name Being and bring it out of concealment.

Heidegger's consideration of technology as the extension of metaphysics by other means results in his failure to comprehend and appreciate the manner in which technology is entrenched in modern social and historical framework. As Tom Rockmore observed; "his [Heidegger's] approach to technology through the

---

7 HEIDEGGER, M., "Only a God can Save Us", *Der Spiegel's Interview with Martin Heidegger*. In RICHARD WOLIN, (ed.), *The Heidegger Controversy, A Critical Reader*, p. 105–106, (pp. 91–116).

8 FEENBERG A., *Questioning Technology*, p. 185.

9 HEIDEGGER, M., "Only a God can Save Us", *Der Spiegel's Interview with Martin Heidegger*, p. 107.

10 HEIDEGGER M., *The Question Concerning Technology*, p. 33.

problem of metaphysics sees technology not as an end in itself but as a means to a further end, in this case as part of his effort to destroy the history of metaphysics."[11]

Overall, Heidegger's critique of technology paints a conspicuously gloomy picture of the technological age. He outlined two pictures, but approved neither of them: "In the first, technology destroys itself and perhaps humankind as well. Little imagination is required to envision how this scenario could be realized [...]. In the second scenario, an accidental opening to Being is discovered (or revealed) and then taken. In other words, a god comes to save humankind and humankind pays heed. In either case, a return to the concerns and activities of the poet will serve as preparation for what might come."[12]

Again, Heidegger's notion of poetic dwelling and thinking negates the intellectualistic approach to the question of being given by St. Thomas Aquinas. Dwelling thinking as thanking, and this thanking hidden within the reality of sacrifice, opens up the ascetical life of the shepherd of being. It is therefore difficult to evade the conclusion that the qualities of authentic Dasein are really Christian virtues in secularized form.

The tasks facing our historical era are for Heidegger: that of recovering the meaning of Being and the arousal of the sense of the holy. Heidegger's philosophical attempt at the question of God failed and became insoluble because philosophy had not named Being. The failure of the philosopher to name Being, the failure of the poet to name the Holy means that the essence of divinity cannot be named. This further means that the question of God cannot be posed in the time of want. But a philosophy of being which does not face the question of God cannot be regarded seriously. An elucidation of man either in terms of phenomenology or the metaphysics of morals which ignores the question of God's existence does not merit an accolade of wisdom.

But more sinister in this scenario is Heidegger's declaration that man is essentially helpless to stop the technological danger that threatens his very humanity. His later philosophy is as amoral as it is apolitical, persistently maintaining that the very attempt to conceive an ethics and politics is a contributing factor to the contemporary crisis of the West. As Andrew Feenberg observed, "[A]lthough Heidegger means his critique to cut deeper than any social or historical facts about our times, it is by no means irrelevant to a modern world

---

11 ROCKMORE, T., Heidegger on Technology and Democracy, in A. FEENBERG & A. HANNAY, *Technology and the Politics of Knowledge,* Bloomington & Indiana: Indiana University Press, 1995, p. 140.

12 HICKMAN, A. L., *Philosophical Tools for Technological Culture: Putting Pragmatism to Work*, Bloomington and Indianapolis: Indiana University Press, 2001, p. 152.

armed with nuclear weapons and controlled by vast technologically based organizations."[13]

There is also a factual complexity here, both in understanding a clear-cut thrust of Heidegger's argument and in connecting his proposal to what we might consider as the necessities of the present. "Heidegger, who scrutinizes technology only within the context of his deeper concern with metaphysics, seems incapable of grasping the relation of technology to society and human being."[14] But then Heidegger himself borrowed Hölderlin's line: "*But where danger is, grows the saving power also*." This mean then that, if technology itself is creating a danger for humanity and for the whole bio-sphere, only science and technology can save us. Indeed, isn't it geographers, ecologist, chemists, meteorologists, physicists and other scientists who have done most to alert us to the catastrophic potential of many current industrial practices or even natural disasters? As Pattison observed, isn't it precisely a better scientific understanding of what is going on that will best prepare us for the most fitting technological response? Solar panels, wind energy, insulation systems, cleaner cars and other 'green' initiatives all depend on the application of science, rather than its abandonment. Surely the further development of such technologies is more important than musing about the metaphysical foundation of enframing.[15] Looking at it this way, we might conclude that Heidegger's approach is, bluntly, one of prospective stalemate, an intellectual surrender, a failure to engage with what is most existentially pressing in the tangible reality of contemporary destiny.

This is why we showed in chapter five of this thesis that Hans Jonas, contrary to Heidegger's persimism, demonstrates that technology and the science that made it possible gives us powers that require a new dimension of responsibility. Since the consequences of our actions can reach far in time and space, new moral rules have to be added to the "neighbourhood ethics". Jonas's theory can thus be considered as a response to Heidegger's interpretation of technology. Technological power demands moral direction at the same time that, according to Jonas, its scientific foundations have eroded confidence in the foundations of moral norms: "We need (moral) wisdom most when we believe in it least," or, more explicitly, "Now we shiver in the nakedness of a nihilism in which near-onmipotence is paired with near-emptiness, greatest capacity with knowing least for what ends to

13 Feenberg, A., *Questioning Technology*, p. 185.

14 Rockmore, T., "Heidegger on Technology and Democracy," p. 141.

15 Pattison, G., *The Later Heidegger*, London: Routledge Publication, 2000, p. 206.

use it."[16] This nihilism associated with criticism of contemporary society demands a new concept of responsibility to address its consequences. This provides an alternative to Heidegger's ingenuous dissent concerning the dangers of technology, a set of concrete practices through which we are able to focus the "saving power" that technology presents and, at the same time, to diffuse the "supreme danger" that it continuously represents.

Jonas assessed the chances of controlling technological dangers and the ability of the political systems of his time to forestall a disaster for human race caused by the domination of the technological push. Although he highlighted ethical responsibility as the only possible way of solving the global challenges, ethics does not tell us how to get there. The emergence of the challenges we are facing today requires setting up a democratic political project, yet to be invented.

Lastly, it is still possible to question the need for such an integral, complete ethics of technology as Jonas proposed. From the political standpoint, it might be objected that artificial overviews of technological development and human action are undemocratic. Today the only thing constant in our technological world is change. Each new generation builds on the work of the previous one, gaining new perspective. We humans are good at coming up with new ideas and technologies, but we sometimes have cognitive narrow-mindedness that makes it hard for us to understand their interactions and predict where hazards may arise. A catastrophe like the Gulf of Mexico oil spill or the Fukushima Diiachi nuclear disaster might be geographically limited, biological catastrophe could possibly infect the entire world. One need not be a Luddite to ask that the risks of this technology, and the difficulties that humans have coping with risk of this complexity, be taken very seriously and the solution redefined in a more reflexive democratic means.

7.3. A New Modernity

On account of those shortcomings associated with modern technology, we suggested and directed our view of modernity to the social science that is increasingly

16 Hans, J., *The Imperative of Responsibility*, p. 21, 23. According to Bernstein's observation, "Heidegger, too, seeks to make us aware of the supreme danger of the essence of modern technology, Gestell, which turns everything into 'standing reserve' and even human beings into'human resources'. What is lacking in Heidegger's analysis—from Jonas perspective—is an'ethics of responsibility'. Jonas affirms as primordial and central precisely what Heidegger denies—that new ethics can be metaphysically grounded, indeed grounded in a more adequate theory of being where nature itself affirms the ought-to-be of life." CF. Bernstein, R. J., "Rethinking Responsibility" in *The Legacy of Hans Jonas*. Hastings Centre Report, 1995, No. 7, p. 17 (pp. 13–20).

being called a second or new modernity. It is marked first of all by a reflexivity that has tried to learn from the shortcomings of high modernity, with its sometimes conceited self-confidence, its unexamined belief in innovation and progress, its tendency to believe in the efficiency of its utopian technological advancement.

Following the arguments from Ulrich Beck and Anthony Giddens, we showed that the transition from 'industrial society' to risk society' is described as reflexivity: modernity is undergoing a process of (unseen) 'self-confrontation'. Although Beck and Giddens initially developed their theories of risk and late modernity largely separately, their writings have much in common, e.g., concepts like "produced uncertainty", "individualization", "reflexivity" etc. By their insistence on the process of individualization, they send the social actor adrift into a free-floating, reflexive existence.

We showed that according to Beck, the theory of world risk society addresses the growing recognition of the uncontainable prevalence of sweeping uncertainty and improbability in the modern world. "The basic institutions, the actors of the first modernity—science and expert systems, the state, commerce and the international system, including the military—responsible for calculating and controlling manufactured uncertainties are undermined by growing awareness that they are inefficient, their actions even counter-productive."[17]

Giddens makes a distinction between *external risk* which is "risk of events that may strike individuals unexpectedly (from the outside, as it were) but that happens regulary enough and often enough in a whole population of people to be broadly predictable, and so insurable", from *manufactured risk*, which is "risk created from the very progression of human development, especially by the

17 Beck, U., "Living in the World Risk Society", in *Economy and Society*, Vol. 35, No. 3, August, 2006, p. 338, (329–345). Here we can highlight some of the similarities between Beck and Giddens: (i) For both of them, risk emerges from modernization and globalization; (ii) For both, risk is qualitatively different in late modern societies where risk has greater impact across space and time; (iii) Both of them are interested in the political ramification of risk and they claim that expert discourses have been undermined by concerns about risk; (iv) Both are interested in the ways that increased reflexivity results as a response to risk and uncertainty in late modernity; On divergent issues, (i) While Beck claims that increased risk reflexivity is the outcome of a greater number of risks and hazards being produced, Giddens claims that risks are merely thought to be greater because human subjectivity is now more sensitive to risk; (ii) While Giddens is more interested in self-reflexivity, i.e., reflexivity directed towards the body and self, Beck is more interested in our reflexive critique of the social and institutional.

progression of science and technology. Manufactured risk refers to new risk environments for which history provides us with little previous experience."[18]

These risks have trickled into our way of thinking and acting in the world—they cannot be ruled out, and notwithstanding the finest efforts of the various institutions generated to controle them, they escape such systematic rationalization in dramatically destructive ways. We saw how this process is distinguished as "reflexive modernization", which is inclusive of, but not limited to the concept of the same name promoted by Giddens and Lash.

Reflexivity signifies the reaction of modern man to risk and/or to conditions that gives rise to fear or anxiety. Reflexivity entails a persistent monitoring of action and its content. It involves critical appraisal of actors and stakeholders, as well as the specialized risk experts. The production of wealth is accompanied by production of risks as an outcome of modernization.

Risk as a political and moral problem carries great practical connotations. According to Giddens, for instance, whether or not we should construct a nuclear power plant cannot be decided by a neutral technical assessment of benefit and risk alone, but must include issue of "political prudence."[19] Beck asserts that "risks become the motor of the *self-politicization* of modernity in industrial society."[20] This directs us to examine and search the moral argument of an alternative policy to the rush-to development. Certainly, our future depends upon our aptitude to fashion an alternative vision of development by reinvigorating the normative potentials latent in tradition in favour of mutual human relationships, citizens' participation and dialogic democracy. This is because the techno-economic sphere has lost its hitherto 'non-political' character. Yet it does not become political in the strict sense of the word. It must be regarded as neither political nor non-political, but *subpolitical.*

The final movement of this thesis begins with a discussion on reflexive political modernization citing subpolitics as a step towards addressing the problem of modern technology as it affects the citizenry within a particular context. The significance of the 'subpolitical' arena for the pursuit of politics is not new. But global warming, oil spills, chemical or nuclear plant explosions, mining accidents, the effects of genetically modified organisms, living in fear of nuclear or bioterrorism, building or bridge collapses, transportation crashes etc provides a new arena in which these concerns have been articulated, contested and reconstructed. One

18 GIDDENS, A., "Risk and Responsibility" in *The Modern Law Review, 62*(1), January, 1999, p. 4, (pp. 1–10).

19 GIDDENS, A., *Beyond Left and Right: The Future of Radical Politics*, p. 79.

20 BECK, U., *Risk Society: Towards a New Modernity*, p. 183.

of the result of these uncertainties is that responsibilities previously linked with conventional politics stand at the moment occupied by new bodies, sometimes referred to as epistemic community such as NGOs, business organizations, experts, professionals etc. who through subpolitical activities, devoid of any formal political party, question the decisions of institutional politics as it affects modern society. "Subpolitcs, then, is an indication that there has emerged a new political praxis based on everyday life and, along with it, a new kind of non-institutional politics."[21]

So we have argued that studies of science and technology have taken a deliberative turn from Heidegger's ontological interpretation in recent decades. Having established that policymakers and politicians have lost their primary responsibility in the midst of defining both problems and solutions in the sphere of decision making on environmental and technological development, we suggested that, although subpolitics might be a substantial menace to the traditional institutions of representative democracy, it also signifies a potential for a new political deliberation and for new ways of dealing with decision making on the complicated issues involved in the environmental and technological realms. Thus, technologically induced risks and environmental degradations lead to calls for the de-monopolization of technological and scientific know-how; its subjection to public analysis, and extension of democratic responsibility or accountability to science, technology and government beyond the formal representative institutions of the political system.

Critics have raised concerns with much of Gidden's and Beck's theory regarding the nature of risk and society. For instance, the risk society has been seen as emanating from particular circumstances of post-war Germany, and its appropriateness for describing the politics of nature and society outside Western Europe, or even Germany, is questionable.[22] They see Beck's risk society to be inundated with the same sorts of universalizing tendencies and far-reaching declarations that are not backed up by, or incompatible with experiential details, with excessive emphasis on the threats and significance of risk.

Gandy is of the view that throughout the modern period, "ecological risk and public health scares have been repeatedly translated into new regulatory regimes"

---

21 Sørensenm, P. Mads & Christiansen, A., *Ulrich Beck: An Introduction to the Theory of Second Modernity and the Risk Society,* New York: Routledge, 2013, p. 99.

22 See Marshall, B. K., "Globalization, Environmental Degradation and Ulrich Beck's Risk Society," in *Environmental Values*, Vol. 8, 1999, p. 267 (pp. 253–275); Dingwall, R., "'Risk Society': The Cult of Theory and the Millennium," in *Social Policy and Administration*, Vol. 33, 1999, p. 475 (pp. 474–491); Rustin, M., "Incomplete Modernity: Ulrich Beck's Risk Society," in *Radical Philosophy*, Vol. 67, 1994, p. 3 (pp. 3–12).

which question the *status quo*, although he argues that the natures of contemporary risks do generate unique challenges. Any provisional distinction between a 'industial society' and a 'risk society' is therefore perhaps uncalled-for. In its place, it is possible to argue that (environmental) risk has repeatedly produced institutional crises, and that some modern-day risks pose particular challenges for the institutions which structure modernity.[23] Although Beck and Giddens are right to point out that notions of risk have always existed but now they are manmade, yet we can say without doubt that historically, man has been seen as responsible for being the architect of his own destruction. Risk provoked by immoral and decadent behaviour preceded notions of risk engendered by technological advancement.

This work has not sought to engage critically with risk society's theoretical edifice, and did not do justice to the complexities to be found therein, but aimed at using its offshoot to provide a cutting edge to the ontological undertone of Heidegger's question concerning technology. We would argue, as Mythen does, that "risk society thesis is assembled in the spirit of exploration and adventure: it is not driven by empirical validity, but by invigorating sociology and providing thought-provoking reflections on the modern condition."[24] Also, with their focus on risk and with their analyses of institutional responses to risk, they provide an organizing perspective and a theoretical vocabulary for understanding the trajectory of second modernity.

Meanwhile our tacit concern remains as follows: Is there no likelihood of subpolitics leading to fundamentalism, and thus accessible to anti-modernist and anti-democratic elements who can capitalize on it to cause mobocracy? Even if the room is created for debate before any technological step that affects the public is taken, is there no chance that the debate "will not necessarily lead to agreement, and policy-makers will in the end have to make the leap one way or the other."[25] Will such a leap be acceptable to all the actors involved and if no, what then could be the best way out of the present modern technological predicament?

7.4. A Better Environment

Lastly, we showed how the modern technological growth, with its excessive exploitation of nature, has resulted in environmental decay and degradation. Heidegger

23 See GANDY, M., "Rethinking the Ecological Leviathan: Environmental Regulation in an Age of Risk," in *Global Environmental Change*, Vol. 9, 1999, p. 66 (pp. 59–69).

24 MYTHEN GABE, "Reappraising the Risk Society thesis: Telescopic Sight or Myopic Vision?" in *Current Sociology*, 2007, 55(6), 793–813, p. 803.

25 GIDDENS, A., *The Politics of Climate Change*, Cambridge, UK: Polity Press, 2009, p. 60.

asks questions of the modern paradigm that may permit us to go beyond the technological disaffection and estrangement that industrial production and modern philosophy, as its cultural superstructure, have created. Climate change is setting a limit on that world view with a finality that has never before been encountered. In Heidegger's understanding, the calculative thinking has turned the environment and humans themselves into standing reserve, at hand for our projects, and we are in a widespread battle from thinking of the meditative, thankful type, which alone can set up our dwelling on the earth. Hence, the irresistible sense of alienation that follows our stride towards technological expertise. The way back, the *Schritt Zurück*, to thinking as dwelling is through arts, but only insofar as the arts are rooted in and brings us to meditative dwelling, thus coming back into our essence.

But out from Heidegger's poetic world to our present day reality, the existence of nuclear reactors is a matter of great concern for those who care for the earth and not an issue of poetizing. They produce economical and abundant energy, but the problem is with nuclear waste. It is extremely radio active, and to date no secure method has been found to dispose of it as we saw in the less harmful *Brent Spar* buoy case. Nuclear wastes from some industrialized countries have been discarded into the sea or soil of some underprivileged nations after inducement of their political leaders. Also the ultimate threat from technology comes from extremely sophisticated nuclear arsenals in military technology. Vast quantities of deadly weapons are stacked upon the earth.

Beck, in second modernity, took the environmental issue as one of the biggest problems to illustrate 'risk society.' In a book, *The Cosmopolitan Vision*, Beck distinguishes between three different axes of conflict in world risk society: "first, *ecological* interdependency crises, which have their own global dynamics; second, *economic* interdependency crises, which are initially individualized and nationalized; and third, the threat produced by *terrorist* interdependency crises."[26] Despite their differences, all of these three "cannot be construed as external environmental crises but must be conceived as culturally manufactured actions, effects and insecurities."[27] He called them "civilizational risks" that "can sharpen global

---

26 Beck, U., *The Cosmopolitan Vision*, trans. Ciaran Cronin, Cambridge: Polity Press, 2006, p. 22; First published in German as *Der Kosmopolitische Blick oder: Krieg ist Frieden*, Frankfurt am Main: Suhrkamp Verlag, 2004.

27 *Ibid*., p. 23.

normative consciousness generate global publics and promote a cosmopolitan outlook."[28]

Lending credence to this, Al Gore, who has made the environment his signature issue, has reminded America and rest of the world of our sacred duty to act as responsible stewards to protect our environment.

> Modern industrial civilization, as presently organized, is colliding violently with our planet's ecological system. The ferocity of its assault on the earth is breathtaking, and the horrific consequences are occurring so quickly as to defy our capacity to recognize them, comprehend their global implications, and organize an appropriate and timely response. [...]. I have come to believe that we must take bold and unequivocal action: we must make the rescue of the environment the central organizing principle for civilization. Whether we realize it or not, we are now engaged in an epic battle to right the balance of our earth, and the tide of this battle will turn only when the majority of people in the world become sufficiently aroused by a shared sense of urgent danger to join an all-out effort. [...] Politics, broadly defined, is the means by which we make collective decisions and choices.[29]

In another work, *An Inconvenient Truth*, Al Gore provided us with technological answers to decipher the problem of global warming: "solar panels, geothermal power stations, fluorescent Lightbulbs, Green roofs, hybrid cars, and wind power."[30] He has confidence in the new technology to match the consequences of technological advancements such as climate change or global warming. Although Al Gore is not a scientist, he does rely heavily on the authority of science in this book which is why scientists are sensitive to its details and claims. Nonetheless, critics have pointed out that nearly every significant statement Al Gore makes regarding climate science and climate policy is either one sided, misleading, exaggerated, speculative, or completely wrong.

Furthermore, since environmental policy has been unproductive in changing behaviour and bringing about societal transformations, it is essential that we make decent choices to reduce undesirable environmental and socio-economic impacts of thoughtless application of technology. Therefore, environmental policies need

---

28 *Ibid.* Elsewhere in an earlier work, we find almost the same argument as this: "... a cosmopolitan everyday consciousness which transcends even the borders between man, animal and plants." See Beck Ulrich, *What is Globalization*, Cambridge: Polity Press, 2000, p. 38.

29 Al Gore, *Earth in the Balance: Ecology and the Human Spirit* with a New Forward, Boston: Houghton Mifflin Company, 2000 (1992), pp. 269–270.

30 Al Gore, *An Inconvenient Truth: The Crisis of Global Warming*, New York: Viking, 2006, p. 166.

to be aimed at inspiring vibrant efficiency, and ought to be incentive-based and flexible. Well-informed ethical decision making and preemptive assessment of emerging technology is essential.

The growing difficulty and costs of Command and Control environmental regulations that seek to control pollution subsequent to its production have shifted the interest of environmental regulators and firms towards flexible environmental strategies that aim at reduction of pollution at source and prevent its generation. Thus, in recent years, environmental agencies have acknowledged the need for flexible regulation with incentives for pollution prevention. Flexible and market-based policy instruments are preferred more than command-and-control methods. Besides, the regulatory agencies have recognized that many industries, especially small ones, require information about new pollution prevention technologies, markets for recycled products, and toxicity of materials. Information and technical support are essential parts of more flexible regulation. This assists firms to take up the most cost effective technologies and methods suitable to their activities in meeting government environmental goals.

Pollution prevention "involves waste minimization, source reduction, cleaner production, design for environment and preventive services."[31] Shen defines source reduction as "any practice which reduces the amount of any hazardous substance, pollutant, or contaminant entering any waste stream or otherwise released into the environment prior to recycling, treatment, or disposal [...]. The term includes: equipment or technology modifications, process or procedure modifications, re-formulation or re-design of products, substitution of raw materials, and improvements in house-keeping, maintenance, training and inventory control."[32] It also incorporates those undertakings that increase durability,

31 Shen, T. Thomas, *Industrial Pollution Prevention*, 2nd Edition, Berlin: Springer-Verlag, 1999, p. 18.

32 *Ibid.*, p. 25. According to Tchobanoglous G. & Kreith, F., "Consumers can participate by buying less or using products more efficiently. The public sector (government entities at all levels: local, state and federal) and the private sector can also be more efficient consumers. They can reevaluate procedures which needlessly distribute paper (multiple copies of documents can be cut back), initiate procedures which require the purchase of products with longer lifespans, and cut down on the purchase of disposable products. The private sector can redesign its manufacturing processes to reduce the amount of waste generated in manufacturing. Reducing the amount of waste may require the use of closed-loop manufacturing processes, different raw materials, and/or different production processes. Finally, the private sector can redesign products by increasing their durability, substituting less toxic materials, or increasing product effectiveness."

reusability and repairability. Everybody, private or public, can practice source reduction and this call again for our change of life style.

Finally, we make it to say that by the adoption of precautionary principle, especially in the absence of scientific accord, we can reduce responsibility that emanates from manufactured risk through discretionary decisions. Today, than ever, precaution or participation are argued to provide necessary elements of a more robust approach to the governance of technology. According to François Ewald;

> The precautionary principle requires an active use of doubt, in the sense Descartes made canonical in his Meditations. Before any action, I must not only ask myself what I need to know and what I need to master, but also what I do not know, what I dread or suspect. I must, out of precaution, imagine the worst possible, the consequence that an infinitely deceptive, malicious daemon could have slipped into an apparently innocent enterprise.[33]

In the United Nations Framework Convention on Climate Change which seeks to deal with the threat of global warming, the precautionary principle is included in the following terms: "The parties should take precautionary measures to anticipate, prevent or minimize the causes of climate change and mitigate its adverse effects. Where there are threats of serious or irreversible damage, lack of full scientific certainty should not be used as a reason for postponing such measure, taking into account that policies and measures to deal with climate change should be cost-effective so that to ensure global benefits at the lowest possible cost."[34]

Furthermore, according to Sonja, the principles empower the state to pursue environmental precaution by four types of actions, the goal being that damage must be avoided: (a) Dangers must be detected early, hence research is essential; (b) When irreversibility is feared, there should be action before there is full understanding, i.e. proof of damage is not required; (c) Technical developments to reduce current levels of discharge of pollutants must be promoted, and (d) The state must contribute to the introduction of cleaner processes/technologies into the private sector.[35]

Cf. Tchobanoglous G. & Kreith F., (eds.) *Handbook of Solid Waste Management*, 2nd Edition, Introduction, New York: McGraw-Hill Companies Inc., 2002, pp. 8–9.

33 Ewald, F., The Reture of Descartes' Malicious Demon: An Outline of a Philosophy of Precaution, in T. Baker & J. Simon, (eds.), Embracing Risk: The Changing Culture of Insurance and Responsibility, Chicago: University of Chicago Press, 2002, (pp. 273–301), p. 285.

34 *United Nations Framework Convention on Climate Change*, 1994, Art. 3.

35 Sonja Boehmer-Christiansen, "The Precautionary Principle in German-Enabling Government," in Timothy O'Riordan and James Cameron (eds.), *Interpreting the Precautionary Principle,* London: Earthsea, 1994, p. 37.

Given the magnitude of the possible consequences of our technological choices, it is an absolute obligation for us to try and anticipate those consequences, assess them, and ground our choices on this assessment. Thus, it is imperative to note that where assessments fall short to specify the level of impacts on the environment and there is no scientific consensus, it is morally and ethically worthwhile to get it wrong on the side of caution by suspending action. This is the precautionary principle, and there is no way around it.

Although precautionary principle is widely accepted as a framework for policy, our foundamental concerns still are as follows: if precautionary principle is to be followed to its letters, will it not be blocking the way for human initiatives and adventures? Is human life itself not partly a risky affair in its profundity? After all, "risk-taking is essential to new thinking in all spheres, to scientific progress and to wealth-creation."[36] Above all, will procastination not be potentially paralyzing to the society since precautionary principle will definitely lead to no particular direction? What perhaps could be the best possible way out when taking action seems dangerous and inaction seems risky? Although religion is being pushed more and more into a corner especially in most technologically advanced societies, we think it could play a major role in our present environmental predicament. Political decisions and policies are often controlled by economic consequences and benefits. But the inspiration from religion motivates people to cultivate a new life style and a new understanding of development.

Conclusion: It is laudable to study Heidegger's thought about Being, Dasein, and how technology fits into this situation. When Heidegger examines "the question concerning technology" he is concerned with the essence of modern technology, not merely any technology; for it is modern technology that poses the problem. He sees a problem with technology; not technology '*in its perseity*' but how its essence is revealed. The key to our losing our sense of self is to change our perspective. Fundamentally, technology orders in a particular method that limits our outlook of nature. This destining, as a mode of revealing, though in itself not bad, only turn out to be limiting when it closes us off to other possibilities and alienates us.

Due to the complex nature of modernity and the interconnectedness between its varying elements, including reason, freedom, science and technology, and how these elements are related to human social concerns, it is difficult for anyone, including Heidegger, to understand modernity and deal with its various facets,

---

36 GIDDENS, A., *The Politics of Climate Change*, p. 57.

unless its particular elements are taken into consideration within its own context, devoid of any predetermined assumption. Despite Heidegger's insinuations, we cannot logically declare that modern man cannot be capable of bringing technology under control. Such despair constitutes an object of conflicting signal and a clog in the wheel of progress.

The standard critique dismisses Heidegger as technological determinists who had an essentialist understanding of technology. It is evident that his transcendental thinking on technology is impracticable to be "falsified" or otherwise be explicitly appraised within the research community. Such profound metaphysical conjectures may all the same, have their place in philosophy. Technology is not out of control as Heidegger and some others might *submit*. What should change is the *mode* of control and how scientific or technological advancement is carried out in human society. The target should be to anticipate negative consequences in advance and more regularly, to set up policy procedures to inspire reflexivity and learning, and as a result generate *avant-garde* possibility for experimentation. Our proposal is for constructive, practical and productive experimentation. Technologies need to be *fostered, encouraged and promoted* but in a sketch out formula that allows different actors to become engaged. Or as Guardini rightly pointed out, we need more technology and not less. "We need stronger, more considered, more human technology. We need more science, but it must be more intellectual and designed; we need more economic and political energy, but it must be more mature and responsible, able to see the details in the whole contexts to which it belongs."[37]

The point basically is that whatever happens—notwithstanding of past destinies and intentions—social, political and technological actors make decisions all the time, which are today more or less driven by market forces. Such decisions often emerge to have moral and ethical implications for technological development and operation which calls for citizen's participation in decisions making. We argue that those decisions require to be reflected upon reflexively in the best possible way, even though, admittedly, it is not so far entirely obvious what the best way is. The research approach outlined in this work ventures to be a step toward discovering the ground that would constitute such "a best way." And that is a fascinating question that could be the object of future research.

---

37 Guardini, R., *Letters from Lake Como*, p. 83.

General Bibliography

Primary Sources: Works by Heidegger

HEIDEGGER, M., *Being and Time*, trans. John Macquarrie & Edward Robinson, 29th edition, Oxford: Blackwell Publishing Ltd. 2007.

- *Being and Truth*, trans. G. Fried & R. Polt Bloomington, Indiana: Indiana University Press, 2010.
- *The Event*, trans. R. Rojcewicz, Bloomington, Indiana: Indiana University Press, 2013.
- *The Question Concerning Technology and Other Essays*, translated and with an Introduction by William Lovitt, New York: Harper & Row, Publishers, Inc, 1977.
- *On the Essence of Truth* in *Basic Writings*, Revised and Expanded Edition, ed. David Ferrell Krell, New York: HarperCollins Publishers, 1993.
- *On the Way to Language*, trans. Peter D. Hertz, San Francisco: Harper and Row, 1982.
- *Poetry, Language, Thought* trans. Albert Hofstadter, Perennial Classic Edition, New York: Harper & Row, 2001.
- *The Essence of Truth: On Plato's Cave Allegory and Theaetetus*, trans. Ted Sadler, Athlone: Continuum International Publishing Group, 2002.
- *Towards the Definition of Philosophy*, trans. T. Sandler, London: The Athlone Press, 2000.
- *Parmenides*, trans. André Schuwer & Richard Rojcewicz, Bloomington, IN. Indiana University Press, 1998.
- *The Principle of Reason*, trans. Iilly Reginald, Indianapolis: Indiana University Press, 1996.
- *The Basic problems of Phenomenology*, trans. Hofstadter A., Bloomington: Indiana University Press, 1982, First Midland Book Edition, 1988.
- "Letter on Humanism" in *Basic Writings*, Revised and Expanded Edition, ed. Davis Farrell Krell, New York: HarperCollins Publications, 1993.
- "The Thinker As Poet" in *Poetry, Language, Thought*, trans. Albert Hofstadter, New York: Perennial Classics edition, 2001.
- "The Way Back into the Ground of Metaphysics," in *Existentialism from Dostoevsky to Sartre*, ed. W. Kaufmann, New York: New American Library, 1975.
- "*What is Metaphysics?*" trans. David Farrell Krell, In *Basic Writing*, New York: Harper Collins Publishers, 1993.

– “What is Metaphysics”, ‘Postscirpt’ in *Existence and Being*, ed. Werner Brock, Indiana: Regnery/Gateway Inc., 1979, pp. 354–356, 1949 edition.
– *Aristotle's Metaphysics θ 1–3: On the Essence and Actuality of Force*, trans. W. Brogan & P. Warnek, Bloomington and Indianapolis: Indiana University Press, 1995.
– *Contributions to Philosophy (From Enowning),* trans. Parvis Emad & Kenneth Maly, Bloomington: Indiana University Press, 1999.
– *Country Path Conversations*, trans. Bret W. Davis, Bloomington, IN: Indiana University Press, 2010.
– *Discourse on Thinking*, trans., John M. Anderson & Hans, E. Freund, London: Harper Colophon Edition, 1969.
– *Early Greek Thinking,* trans. David F. Krell & Frank Capuzzi, New York: Harper and Row, 1975.
– *Four Seminars*, trans. Andrew Mitchell & François Raffoul, Bloomington: Indiana University Press, 2003.
– *Identity and Difference*, trans. Joan Stambaugh, Chicago: The University of Chicago Press bi-lingual edition, 2002.
– *Introduction to Metaphysics*, trans. Gregory Fried & Richard Polt, New Haven/ London: Yale University Press, 2000.
– *Kant and the Problem of Metaphysics*, 5th Edition, Enlarged, trans. Richard Taft, Indianapolis: Indiana University Press, 1997.
– *On Time and Being*, translated and with Introduction by Joan Stambaugh, Chicago: University of Chicago Press edition, 2002.
– *The Fundamental Concepts of Metaphysics: World, Finitude, Solitude*, trans. W. McNeil & N. Walker, Bloomington and Indianapolis: Indiana University Press, 1995.
– *The Question of Being*, trans. W. Kluback and J.T. Wild, London: Vision Press Ltd. 1956.
– *What is a Thing*, trans. W. B. Barton, Jr. and Vera Deutsch, Indiana: Gateway edition, 1969.
– *What is Called Thinking*, trans. J. Glenngray, New York: HarperCollins Publishers Inc., 2004 edition.
– *What is Philosophy*? trans., William Kluback & Jean T. Wilde, New York: Twayne, 1958.
– *Introduction to Phenomenological Reasearch*, trans. Daniel, O. Dahlstrom, Bloomington, Indiana: Indiana University Press, 2005.
– “Only a God can Save Us,” Der *Spiegel*, “Interview with Martin Heidegger,” trans. Maria Alter & John D. Caputo, in *Philosophy Today* 20 (Winter 1976): 267–284.

– *Sein und Zeit,* 17. Aufl. Tübingen: Max Niemeyeer, 1993.

– *Sein und Wahrheit*, Frankfurt am Main: Vittorio Klostermann, 2001.

– Brief über den Humanismus, *Wegmarken*, (Weg), 3., durchgesehen Aufl. Frankfurt a. M: Vittorio Klostermann, 1996.

– *Identitat und Differenz,* Verlag Günther Neska in Pfullingen, 1957.

– *Die Grundbegriffe der Metaphysik. Welt—Endlichkeit—Einsamkei*t, Frankfurt am Main: Vittorio Klostermann, 1983, 1992.

– *Kant und das Problem der Metaphysik*, 2. Aufl Frankfurt am Main: Vittorio Klostermann, 1951.

– "Nur noch ein Gott kann uns retten," *Der Spiegel*, 31 May 1976.

– *Phenomenologie und Theologie*, Frankfurt am Main: Vittorio Klostermann, 1970.

– *Vom Wesen der Wahrheit*, First published in 1943, Frankfurt a. M, in *Wegmarken (Weg)* 3., durchgesehene Aufl. Frankfurt a. M.: Vittorio Klostermann, 1996.

– *Was ist Metaphysik*? Frankfurt am Main: Vittorio Klostermann, (1975) 1996.

– *Was Heisst Denken*? Tübingen: Max Niemeyer Verlag, 1954.

– *Zur Sache des Denkens,* 2. Auflage, Tübingen: Max Niemeyer, 1976.

– *Bauen Wohnen Denken* (1951) in *Vorträge und Aufsätze*, 8. Aufl. Stuttgart: Günter Neske, 1997, 139–156.

– *Aus der Erfahrung des Denken*, 4 Aufl. Pfullingen: Neske, 1977.

– *Der Feldweg*, 3rd ed. Frankfurt am Main: Klostermann, 1962.

– *Der Satz vom Grund*, 5. Aufl. Pfullingen: Neske, 1978.

– *Einführung in die Metaphysik*, Tübingen: Max Niemeyer Verlag, 6. Aufl. 1998.

– *Erläuterungen zu Hölderlins Dichtung*, 5. Aufl. Frankfurt am Main: Vittorio Klostermann, 1981.

– *Gelassenheit*, 11. Aufl. Stuttgart: Günther Neske, 1999.

– *Holzweg*, 7. Durchgesehen Aufl. Frankfurt a. M.: Vittorio Klostermann, 1994.

– *Was ist das—die Philosophie*? 5 Aufl. Pfullingen: Neske, 1972.

– Gesamtausgabe, B. 2, *Sein und Zeit*, Frankfurt am Main: Klostermann, 1977.

– Gesamtausgabe, 12, *Unterwegs zur Spache*, 11., durchgesehene Aufl., Stuttgart: Günther Neske, 1997. Originally published by Verlag Günther Neske, Pfullingen, 1959.

– Gesamtausgabe, 71: *Das Ereignis*. Edited by Friedrich-Wilhelm v. Herrman, Frankfurt am Main: Vittorio Klostermann, 2009.

– Gesamtausgabe, 79: 3–4 *Bremer und Freiburger Vorträge*, Ed. Petra Jaeger, Frankfurt: Klostermann, 1994.

- Gesamtausgabe, 54, (*Parmenides*), Frankfurt an Main: Vittorio Klostermann, 1982.
- Gesamtausgabe, 16, „*Reden und andere Zeugnisse eines Lebensweges*“: 1940–1976. Ed. Hermann Heidegger, Frankfurt: Klostermann, 2000.
- Gesamtausgabe, 9, “eine Wandlung des Denkens”: “*Vom Wesen der Wahrheit,*” 187.21–22.
- Gesamtausgabe, 15, *Seminare*, Frankfurt am Main: Vittorio Klostermann, 1986.
- Gesamtausgabe, 65, *Beiträge zur Philosophie, (Vom Ereignis)* ed. Friedrich-Wilhelm von Herrmann, 2nd, rev. ed. Frankfurt am Main: Vittorio Klostermann, 1994.
- Gesamtausgabe, 58: *Grundproblem der Phänomenologie*, Frankfurt a, M: Vittorio Klostermann, 1993.
- Gesamtausgabe, Abt. II, Bd. 26, *Metaphysiche Anfangsgründe der Logik im Ausgang von Leibniz,* Frankfurt am Main: Vittorio Klostermann, 1978.
- Gesamtausgabe, 17, *Einführung in die phänomenologische Forschung*, Frankfurt am Main: Vittorio Klostermann, 1994.

Secondary Sources

Ackermann, R., *Data, Instruments, and Theory: A Dialectical Approach to Understanding Science,* Princeton: Princeton University Press, 1985.

Adam C. Tannery P. (eds.), *Œuvres de Descartes*, vol. 111, Paris: Vrin, 1996.

Adam, B., Beck, U. & Joost van Loon, *The Risk Society and Beyond: Critical Issues for Social Theory*, London: Sage Publications, 2000.

Adam, B., *Timescapes of Modernity: The Environment and Invisible Hazards*, London: Routledge, 1998.

Adornor, T. W. & Horkheimer, M., *Dialectics of Enlightenment*, trans. E. Jephcott, Stanford, Calif: Stanford University Press, 2002.

Anderson, A., “Environmental Pressure Politics and the ‘Risk Society’” in Stuart Allen, Barbara Adam and Cynthia Carter (eds.), *Environmental Risks and the Media*, London: Routledge, 2000.

Alford, F., Levinas, the Frankfurt School and Psychoanalysis, London: Continuum, 2002.

Aquinas, T., *Summa Theologiae*, English translation, Introduction, Notes, Appendices & Glossary by Thomas, Gilby, Cambridge: Blackfriars, 1967.

- *Truth (Quaestiones Disputatae De Verite)*, q. 10, art. 8, trans. By Robert W. Mulligan, J.V. McGlynn, and R.W. Schmidt, Chicago, IL: Henry Regenery Company, 1952–54.

Aristotle, *Metaphysics*, in *The Complete Works of Aristotle: The Revised Oxford Translation*, ed. Jonathan Barnes, New Jersey: Princeton University Press, 1984.

Ballard, B.W. *The Role of Mood in Heidegger's Ontology*, Lanham, Maryland: University Press of America, 1991.

Barash, J. A., *Martin Heidegger and the Problem of Historical Meaning*, New York: Fordham University Press, 2003.

Barker, R., *The Hiroshima Maidens: A Story of Courage, Compassion and Survival*, New York: Viking, 1985.

Barnett, C., Cloke, P., Clarke, N. & Malpass, A., *Globalizing Responsibility: The Political Rationalities of Ethical Consumption*, West Sussex, UK: Blackwell Publishing, 2011.

Beck, U., "Reflexive Governance: Politics in the Global Risk Society," in Von Jan-Peter Voss, Dierk Bauknecht, René Kemp, editors, *Reflexive Governance for Sustainable Development*, Cheltenham, UK: Edward Elgar Publishing Ltd., 2006.

– "Risk Society and the Provident State", trans. Martin Chalmer, in Scott Lash, Bronislaw Szerszynski & Brian Wynne, (eds.), *Risk, Environment and Modernity: Towards a New Ecology*, London: SAGE Publications, 1996.

– "The Politics of Risk Society" in *The Politics of Risk Society*, ed. Jane Franklin, London: Polity Press, 1998.

– "The Reinvention of Politics: Towards a Theory of Reflexive Modernization" in U. Beck, A. Giddens, S. Lash, *Reflexive Modernization: Politics, Tradition and Aesthetics in the Modern Social Order*, Cambridge: Polity Press, 1994.

– *Democracy Without Enemies*, Cambridge: Polity Press, 1997.

– *Risk Society: Towards a New Modernity*, trans. Mark Ritter, Los Angeles & London: Sage Publications, 1992. First published in German as *Risikogesellschaft: Auf dem Weg in eine andere Modern*, Frankfurt am Main: Surbkamp Verlag, 1986.

– *The Brave New World of Work*, Cambridge: Polity Press, 2001.

– *The Cosmopolitan Vision*, trans. Ciaran Cronin, Cambridge: Polity Press, 2006. First published in German as *Der Kosmopolitische Blick oder: Krieg ist Frieden*, Frankfurt am Main: Suhrkamp Verlag, 2004.

– *What is Globalization*, Cambridge: Polity Press, 2000.

– *World at Risk*, Cambridge, UK: Polity Press, 2009.

– *World Risk Society*, Cambridge, UK & Massachusetts, USA: Polity Press in association with Blackwell Publishers, 1999.

– *Risk Society: Towards a New Modernity*, London: Sage, 1992.

– *The Reinvention of Politics: Rethinking Modernity in the Global Social Order*, trans., Mark Ritter, Cambridge: Polity Press, 1997; First published in German as *Die Erfindung des Politischen*, Frankfurt am Main: Suhrkamp Verlag, 1993.

- *Ecological Enlightenment: Essays in the Politics of the Risk Society*, Atlantic Highlands, NJ: Humanities Press, 1994.

Beck, U, Giddens, A. & Lash, S., *Reflexive Modernization: Politics, Tradition and Aesthetics in the Modern Social Order,* Cambridge: Polity Press, 1994.

Beck, U & Beck-Gernsheim, E., "Individualization and 'Precious Freedom': Perspectives and Controversies of a Subject-Orientated Sociology", in P. Heelas, S. Lash & P. Morris (eds.), *Detraditionalization*, Oxford: Blackwell Publishers, 1996.

- *The Normal Chaos of Love*, Cambridge: Polity Press, 1995.

- *Individualization: Institutionalized Individualism and its Social and Political Consequences,* London: SAGA Publications Ltd., 2002.

Beck, U. & Williams, J., *Conversation with Ulrich Beck,* Cambridge: Polity press, 2004.

Behler, E., *Irony and the Discourse of Modernity*, Seattle & London: University of Washington Press, 1990.

Benhabib, S., "Towards a Deliberative Model of Democratic Legitimacy", in Benhabib, S., (ed.), *Democracy and Differences: Contesting the Boundaries of the Political,* Princeton, New Jersey: Princeton University Press, 1996.

Benton, T., "Social Theory and Ecological Politics: Reflexive Modernization or Green Socialism," in Von Riley E. Dunlap, Frederick H. Buttel, Peter Dickens, August Gijswijt, (eds.) *Sociological Theory andthe Enviromental: Classical Foundations, Contemporary Insights,* Maryland, USA: Rowman & Littlefield Publishers, Inc., 2002.

Bernard. A. O. Williams, *Descartes: The Project of Pure Inquiry*, New York: Routledge, 2005.

Biemel, W., "Heidegger's Concept of Dasein" in *Heidegger's Existential Analytic*, Frederick Elliston, (ed.), The Hague: Mouton Publishers, 1978.

- *Le Concept de Monde chez Heidegger*, Louvain: E. Nauwelaerts, 1950.

- *Martin Heidegger*, Reinbek bei Hamnurg: Rowohlt, 1973.

- "Poetry and Language in Heidegger," in Kockelmans J. Joseph, *On Heidegger and Language*, Evanston: Northwestern University Press, 1972.

Blattner, W. *Heidegger's Temporal Idealism*, Cambridge: Cambridge University Press, 1999.

Bochenski, I. M., *Contemporary European Philosophy*, trans. Donald Nicholl & Karl Aschenbrenncr, Berkeley and Los Angeles: University of California Press, 1957.

Bohman, J. & Rehg, W., eds., *Deliberative Democracy: Essays on Reason and Politics*, Cambridge, MA: MIT Press, 1997.

Bohman, J., *Public Deliberation: Pluralism, Complexity, and Democracy*, Cambridge, MA: MIT Press, 2000.

Boedeker, E., "Phenomenology," in *A Companion to Heidegger*, Dreyfus, H. L & Wrathall, M. A., (eds.), Malden, MA: Blackwell Publishing, 2005.

Borgmann, A., "Technology" in *A Companion to Heidegger*, Dreyfus, H. L & Wrathall, M. A., (eds.), Malden, MA: Blackwell Publishing, 2005.

– *Technology and the Character of Contemporary Life*, Chicago: University of Chicago Press, 1984.

– *Power Failure: Christianity in the Culture of Technology*, Michigan: Brazo Press, 2003.

Braithwaite, J. & Drahos, P., *Global Business Regulation*, Cambridge: Cambridge University Press, 2000.

Brentano, F., *Psychology from an Empirical Standpoint,* trans. A.C. Rancurello, D.B. Terrell, & L. McAlister, London: Routledge, 1995.

Bretschneider, W., *Sein und Wahrheit: Über die Zusammengehörigkeit von Sein und Wahrheit im Denken Martin Heideggers,* Meisenheim am Glan: Verlag Anton Hain, 1965.

Brey, P., "Theorizing Modernity and Technology," in Thomas J. Misa, Philip Brey, and Andrew Feenberg, *Modernity and Technology*, Cambridge, Mass.: The MIT Press, 2003.

Brisart, R., 'L'intentionalité comme "titre d'un problème central" selon Heidegger,' in *Cahiers du centre d'études phénoménologiques* (CEP 2) Louvain-la-Neuve: Cabay, 1982.

Brown, M. L., "Policymaking in International Organizations: The European Union and Ozone Protection" in Robert V. Bartlett, Priya A, Kurian, Madhu Malik (eds.), *International Organizations and Environmental Policy*, Westport, CT: Greenwood Press, 1995.

Bucchi, M. & Trench, B., (eds.), *Handbook of Public Communication of Science and Technology*, London & New York: Routledge, 2008, chaps. 10, 12 & 14.

Buizer, M., "Local Initiatives Challenging Mainstream Policies: Signs of Sub-Politics?" in Henk de Haan, René van der Duim, (eds), *Landscapes, Leisure and Tourism*, Delft, The Netherlands: Uitgeverij Eburon, 2008.

Burggraeve, R., *The Wisdom of Love in the Service of Love*, trans. Jeffery Bloechl, Milwaukee, WI: Marquette Univ. Press, 2002.

Cahoon, L., (ed.), *From Modernism to Postmodernism: An Anthology*, Oxford: Blackwell, 1996.

Callon, M., "Techno-economic networks and irreversibility," in Law, J. (ed.), *A Sociology of Monsters: Essays on Power, Technology and Domination*, London: Routledge, 1991.

Callon, M., Lascoumes, P. & Barthe, Y., *Acting in an Uncertain World: An Essay on Technical Democracy*, trans. Graham Burchell, Cambridge, MA: The MIT Press, 2009, First published as *Agir dans un monde incertain: Essai sur la democratie technique* by Editions du Seuil, 2001.

Caputo, D. J., *The Mystical Element in Heidegger's Thought*, New York: Fordham University Press, 1986.

– *Heidegger and Aquinas: An Essay on Overcoming Metaphysics*, New York: Fordham University Press, 1982.

– "Heidegger and Theology" in Guignon, Charles B., (ed.). *The Cambridge Companion to Heidegger*, Cambridge and New York: Cambridge University Press, 2000.

– *Demythologizing Heidegger*, Bloomington, Indianapolis: Indiana University Press, 1993.

– *The Mystical Elements in Heidegger's Thought*, Fordham University Press, 1986.

– "Being and the Mystery of Person," in *The Universe as Journey: Conversation with W. Norris Clark, S. J.*, Bronx, N.Y: Fordham Univ. Press, 1988.

Carmen, T., *Heidegger Analytic: Interpretation, Discourse and Authenticity in Being and Time*, New York: Cambridge University Press, 2003.

CERBONE, R. David, *Understanding Pheomenology*, Routledge, 2014.

Chabot, P., *La Philosophie de Simondon*, Paris: Librairie Philosophique J. Vrin, 2003.

Chalier, C., "Emmauel Levinas: Responsibility and Election," in *Ethics*, P. Griffiths, (ed.), Cambridge: Cambridge University Press, 1993.

Chanter, T., *Time, Death, and the Feminine: Levinas with Heidegger*, Stanford: Stanford University Press, 2001.

Chapman, A., *Democratizing Technology: Risk, Responsibility and the Regulation of Chemicals*, London: Earthscan, 2007.

Ciaramelli, F., "Levinas's Ethical Discourse Between Individuation and Universality," in Bernasconi. R., & Critchley, S., eds., *Reading Levinas*, Bloomington: Indiana University Press, 1991.

Clifford, G., Christians, John P. Ferre & Fackler, P. Mark, *Good News: Social Ethics and the Press*, New York: Oxford University Press, 1993.

Cohen, A. R., *Face to Face with Levinas*, New York: State University of New York Press, 1986.

– *Ethics, Exegesis, and Philosophy: Interpretation after Levinas*, Cambridge: Cambridge University Press, 2001.

Cohen, J., "Deliberation and Democratic Legitimacy," in Bohman J. & Rehg W., *Deliberative Democracy: Essays on Reason and Politics,* Massachusetts: The MIT Press, 1997.

Colin, D., *Levinas: An Introduction*, Notre Dame: University of Notre Dame Press, 1996.

Copleston, F. C., *A History of Medieval Philosophy*, Notre Dame & London: University of Notre Dame Press, 1990.

Cornelio, F., *God in Exile: Modern Atheism From its Roots in the Cartesian Cogito to the Present Day*, trans. Arthur Gibson, Westminster, Maryland: Newman Press, 1968.

Critchley, S. & Bernasconi, R., (eds.), *The Cambridge Companion to Levinas*, Cambridge: Cambridge University Press, 2002.

Critchley, S. & Schürmann, R., *On Heidegger's Being and Time*, ed. Steven Levine, London: Routledge, 2008.

Critchley, S., "Introduction," in Bernasconi R., & Critchley S., (eds.), *Cambridge Campanion to Levinas*, Cambridge: Cambridge University Press, 2004.

– *Infinitely Demanding: Ethics of Commitment, Politics of Resistance,* New York: Verso Press, 2007.

– Leaving the Climate of Heidegger's Thinking. In Hansel, J., (ed.), *Levinas in Jerusalem: Phenmenology, Ethics, Politics, Aesthetics,* (Amsterdam Studies in Jewish Philosophy), Springer, 2009.

Craig, M., *Levinas and James: Towards a Pragmatic Phenomenology*, Bloomington, Indiana: Indiana University Press, 2010.

Crowell, S. & Malpas, J., (ed.), *Transcendental Heidegger*, Stanford, California: Stanford University Press, 2007.

Dalton, D. *Longing for the Other: Levinas and Metaphysical Desire*, Pittsburg: Duquesne University Press, 2009.

Dalle Pezze, B., *Martin Heidegger and Meister Eckhardt: A Path Towards Gelassenheit*, The Edwin Mellen Press, 2009.

Daniel, G. "Semel in Vita: The Scientific Background to Descartes' Meditations" in Amélie Oksenberg Rorty (ed.) *Essays on Descartes' Maditations* Los Angeles: University of California Press, 1986.

Dastur, F., "The Reception and Non-Reception of Heidegger in France," in *French Interpretation of Heidegger,* Albany, New York: Sunny Press, 2008.

– La Question de la Technique: Le Dialogue Entre Patočka et Heidegger. In Frogneux, N., (ed.), *Jan Patočka : Liberté, Existence et Monde Commun*, Le Cercle Herméneutique Editeur, Argenteuil, 2012.

Dawson, C., *Progress and Religion*, New Jersey: Doubleday Image, 1960.

Dedeurwaerdere, T., "La dimension éthique dans la sphère de la rationalité technologique: normes, contextes et arrière-plans," *Les Carnets du Centre de Philosophie du Droit*, N° 82, 2000.

– "The Contribution of Network Governance to sustainable Development", Institute du Développment et des Relations Durable Internationales (Iddris), *Les seminaires de l'Iddri*, Paris, no 13, 2005.

Depré, O., *Hans Jonas*, Paris: Ellipses Édition, 2003.

– *Philosophy Moral*, Louvain-la-Neuve, Beligum: Academia Bruylant, 1999.

Depre, O. & Lories, D., *Lire Descartes Aujourd'hui*, Louvain-Paris: Édition Peeters, 1997.

Derek, R. M., *Heidegger's Philosophy and Theory of the Self*, Aldershot, England: Ashgate Publishing Limited, 2001.

Dermot, M. & Mooney, T., (eds.), *The Phenomenology Reader*, London: Routledge, 2002.

Dicker, G., *Descartes: An Analytical and Historical Introduction*, New York, Oxford: Oxford University Press, 1993.

Derrida, J., *Adieu a Emmanuel Levinas*, Paris: Galilée, 1997.

– *Meaning of Philosophy*, trans. Alan Bass, Chicago: The University of Chicago Press, 1982.

Descartes, R., *Meditations on First Philosophy with Selections from Objections and Replies*, trans. John Cottingham, with Introduction by Bernard Williams, Cambridge: Cambridge University Press, 1986.

Deuze, M., *Media Work*, Cambridge: Polity Press, 2007.

Didier, F., *Heidegger et le probleme de l'espace*, Paris: Editions de Minuit, 1986.

Dominique, J., *Power of the Rational*, trans. Peg & Elizabeth, Birmingham, Indianapolis: Indiana University Press, 1994.

Donald, R. Wehrs & David P. Haney, (eds.), *Levinas and Nineteenth-Century Literature: Ethics and Otherness from Romanticism through Realism*, (Introduction), University of Delaware Press, 2009.

Douglas, M. & Wildavsky, A., *Risk and Culture: An Essay in the Selection and Interpretation of Technological and Environmental Dangers*, Berkeley, CA: University of California Press, 1982.

DRYFUS, L. H., *Being-in-the-world: A Commentary on Heidegger's Being and Time, Division I*, Masschusetts: MIT Press, 1991.

– "Heidegger on Gaining a Free Relation to Technology" in *Readings in the Philosophy of Technology*, ed. David M. Kaplan, Lanham, U.S.A: Rowman & Littlefield Publishers, Inc., 2004. Originally published in *Technology and the Politics of knowledge*, Feenberg & Hannay, (eds.), Indiana University Press, 1995.

DRYZEK, J., *Deliberative Democracy and Beyond: Liberals, Critics, Contestations*, Oxford: Oxford University Press, 2000.

DUMOUCHEL, P., "Gilbert Simondon's Plea for a Philosophy of Technology", in A. Feenberg & H. Alastair, eds., *Technology and the Politics of Knowledge*, Bloomington & Indianapolis: Indiana University Press, 1995.

DE BOER, T., An Ethical Transcendental Philosophy, in *Face to Face with Levinas*, Cohen A. Richard, (ed.) New York: State University of New York Press, 1986.

DILTHY, W., *Selected Works: Poetry and Experience*, vol. 4, Rudolf Makkreel & Frithjof Rodi (eds.) Princeton, NJ: Princeton University Press, 1996.

DOUGLAS, M., *Risk and Blame: Essays in Cultural Theory*, London: Routledge 1992.

Duns Scotus, J., *Ordinatio*, In *Philosophical Writings*, edited and translated by A. B. Wolter, Edinburgh: Nelson's Philosophical Texts, 1962.

DUSEK, V., *Philosophy of Technology: An Introduction*, Malden, MA: Blackwell Publishing Ltd., 2006.

EDITA, M., *Les Fleurs D'Hiroshima*, trans. Suzanne Lipinska, Paris: René Julliard, 1961.

EDWARDS, J. C. The Thinging of the Thing, in *A Companion to Heidegger*, Dryfus, H. L & Wrathall, M. A. (eds), Malden, MA: Blackwell Publishing, 2005.

ELI GORDON, P., "Fidelity as Heresy: Levinas, Heidegger, and the Crisis of the Transcendental Ego," in *Heidegger's Jewish Followers*, edited by Samuel Fleischacker, Pittsburgh, Pennsylvania: Duquesne University Press, 2008.

ELLIOT, A., *Contemporary Social Theory: An Introduction*, New York; Routledge, 2009.

ELLUL, J., *The Technological Society*, trans. John Wilkinson, New York: Knopf, 1964.

ERICKSON, A. S., *Language and Being: An Analytic Phenomenology*, New Haven: Yale University Press, 1970.

ESTRIN, B. L., *The American love lyric after Auschwitz and Hiroshima*, New York: Palgrave, 2001.

EWALD, F., "The Return of Descartes' Malicious Demon: An Outline of a Philosophy of Precaution", in T. Baker & J. Simon, (eds.), *Embracing Risk: The Changing Culture of Insurance and Responsibility*, Chicago: University of Chicago Press, 2002.

Falk, R., "The Making of Global Citizenship", in B. van Steenbergen (ed.), *The Conditions of Citizenship*, London: Sage, 1994.

Fasching, D. J., *The Thought of Jacques Ellul: A Systematic Exposition*, Toronto: Edwin Mellen, 1981.

Fedier, F., "Heidegger et Dieu" in Kearney R. & O'Leary J., eds., *Heidegger et la Question de Dieu*, Paris: Bernard Grasset, 1980.

Feenberg, A., *Critical Theory of Technology*, New York: Oxford University Press, 1991.

- *Questioning Technology*, London & New York: Routledge, 1999.

- *Transforming Technology: A Critical Theory Revisited*, Oxford: Oxford University Press, 2002.

- *Heidegger and Marcuse: The Catastrophe and Redemption of History*, N.Y.: Routledge, 2005.

- *Between Reason and Experience: Essays in Technology and Modernity*, Cambridge, Massachusetts: The MIT Press, 2010.

Ferry, L. & Renaut, A., *Heidegger and Modernity*, trans. Franklin Philip, Chicago & London: University of Chicago press, 1990. Originally, published as *Heidegger et les Modern*, Paris: Grasset & Fasquelle, 1988.

Fischer, F., *Citizen, Experts, and the Environment*, Durham: Duke University Press, 2005.

Foltz, V. C., *Inhabiting the Earth: Heidegger, Environmental Ethics and the Metaphysics of Nature*, Atlantic Highlands, NJ: Humanities Press International, Inc., 1995.

Foucault, M., *The Order of Things*, New York: Vintage Books, 1973.

Frank, P., "Ontology and Ethics: Questioning First Philosophy in Levinas, Heidegger, and Fichte," in *Heidegger's Jewish Followers*, edited by Samuel Fleischacker, Pittsburgh, Pennsylvania: Duquesne University Press, 2008.

Frede, D., "The Question of Being," In Guignon, C. (ed.), *The Cambridge Companion to Heidegger*, Cambridge, UK: Cambridge University Press, 1993.

Friedrich-Wilhelm von Herrmann, *Subjekt und Dasein: Interpretationen zu "Sein und Zeit,"* 2. Stark erw. Aufl. Frankfurt a. M.: Vittorio Klostermann, 1985.

Frogneux, N., *Hans Jonas ou la vie dans le monde,* Bruxelles: De Boeck Université, 2001.

Froman Wayne, J. Levinas and Heidegger: a strange conversation. In Dahlstrom, D.O. editor, *Interpreting Heidegger: Critical Essays*, Cambridge: Cambridge University Press, 2011.

Fryer, R. D., *The Intervention of the Other: Ethical Subjectivity in Levinas and Lacan*, New York: Other Press LLC, 2004.

GADAMER, H-G., "Hermeneutics and Logocentism," in Diane P. Michelfelder & Richard E. Palmer, (eds.) *Dialogue and Deconstruction: The Gadamer-Derrida Encounter*, Albany: State University of New York Press, 1989.

GADAMER, H-G., *Truth and Method*, 2nd ed., trans. Joel Weinsheimer & Donald G. Marshall, New York: Continuum, 2002.

GARBER, D., *Descartes Embodied: Reading Cartesian Philosophy Through Cartesian Science*, Cambride: Cambridge University Press, 2001.

GELVEN, M., *A Commentary on Heidegger's Being and Time,* Chicago: Northern Illinois University Press, 1989.

GIDDENS, A., *Runaway World: How Globalization is Reshaping our Lives*, 2nd Edition, London: Profile Books, 2002.

– *The Politics of Climate Change*, Cambridge, UK: Polity Press, 2009.

– "Risk Society, the Context of British Politics," in Franklin, J. (ed.) *The Politics of Risk Society,* Oxford: Polity Press, 1998.

– *Beyond Left and Right: The Future of Radical Politics*, Cambridge, UK: Polity Press, 1994.

– *Modernity and Self-Identity: Self and Society in the Late Modern Age*, Cambridge: Polity Press, 1991.

– *The Consequences of Modernity*, Stanford, California: Stanford University Press, 1990.

– *The New Rules of Sociological Method*, London: Hutchinson, 1976.

– *Sociology*, 4th edition, Cambridge: Polity Press, 2001.

– Industrialization, Ecology, and the Development of Life Politics, in W. V. D'Antonio, M. Sasaki & Y. Yonebayashi, (eds.), *Ecology, Society & The Quality of Social Life*, New Brunswick, New Jersey: Transaction Publishers, 1994.

GIDDENS, A., & PIERSON, C., *Conversations with Anthony Giddens: Making Sense of Modernity*, Stanford, California: Stanford University Press, 1998.

GILBRAITH, J. K., *The New Industrial State*, New York: New American Library, 1967.

GILLESPIE, M. A. *The Theological Origins of Modernity*, Ill, Chicago: The University of Chicago Press, 2008.

GLAZEBROOK, T., *Heidegger's Philosophy of Science*, New York: Fordham University Press, 2000.

GLENDINNING, S., *In the Name of Phenomenology*, London and New York: Routledge, 2007.

GOLEE, J. M., "Heidegger's 'From the Dark Opening…'" in Boetzkes, A & Vinegar, A. (eds.), *Heidegger and the Work of Art History*, Surrey, England: Ashgate Publishing Limited, 2014.

Gordon, J-S., Burckhart, H. & Segler, P., (Introduction), *Global Ethics and Moral Responsibility: Hans Jonas and his Critics,* Gordon, J-S., Burckhart, H. (eds.), Surrey, England: Ashgate Publishing Limited, 2014.

Gore, Al., *Earth in the Balance: Ecology and the Human Spirit,* Boston: Houghton Mifflin Company, 2000 (1992).

– *An Inconvenient Truth: The Crisis of Global Warming,* New York: Viking, 2006.

Gorner, P., *Heidegger's Being and Time An Introduction,* Cambridge: Cambridge University Press, 2007.

Graeme, N., *Illustrations of Being: Drawing upon Heidegger and upon Metaphysics,* London: Humanities Press International, Inc., 1992.

Graham, S., *Deliberative Democracy and the Environment*, London: Routledge, 2003.

Gray, J., *False Dawn: The Delusion of Global Capitalism,* London: Granta, 1998.

Greaves, T., *Starting with Heidegger*, London: Continuum International Publishing Group, 2010.

Grondin, J., *Introduction to Philosophical Hermeneutics*, New Haven: Yale University Press, 1991.

– *Le tournant dans la pensée de Martin Heidegger*, Paris: Presses Universitaires de France, 1987.

– *Sources of Hermeneutics*, Albany: University of New York Press, 1995.

Guardini, R., *The End of the Modern World*, Chicago: Henry Regnery Co. 1956; Delaware: ISI Books, Revised edition, 1998.

– *The Essential Guardini: An Anthology of the Writings of Romano Guardini*, selected and with an Introduction by Heinz R. Kuehn, Chicago: Liturgy Training Publications, 1997.

– *The Letters from Lake Como: Explorations in technology and Human Race*, trans. Geoffrey W. Bromiley, Grand Rapids, Mich.: Eerdmans Pub. Co., 1994. Originally published in German as Romano Guardini, *Die Technik und der Mensch: Briefe vom Comer See*, mit einem Vorwort von Walter Dirks, Mainz: Matthias-Grünewald-Verlag, (1927), 1981.

Gueroult, M., *Descartes' Philosophy Interpreted According to the Order of Reason, vol. 1: The Soul and God,* trans. Roger Ariew, Minneapolis: University of Minnesota Press, 1984. Originally published as *Descartes selon l'ordre des raisons*, vol. 1, l'âme et Dieu, 1952, 1968.

Guignon, C., *Heidegger and the Problem of Knowledge*, Indianapolis: Hackett Publishing Co., 1983.

– "The Twofold Task: Heidegger's Foundational Historicism" *in "Being and Time", Tulane Studies in Philosophy: The Thought of Martin Heidegger*, vol. 32, Zimmerman, Michael, E., ed., New Orleans: Tulane University Press, 1984.

– "The History of Being," in *A Companion to Heidegger*, (ed.), Dreyfus and Wrathall, Oxford, UK: Blackwell Publishing Co., 2005.

– "Authenticity and the Question of Being," in McManus, D., ed., *Heidegger, Authenticity and the Self: Themes From Division Two of Being and Time*, New York: Routledge, 2015.

Guilead, R., *Être et Liberté: Une Etude sur le Dernier Heidegger*, Louvain: Nauwelaerts, 1965.

Gutmann, A, & Thompson, D., *Why Deliberative Democracy?* Princeton: Princeton University Press, 2004.

Gutting, G., *French Philosophy in the Twentieth Century*, Cambridge: Cambridge University Press, 2001.

Haas, M. P., *Saving the Mediterranean: The Politics of International Environmental Cooperation*, New York, Oxford: Columbia University Press, 1990.

Habermas, J., *The Philosophical Discourse of Modernity: Twelve Lectures*, Cambridge, Mass.: MIT Press, 1987.

– *The Postnational Constellation: Political Essays*, trans. by Max Pensky, Cambridge & Oxford: Polity Press and Blackwell Publishers, 2001 (1998).

– *The Structural Transformation of the Public Sphere*, Cambridge, Mass: MIT Press, 1989, 1991.

– *Theory and Practice,* Boston: Beacon Press, 1973.

– *Toward a Rational Society: Student Protest, Science and Politics*, Boston: Beacon Press, 1971.

– "Questions and Counterquestions" in *Habermas and Modernity*, Bernstein, R. J., ed., Cambridge, Mass.: MIT Press, 1985.

– "Modernity—An Incomplete Project," in Hal Foster, *The Anti-Aesthetic: Essays on Postmodern Culture*, New York: The New Press, 2002, (3–15).

– "Technology and Science as Ideology," in *Towards a Rational Society*, trans. J. Shapiro, Boston: Beacon Press, 1970.

– *Theory of Communicative Action, Vol. I*, Cambridge, Mass.: MIT Press, 1983.

Halliburton, D., *Poetic Thinking: An Approach to Heidegger*, Chicago: University of Chicago Press, 1981.

Hammond, M., Howarth, J. & Keat, R., *Understanding Phenomenology*, Oxford: Basil Backwell, 1991.

Hand, S., *Emmanuel Levinas*, New York: Routledge, 2009.

Hanks, C., ed. *Technology and Values (Essential Readings),* West Sussex, UK: Wiley-Blackwell Edition, 2010.

Hans Urs von Balthasar, *Romano Guardini: Reform from the Source*, New Edition, San Francisco: A Communion Book, Ignatius Press, 2010. German Original: *Romano Guardini: Reform aus dem Ursprung,* Einsiedeln, Freiburg: Johannes Verlag, 1995.

Hans, J., "Toward a Philosophy of Technology," in *Technology and Values: Essential Readings*, ed. Craig Hanks, West Sussex, UK: Wiley-Blackwell Edition, 2010.

– *Philosophical Essays: From Ancient Creed to Technological Man*, Chicago: University of Chicago Press, 1980; Atropos Press, 2010.

– *Mortality and Morality: A Search for the Good after Auschwitz,* edited with an introduction by Lawrence Vogel, Evanston, Ill.: Northwestern University Press, 1996.

– "Zur ontologischen Grundlegung einer Zukunftsethik", in *Philosophischen Untersuchungen und metaphysische Vermutungen*, Frankfurt am Main: Insel Verlag, 1992.

– *The Imperative of Responsibility: In Search of an Ethics for the Technological Age*, Chicago: The University of Chicago Press, 1984.

– *On Faith, Reason and Responsibility*, Claremont, California: The Institute for Antiquity and Christianity Studies, 1981.

– *Das Prinzip Verantwortung: Versuch einer Ethikfür die technologische Zivilisation*, Frankurt am Main: Insel Verlag, 1979.

– *The Phenomenology of Life: Toward a Philosophical Biology*, Chicago & London: The University of Chicago Press, 1966.

– *Gnosis und spätantiker Geist I–II* (1934 and 1954), Göttingen: Vandenhoeck & Ruprecht, 1964–1988, 1993. An English abridgment of the two volumes, under the title *The Gnostic Religion: The Message of the Alien God and the Beginnings of Chiristianity,* Boston: Beacon Press, 1958.

Hans, W. C., *Heidegger and the Roots of Existential Therapy*, London: Continuum, 2002.

Harman, G., *The Quadruple Object*, Alresford, UK: Zero Books, 2011.

– *Tool-being: Heidegger and the Metaphysics of Objects*. Peru, Illinois: Open Court Publishing, 2011.

Haynes, J. D., *Perspectival Thinking For Inquiring Organizations,* Santa Rosa, California: Informing Science Press, 2000 (2007).

Hegel, G.W. F., *Phenomenology of Mind*, trans. J.B. Baillie, New York: Dover Publications, Inc., 2003.

Herman, P., *Heidegger's Philosophy of Being: A Critical Interpretation*, Princeton, New Jersey: Princeton University Press, 1998.

Hickman, A. L., *Philosophical Tools for Technological Culutre: Putting Pragmatism to Work.* Bloomington and Indianapolis: Indiana University Press, 2001.

Hoedemaker, B., *Secularisation and Mission*, Harrisburg: Trinity Press International, 1997.

Hofmeyr, B., (ed.), *Radical Passivity: Rethinking Ethical Agency in Levinas*, Springer, 2009.

Hölderlin, F., *Poems and Fragments*, trans. Hamburger, M., Ann Arbor: The University of Michigan Press, 1966.

Holzer, B., *Transnational Subpolitics and Corporate Discourse: a Study of Environmental Protest and the Royal Dutch/Shell group*, Department of Sociology, London: London School of Economics, 2002.

Hordern, W., *Speaking of God: The Nature and Purpose of Theological Language*, New York: The Macmillian and co., 1964.

Hottois, G. & Pinsart, M-G., *Hans Jonas: Nature et Responsabilité*, Paris: Vrin, 1993.

– *Aux Fondements d' une Èthique Contemporaine: H. Jonas et H.T. Engelhardt en Perspective*, Paris: J. Vrin, 1993.

– *La Philosophie des technosciences*, Abidjan: Presses des Universités de Cote d'Ivoire, 1997.

– *La Science entre valeurs modernes et postmodernité*, Paris: Vrin, 2005.

– *Philosophies des sciences, philosophies des techniques, sous l'égide du college de France*, Paris: Odile Jacob, 2004.

Hoy Couzens, D., "Heidegger and the Hermeneutic Turn," in Guignon B. Charles, ed., *The Cambridge Companion to Heidegger*, Cambridge: Cambridge University Press, 1993.

Hughes, T. P., *Human-built World: How to Think about Technology and Culture*, Chicago: University of Chicago Press, 2004.

Hunter, J. D., "What is Modernity? Historical Roots and Contemporary Features" in P. Sampson, V. Samuel & C. Sugden, *Faith and Modernity*, Oxford: Regnum Books, 1994.

Husserl, E., *Ideas Pertaining to a Pure Phenomenology and a Phenomenological Philosophy*, trans. F. Kersten, The Hague: Nijhoff, (1982) 1913.

– *Logical Investigations*, Ed. Dermot Moran, 2nd ed. 2 vols, London: Routledge, [1900/1901] 2001.

– *Psychological and Transcendental Phenomenology and the Confrontation with Heidegger* (1927–1931) Dordrecht: Kluwer, 1997.

– *Cartesian Meditations: An Introduction to Phenomenology*, Dordrecht: Kluwer Academic Publishers, 1995.

Ihde, D., *Heidegger's Technologies: Postphenomenological Perspectives*, New York: Fordham University Press, 2010.

– *Listening and Voice: Phenomenologies of Sound*, 2nd Edition, Albany, New York: State University of New York Press, 2007.

– *Philosophy of Technology: An Introduction,* New York: Paragon Press, 1993.

– *Technology and the Lifeworld: From Garden to Earth*, Bloomington: Indiana University Press, 1990.

– *Existential Technics*, Albany: SUNY Press, 1983.

– *Technics and Praxis*, Dordrecht: Reidel, 1979.

Inwood, M., *Heidegger, A Very Short Introduction*, Oxford: Oxford University Press, 2000.

Iroegbu, O. P., *Metaphysics the Kpim of Philosophy,* Owerri, Nig.: International Universities Press, Ltd. 1995.

– *Enwisdomization and African Philosophy*, Owerri, Nig: International Universities Press, 1994.

– *Kpim of Time – Eternity*, Ibadan: Hope Publications Ltd., 2004.

Isaac, N., *Mathematical Principle of Natural Philosophy and His System of the World,* trans. Andrew Motte, 1729; revised translation, Florian Cajori, Berkeley: University of California Press, 1946.

Jameson, F., *A Singular Modernity: Essays on the Ontology of the Present,* London: Verso, 2002.

Jamieson, D., (ed.), *A Companion to Environmental Philosophy*, Malden, MA: Blackwell Publishing, 2003.

Janicaud, D., 'Overcoming Metaphysics?' in Dominique Janicaud and Jean-Francois Mattéi, *Heidegger: From Metaphysics to Thought*, trans. Micheal Gendre, New York: Sunny, 1995.

– *Heidegger in France*, Paris: Albin Michel, 2000.

Jay Lifton, R., *Death in Life: Survivors of Hiroshima*, New York: Random House, 1967.

John, H. Van Engen, *Sisters and Brothers of the Common Life: The Devotio Moderna and the World of the Later Middle Ages*, Philadelphia, Pennsylvania: University of Pennsylvania Press, 2008.

Jordan, G. S., *Greenpeace and the Brent Spar*, Basingstoke, England: Palgrave, 2001.

Kant, I., *Kritik der praktischen Vernunft. Die Kritiken*, second edition, (first edition 1974), Frankfurt am Main: Suhrkamp, 1996.

Kaufmann, P., *L'Experience emotionnelle de l'espace,* Paris: Vrin, 1967.

Keith, S., *Environmental Hazards: Assessing Risk and Reducing Disaster*, 3rd Edition, New York: Routledge, 2001.

Kellner, D., *Critical Theory, Marxism and Modernity*, Baltimore, Md.: John Hopkins University Press, 1989.

KEMP, P., "Ricœur Between Heidegger and Levinas: Original Affirmation between Ontological Contestation and Ethical Injunction," in Kearney, R., (ed.), *Paul Ricœur: The Hermeneutics of Action*, London: Sage, 1996.

Kettering, E., *NÄHE: Das Denken Martin Heidegger*, Pfullingen: Günther Neske, 1987.

Kimaid, M., *Modernity, Metatheory, and the Temporal-Spatial Divide: From Mythos to Techne*, New York: Routledge, 2015.

Kisiel, T., *Heidegger's Way of Thought*, London: Continuum Books, 2002.

– *The Genesis of Heidegger's Being and Time*, Los Angeles: University of California Press, 1993.

Kockelmans, J. J, *On the Truth of Being: Reflection on Heidegger's Later Philosophy*, Bloomington: Indiana University Press, 1984.

– "Ontological Difference, Hermeneutics, and Language," In *On Heidegger and Language*, Evanston: North-western University Press, (1972), 1980.

– *Edmund Husserl's Phenomenology*, West Lafayette, Indiana: Purdue University Press, 1994.

– *Heidegger on Art and Artworks*, Dordrecht and Boston: Martinus Nijhoff, 1985.

– *Martin Heidegger, A First Introduction to his Philosophy*, Pittsburgh, Pa: Duquesne University Press, 1965.

– *Heidegger and Science*, Washington DC: University Press of America, 1985.

Kort, M., *The Columbia guide to Hiroshima and the bomb*, New York: Columbia University Press, 2007.

Kovacs, G., *The Question of God in Heidegger's Phenomenology*, Evanston, Illinois: Northwestern University Press, 1990.

Kumar, K., *From Post-industrial to Post-modern Society. New Theories of the Contemporary World*, Oxford/Cambridge: Blackwell, 1995.

Ladriere, J., *L'éthique dans l'univers de la rationalité*, Namur: Artel/Fides, 1997.

– *Les Enjeux de la rationalité*, Paris: Aubier-Montaigne, 1977.

Langan, T. *The Meaning of Heidegger*, New York, 1959.

Langdon, W., *The Whale and the Reactor: A Search for Limits in an Age of High Technology*, Chicago: University of Chicago Press, 1986.

Lash, S. "Forward: Individualization in a Non-Linear Mode" in Ulrich Beck & Elisabeth Beck-Gernsheim, *Individualization*, Sage Publication, 2003.

– “Reflexivity and its Doubles: Structure, Aesthetics, Community” in *Reflexive Modernization: Politics, Tradition and Aesthetics in the Modern Social Order,* Cambridge: Polity Press, 1994.

Latour, B., “Technology is Society made Durable”, in Law John. (ed.), A *Sociology of Monsters: Essays on Power, Technology and Domination*, London: Routledge, 1991.

– *Politics of Nature: How to Bring the Sciences into Democracy*, trans., Catherine Porter, Cambridge, Mass.: Harvard University Press, 2004, originally published as *Politique de la Nature*, Paris: Editions La Découverte, 1999.

– *Reassembling the Social: An Introduction to Actor-Network-Theory,* Oxford: Oxford University Press, 2005.

– *Science in Action*, Cambridge: Harvard University Press, 1987.

– *Science in Action: How to Follow Scientists and Engineers through Society*, Cambridge: Harvard University Press, 1987.

– *We Have Never Been Modern*, London: Harvester Wheatsheaf, 1993.

Leach, M., Scoones, I. & Wynne, B., (eds.) *Science and Citizens: Globalization and the Challenge of Engagement,* London: Zed Books, 2005.

Leighton, D., ‘Searching for Politics in an Uncertain World: Interview with Zygmunt Bauman’, *Archive*, 10, 1, 2001.

Lenoble, J. & Maesschalck, M., *Towards a Theory of Governance: The Action of Norms*, trans. John Paterson, The Hague: Kluwer Law International, 2003.

Leroy, P., Tatenhove, J. van., “Political Modernization Theory and Environmental Politics,” in Von Gert Spaargaren, Arthur. P. J. Mol, Frederick H. Buttel, (eds.), *Environment and Global Modernity*, London: Sage Publications, 2000.

Lescoe, J. F. *Existentialism: With or Without God*, New York: Alba House, 1972.

Leslie, P. Thiele, *Timely Meditations: Martin Heidegger and Postmodern Politics,* Princeton, New Jersey: Princeton University Press, 1995.

Levinas, E. & Kearney, R., “Dialogue with Emmanuel Levinas” in *Face to Face with Levinas*, Cohen, A. R., ed. New York: State University of New York Press, 1986.

Levinas, E. “Ideology and Idealism” in *Modern Jewish Ethics*, ed. Marvin Fox, Columbus: Ohio State University Press, 1975.

– *On Escape De l’invasion*, trans. Berttina Bergo, Stanford, Calif: Stanford University Press, 2003.

– “Is Ontology Fundamental?” *Emmanuel Levinas: Basic Philosophical Writings*. Eds. Adriaan T. Peperzak, Simon Critchley, and Robert Bernasconi, Bloomington: Indiana University Press, 1996.

– “Phenomenology and Enigma,” in *Collected Philosophical Papers*, trans. Lingis Alphonso, Dordrecht, The Netherland: Martinus Nijhoff, 1987.

– “Substitution,” in Peperzak, A., Critchley S., and Bernasconi R., (eds.) *Basic Philosophical Writings*, Bloomington, Illinois: Indiana University Press, 1996.

– *De Dieu qui vient a l'idee*, Paris: J. Vrin, 1982.

– *Discovering Existence with Husserl*, trans. and eds. Richard A. Cohen and Michael B. Smith, Evanston: Northwestern University Press, 1998.

– *En decouvrant l'existence avec Husserl et Heidegger*, Paris: J. Vrin, 1982.

– *Entre Nous: On Thinking-of-the-Other*, trans. Michael B. Smith and Barbara Harshav, New York: Columbia University Press, 1998.

– Ethics *and Infinity: Coversations with Philippe Nemo*, trans. Richard A. Cohen, Pittsburg: Duquesne University Press, 1985.

– *Existence and Existents*, trans. Alphonso Lingis, Pittsburgh: Duquesne University Press, 2001.

– *Otherwise than Being or Beyond Essence*, trans. Alphonso Lingis, Pittsburg: Duquesne University Press, 2000.

– *The Levinas Reader*, Sean Hand, (ed.) Cambridge, Mass: Basil Blackwell, 1989.

– *The Theory of Intuition in Husserl's Phenomenology*, second edition, trans. André Orianne, Evanston, Illinois: Northwestern University Press, 1995.

– *Time and the Other*, trans Richard A. Cohen, Pitsburgh, PA: Duquesne Unversity Press, 1987.

– *Theory of Intuition in Husserl's Phenomenology*, 2nd Edition, Evanson, IL, USA: Northwestern University Press, 1995.

– *Totalité et infini, essai sur l' extériorité*, Paris: Le Livre de Poche/Kluwer, Edition 1994.

– *Totality and Infinity: An Essay on Exteriority*, trans. Alphonso Lingis, Pittsburg, Pennsylvania: Duquesne University Press, 1969. First published in French under the title *Totalite et Infini,* The Hague, Netherlands: Martinus Nijhoff, 1961.

Levy, J. D., *Hans Jonas: The Integrity of Thinking*, Columbia & London: University of Missouri Press, 2002.

Lewis, M. & Steahler, T., *Phenomenology: An Introduction*, London: Continnum International Publishing Group, 2010.

Lewis, R. in Ronald, T. Takaki, *Hiroshima: Why America Dropped the Atomic Bomb*, New York, Boston: Little, Brown and Company, 1995.

Lingis, A., Translator's introduction to *Otherwise than Being or Beyond Essence,* Pittsburgh: Duquesne University Press, 1997.

LLEWELYN, J., *Emmanuel Levinas: The Genealogy of Ethics,* New York: Routledge, 1995.

LORI GRUEN, Technology, in Craig Hanks, ed. T*echnology and Values (Essential Readings),* West Sussex, UK: Wiley-Blackwell Edition, 2010.

LOSCERBO, J., *Being and Technology: A Study in the Philosophy of Martin Heidegger*, The Hague: Martinus Nijhoff Publishers, 1981.

LOVITT, W. & LOVITT, H. BRUNDAGE, *Modern Technology in the Heideggerian Perspective: Problems in Contemporary Philosophy,* Vol. I, Lewiston, New York: The Edwin Mellen Press, 1995.

LOY, M. VAIL, *Heidegger and Ontological Difference*, University Park, Pa: The Pennsylvania State University Press, 1972.

LUHMANN, N., *Risk: A Sociological Theory*, New York: Aldine De Gruyter 1993.

LYOTARD, J-F, *Phenomenology,* trans. Brian Beakley, Albany, USA: State University of New York Press, 1991.

MACANN, E. CHRISTOPHER, *Four Phenomenological Philosophers: Husserl, Heidegger, Sartre, Merleau-Ponty*, London: Routledge, 1993 (Taylor and Francis edition, 2005).

MAESSCHALCK, M., *Transformations de l'Éthique de la phénoménologie radicale au pragmatisme social*, Bruxelles: P.I.E. Peter Lang, 2010.

– *Normes et Contextes. Les Fondements d'une Pragmatique Contextuelle*, Zürich/ New York: Olms, Hildescheim, 2001.

– *Le Principe d'Autonomie: Introduction aux auteurs modernes*, Louvain-la-Neuve: Éditions de l'insitut supérieur de philosophie, 1992.

– *raison et pouvoir: les impasses de la pensée politique postmoderne*, Bruxelles: Facultés universitaires Saint-Louis, 1992.

MAGDA, K., *A Guide to Heidegger's Being and Time*, edited by John Llewelyn, New York: State University of New York, 2001.

MALPAS, J., *Heidegger's Topology: Being, Place, World,* Cambridge, MA: MIT Press, 2006.

MALPAS, J., *Heidegger and the Thinking of Place: Explorations in the Topology of Being*, Cambridge, MA: The MIT Press, 2012.

MARCUSE, H., One-*Dimensional Man,* Boston: Beacon Press, 1964.

– *Technology, War and Fascism, Collected Papers of Herbert Marcuse*, Vol. 1, ed. Douglas Keller, London and New York: Routledge, 1998.

– The New Forms of Control, in *Technology and Values: (Essential Readings)*, C., Hanks, (ed.), West Sussex, UK: Wiley-Blackwell Edition, 2010.

Marion, J-L. (ed. & trans.), *Règles utiles et claires pour la direction de l'esprit*, The Hague : Nijhoff, 1977.

–"Heidegger and Descartes" in *Critical Heidegger*, ed. Christopher Macann, London & New York: Routledge, 1996.

– *On Descartes Metaphysical Prism*, trans. Jeffery L. Kosky, Chicago & London: University of Chicago Press, 1999. Originally published as *Sur le prisme métaphysique de Descartes,* Presses Universitaires de France, 1986.

– "On Descartes' Constitution of Metaphysics", in Tom Sorell, ed. *Descartes,* Aldershot, England: Dartmouth Pub. Comp., 1999.

Mark, J., "Ecology and the Urban Aesthetic" in *Ecological Design and Planning,* eds., George F. Thompson and Frederick R.Steiner, New York: John Wiley & Sons Inc., 1997.

Maurice de Gandillac, *Le siècle traverse: Souvenirs de neuf décennies,* Paris: Albin Michel, 1998.

McClellan III, J. E. & Dorn, H., *Science and Technology in World History: An-Introduction*, 2nd Edition, Baltimore, Maryland: The John Hopkins University Press, 2006.

Mckechnie, B. R. & Welsh, I., "When the Global Meets the Local: Critical Reflections on Reflexive Modernization" in E. D. Von Riley, F. H. Buttel, P. Dickens, & A. Gijswijt, (eds.) *Sociological Theory and the Enviromental: Classical Foundations, Contemporary Insights,* Maryland, USA: Rowman & Littlefield Publishers, Inc., 2002.

Mehta, J. Lal, *Martin Heidegger: The Way and the Vision*, Honolulu: University of Hawaii Press, 1976.

Micheletti, M., *Political Virtue and Shopping: Individuals, Consumerism, and Collective Action*, New York: Palgrave Macmillan, 2003.

Micheletti, M., Follesdal, A. & Stolle, D., (eds.), *Politics, Products, and Market: Exploring Political Consumerism Past and Present,* New Brunswick, New Jersey: Transaction Publishers, 2004.

Misa, T. J., "The Compelling Tangle of Modernity and Technology" in *Modernity and Technology*, J. T. Misa, P. Brey, & A. Feenberg, (eds.), Cambridge: The MIT Press, 2003.

Misa, T. J., Brey, P & A. Feenberg, (eds.), *Modernity and Technology*, Cambridge, Massachusetts: The IMT Press, 2003.

Mitchell, A. J., Heidegger among the Sculptors: Body, Space, and the Art of Dwelling, Stanford, California: Standford University Press, 2010.

Moran, D., "The Destruction of the Destruction: Heidegger's Versions of the History of Philosophy" in *Martin Heidegger: Politics, Art and Technology*, eds. Karsten Harries and Christopher Jamme, New York: Homes and Meier, 1994.

– *Introduction to Phenomenology*, London: Routledge, 2000.

Mortley, R., *French Philosophers in Conversation*, New York: Routledge, 1991.

Morton, E. Winston & Edelback, D. Ralph, *Society, Ethics, and Technology*, 4th Edition, Introd. "Children of Invention Revisited" by Morton Winston, Boston, MA: Wadsworth Publishing, 2011.

Mouffe, C., *On The Political*, London: Routledge, 2005.

Mugerauer, R., *Heidegger and Homecoming: The Leitmotif of the Later Writings*, Toronto: University of Toronto Press, 2008.

Mulhall, S., "Human Mortality: Heidegger on How to Portray the Impossible Possibility of Dasein" in *A Companion to Heidegger*eds. Hubert L. Dreyfus and Mark A. Wrathall, Oxford, England: Blackwell Publishing Ltd., 2007.

– *Heidegger and Being and Time*, second Edition, London and New York: Routledge, 2005.

– *Inheritance and Originality*, Oxford: Oxford University Press, 2001.

– *The Routledge Guidebook to Heidegger's Being and Time*, New York: Routledge, 2013.

Mumford, L., *The Myth of the Machine: Technics and Human Development*, New York: Harcourt Brace Jovanovich, 1967.

Neske, G. & Kettering, E., (eds.), *Martin Heidegger and National Socialism*, New York: Paragon House, 1990.

Norman, G., *Systematic Theology, Introduction, Bible*, Vol. 1, Minneapolis: Bethany House, 2002.

Norris, C., *Reclaiming Truth: Contribution to a Critique of Cultural Relativism*, London: Lawrence & Wishart, 1996.

Nuyen, A. T., "Altruism as the Condition of Subjectivity" in Kapur Basant, Kim Chong Chong, (eds.) *Altruistic Reveries: Perspectives from the Humanities and Social Sciences*, Dordrecht, The Netherlands: Kluwer Academic Publishers, 2002.

Oberst, J. *Heidegger on Language and Death: The Intrinsic Connection in Human Existence*, London and NewYork: Continuum International Publishing Group, 2009.

Ogura, T., *Letters from the end of the world: a firsthand account of the bombing of Hiroshima*, New York: Kodansha International, 1997.

Okoro, J. I., *The Earth as a Living Superorganism: From Scientific Gaia (Hypothesis) to the Metaphysics of Nature*, Frankfurt am Main: Peter Lang, 2005.

Ott, Hugo, *Martin Heidegger: unterwegs zur seiner Biographie*, Frankfurt a. M. & New York: Campus Verlag, 1988.

Outhwaite, W., *Habermas: A Critical Introduction*, 2nd Edition, Cambridge: Polity Press, 2009.

Pacey, A., *The Culture of Technology*, Cambridge, MA: MIT Press, 1983.

Palmer, E. R., *Hermeneutics*, Evaston, IL: Northwestern University Press, 1969.

Partenie, C & Rockmore, T. (eds.), *Heidegger and Plato: Toward Dialogue*, Evanston, Illinois: Northwestern University Press, 2005.

Pattison, G., *The Later Heidegger*, London: Routledge Publication, 2000.

– *Thinking About God in an Age of Technology*, Oxford: Oxford University Press, 2005.

Peperzak, T. A., *To the Other: An Introduction to the Philosophy of Emmanuel Levinas*, West Lafayette, Indiana: Purdue University Presss, 1993.

– *Beyond: The Philosophy of Emmanuel Levinas*, Evanston, Illinois: Northwestern University Press, 1997.

Pilot, R., *Heidegger An Introduction*, Ithaca, New York: Cornell University Press, 1999.

Pippin, R. B., *Idealism as Modernism (Hegelian Variation),* Cambridge: Cambridge University Press, 1997.

Pöggeler, O. *Martin Heidegger's Path of Thinking,* Atlantic Highlands, N.J.: Humanities Press, 1991.

– *Der Denkweg Martin Heidegger's*, 2nd Auflage, Pfullingen: Günther Neske, 1983.

Poirié, F., *Emmanuel Levinas, Qui êtes-vous?*, Lyons: La Manufacture, 1987.

Polt, R., "Ereignis," in *A Companion to Heidegger,* Dreyfus H. L. & Wrathall, M. A. (eds.), Oxford, England: Blackwell Publishing Ltd., 2007.

– *The Emergency of Being, On Heidegger's Contributions to Philosophy*, Ithaca and London: Cornell University Press, 2006.

Poolos, J., *The Atomic Bombing of Hiroshima and Nagasaki,* Chelsea House Pub., 2008.

Porter, R., *The Creation of the Modern World: The Untold Story of the British Enlightenment*, New York: W.W. Norton, 2000.

Prauss, G., *Erkennen und Handle in Heidegger 'Sein und Zeit'* Freiburg: Karl Alber, 1977.

Punter, A., *Modernity,* New York: Palgrave Macmillan, 2007.

Puthenpurackal, J. J., *Heidegger Through Authentic Totality to Total Authenticity,* Leuven: Leuven University Press, 1987.

Rae, G., *Ontology in Heidegger and Delueze: A Comparative Analysis*, Palgrave Macmillan, 2014.

Rahner, K., *Theological Investigation*, vol. 6, Baltimore: Helicon Press, 1969.

Raffoul, F., *The Origins of Responsibility*, Blommington, Indiana: Indiana University Press, 2010.

Remmon, E. B., *Heidegger and a New Possibility of Dwelling*, Frankfurt/M: Peter Lang, (European Universities Studies, Series xx), 2003.

Richardson, J. W., *Heidegger, Through Phenomenology to Thought,* New York. Fordham University Press, fourth edition, 2003. Originally, The Hague: Nijhoff, 1963.

Ricœur, P., *Temps et Récit* III, Paris: Seuil, 1985.

Risser, J. (ed.), *Heidegger Toward the Turn, Essays on the Work of the 1930s,* New York: State University of New York Press, 1999.

Robbins, J., (ed.), *Is it Righteous to Be? Interviews with Emmanuel Levinas*, Stanford, Calif: Stanford University Press, 2001.

Robert, J., *Philosophizing the Everyday: Revolutionary Praxis and the Fate of Cultural Theory*, Ann Arbor, MI: Pluto Press, 2006.

Rockmore, T., "Heidegger on Technology and Democracy", in Feenberg. A & Hannay, A., *Technology and the Politics of Knowledge,* Bloomington & Indiana: Indiana University Press, 1995.

Rojcewicz, R., *The Gods and Technology: A Reading of Heidegger*, Albany: State University of New York Press, 2006.

Rorty, R., *Contingency, Irony and Solidarity*, Cambridge, UK: Cambridge University Press, 1989.

Rudd, A., *Expressing the World, Skepticism, Wittgenstein, and Heidegger*, Chicago: Open Court Publishing Company, 2003.

Rychter, E., *(Un)Saying the Other Allegory and Irony in Emmanuel Levinas's Ethical Language*, New York: Peter lang Publishing, 2004.

Ryle, G., "Heidegger's Sein und Zeit", in *Heidegger and Modern Philosophy: Critical Essays*, Murray, M., ed., London: Yale University Press, 1978.

Sallis, J., *Delimitations: Phenomenology and the End of Metaphysics*, Bloomington: Indiana University Press, 1986.

– *The Concept of the World: A Study in the Phenomenological Ontology of Martin Heidegger*, Michigan: University Microfilms International, 1977.

Schmidt, L. K., *Understanding Hermeneutics*, Stocksfield, UK: Acumen Publishing Ltd., 2006.

Schroeder, W. R., *Sartre and His Predecessors: The Self and the Other*, London: Routledge and Kegan Paul, 1984.

Schürmann, R., "Heidegger and Meister Eckhart on Releasement" in Dreyfus, H. & Wrathall, M., (eds.), *Heidegger Reexamined,* New York and London: Routledge, 2003.

– *Wandering Joy: Meister Eckhart's Mystical Philosophy,* Massachussets: Lindisfarne Books, 2001.

Scott, E. Charles, Care and Authenticity, in Davis, W. Bret, (ed.) *Martin Heidegger: Key Concepts*, Durham, UK: Acumen Publishing Limited, 2010.

Sebbah, F-D., *Lévinas et le Contemporain: Les Préoccupations de l'heure*, Besançon: Les Solitaires Intempestifs, Éditions, 2009.

Secada, J., *Cartesian Metaphysics: The Late Scholastic Origins of Modern Philosophy*, Oxford: Oxford University Press, 2000.

Seidel, J.G., *Martin Heidegger and the Pre-Socrates*, Lincoln: University of Nebraska Press, 1976.

Sheehan, T., Facticity and Ereignis, in Dahlstrom, D. O., ed., *Interpreting Heidegger: Critical Essays*, Cambridge: Cambridge University Press, 2011.

– "Kehre and Ereignis: A Prolegomenon to Introduction to Metaphysics," in Polt, R. & Fried, G., (eds.), *A Companion to Heidegger's Introduction to Metaphysics*, New Haven & London: Yale University Press, 2000.

Sheffler Manning, R. J., *Interpreting Otherwise Than Heidegger: Emmanuel Levinas's Ethics as First Philosophy*, Pennsylvania: Duquesne University Press, 1993.

Sheila, J., "Judgment Under Siege: The Three-Body Problem of Expert Legitimacy" *in* Maasen S. & Weingart, P., (eds.), *Democratization of Expertise? Exploring Novel Forms of Scientific Advice in Political Decision-Making,* Dordrecht, The Netherlands: Springer, 2005.

Shen, T. Thomas, *Industrial Pollution Prevention*, 2nd Edition, Berlin: Springer-Verlag, 1999.

Shepherd, A., *The Gift of the Other: Levinas, Derrida, and a Theology of Hospitality,* Cambridge, UK: James Clerke & Co, 2014.

Shildrick, M., "Ethics at the Limits: Levinas, Politics and Responsibility," in Robin May Schott & Kirsten Klercke, (eds.) Philosophy *on the Border,* Copenhagen: Museum Tusculanum Press, 2007.

Simondon, G., *Du Moded' existence des objets techniques,* Méot, 1958; second edition, Paris: Aubier, 1989.

– *L'individu et sa Genèse Physico-Biologique,* Paris: Presses Universitaires de France, 1964.

– *L'individuation Psychique et Collective*, Paris: Aubier, (1989) 2007.

Smith, B. G., *Nietzsche, Heidegger and the Transition to Modernity*, Chicago: University of Chicago Press, 1996.

Sonja, B-C., "The Precautionary Principle in German-Enabling Government," in T. O.'Riordan & J. Cameron (eds), *Interpreting the Precautionary Principle,* London: Earthsea, 1994.

Sørensen, P. Mads, "Den politiske forbruger som subpolitiker", in Fenger-Gron & Kristensen J. E., (eds) *Kritik af den okonomiske fornuft,* Copenhagen: Has Reitzels Forlag, 2001.

Sørensenm, P. Mads & Christiansen, A., *Ulrich Beck: An Introduction to the Theory of Second Modernity and the Risk Society,* New York: Routledge, 2013.

Spaemann, R.,*Grenszen. Zur ethischen Dimension des Handelns,* Stuttgart: Klett-Cotta, 2001.

Spiegelberg, H., *The Phenomenological Movement: A Historical Introduction*, 3rd ed., The Hague: Martinus Nijhoff, 1982.

Stambaugh, J., *The Finitude of Being*, Albany: State University of New York Press, 1992.

Stapleton, T., "Dasein as being–in-the-world" in Bret, W. D. (ed.), *Martin Heidegger: key Concepts*, Durham, UK: Acumen Publishing Limited, 2010.

Stiegler, B., *Technics and Time, 1: The Fault of Epimetheus*, trans. Beardsworth, R., & Collins, G., Stanford, California: Stanford University Press, 1998.

Stirling, A., "Precaution, Foresight and Sustainability: Reflection and Reflexivity in the Governance of Science and Technology," in *Reflexive Governance for Sustainable Development*, Cheltenham, UK: Edward Elgar Publishing Ltd., 2006.

Stirner, M., *The Ego and Its Own*, Cambridge: Cambridge University Press, 1995.

Strhan, A., *Levinas, Subjectivity, Education: Towards an Ethics of Radical Responsibility*, Chichester, UK: Blackwell Publishing Limited, 2012.

Taminiaux, J. "The Presence of *Being and Time* in *Totality and Infinity*", In Hansel J., (ed.), *Levinas in Jerusalem: Phenmenology, Ethics, Politics, Aesthetics,* (Amsterdam Studies in Jewish Philosophy), Springer, 2009.

Tatenhove, J.V., Arts, B. & Leroy, P., (eds.), *Political Modernization and the Environment: The Renewal of Environmental Policy Arrangement,* Dordrecht, The Netherlands: Kluwer Academic Publishers, 2000.

Taylor, C., "Heidegger on Language," in Dreyfus, H. & Wrathall M., (eds.), *A Companion to Heidegger*, Malden, MA, USA: Blackwell Publishing Ltd., 2005.

Tchobanoglous, G. & Kreith, F., (eds.) *Handbook of Solid Waste Management*, 2nd Edition, Introduction, New York: McGraw-Hill Companies Inc., 2002.

THEIS, R., *Jonas: Habiter le Monde*, Paris: Michalon, 2008.

TOULMIN, S., *Cosmopolis: The Hidden Agenda of Modernity*, Chicago: The University of Chicago Press, 1990.

TOURAINE, A., *Critique of Modernity*, Oxford: Blackwell, 1995.

TURNER, D., "Humanity as Shepherd of Being: Heidegger's Philosophy and the Animal Other," in L. MCWHORTER & G. STENSTAD, (eds.), *Heidegger & the Earth: Essays in Environmental Philosophy*, 2nd Edition, Toronto: University of Toronto Press Inc., 2009.

ULRIKE, F. & WYNNE, B., *Taking European Knowledge Society Seriously: Report of the Expert Group on Science and Governance to the Science, Economy and Society Directorate, Directorate-General for Research, European Commission*, 2007.

VAN BUREN, J., "Martin Heidegger, Martin Luther" in T. KISIEL, & J. VAN BUREN, (eds.), *Reading Heidegger from the Start: Essays in is Earliest Thought*, New York: Sunny, 1994.

VAN LOON, J., *Risk and Technological Culture, Towards a Sociology of Virulence*, London: Routledge, 2002.

VANDENBERGHE, F., *A Philosophical History of German Sociology*, trans. Carolyn Shread, USA & Canada: Routledge, 2009.

VANNIER, G., *L'existentialisme: Littérature et philosophie*, Paris: L'Harmattan, 2001.

VENSUS, A. G., *Authentic Human Destiny: The Paths of Shankara and Heidegger*, Washington, D.C.: C.R.V.P. Publication, 1988.

– *From Being-in-the-World to Being –towards- Being: Martin Heidegger's Philosophy of the Authentic Human Person*, Nagpur: SAC Publications, 1996.

– *The Experience of Being As Goal of Human Existence: The Heideggerian Approach*, Washington, D.C: C.R.V.P. Publication, Series IIIB, South Asia, Vol. 2, 2000.

VERBEEK, PETER-PAUL, *Moralizing Technology: Understanding and Desgining the Morality of Things*, Chicago: University of Chicago Press, 2011.

– *What Things do: Philosophical Reflections on Technology, Agency, and Design*, trans. Robert P. Crease, Pennsylvania: The Pennsylvania State University Press, 2005.

VON, G. DELANTY, *Social Theory in a Changing World: Conceptions of Modernity*, Cambridge, UK: Polity Press in association with Blackwell Publishers Ltd., 1999.

VYCINAS, V., *Earth and Gods: An Introduction to the Philosophy of Martin Heidegger*, The Hague: Martinus Nijhoff, 1961.

WALZER, M., "Deliberation and What Else"? in MACEDO, S., (ed.), *Deliberative Politics: Essays on Democracy and Disagreement*, Oxford-New York: Oxford University Press, 1999.

Ward, K., "The Decline and Fall of Reason", in King, U., (ed.), *Faith and Practice in a Postmodern Age,* London: Cassell, 1998.

Wätzold, F., "When Environmentalists Have Power: The Case of the Brent Spar", in Madsen H., and Ulhoi J.P., (eds.) *Industry and the Environment: Practical Applications of Environmental Management Approaches in Business*, Gylling: Naryana Press, 1996.

Weber, M., "The Profession and Vocation of Politics" in *Max Weber: Political Writings*, (1919), ed./trans. Lassman, P. & Speirs, R., Cambridge: Cambridge University Press, 1994.

– *The Protestant Ethic and the Spirit of Capitalism,* New York: Charles Scribner's Sons, 1958.

– *Wissenschaft als Beruf 1917/1919. Politik LA Beruf 1919*, ed. Wolfgang J. Mommsen & Wolfgang Schluchter, Tübingen: Mohr, 1994. Published in English as *Politics as a Vocation*, Philadelphia: Fortress Press, 1965 (1992).

– *Roscher and Knies: The Logical Problems of Historical Economies*, (1903–06), trans. Oakes, G., New York: Free Press, 1975.

Weingart, P., "How Robust is 'Socially Robust Knowledge'"? In *The Challenge of the Social and the Pressure of Practice: Science and Values Revisited,* eds. Carrier Martin, Howard Don., & Kourany Janet, Pittsburgh: University of Pittsburgh Press, 2008.

Werner, M., *Heidegger and Tradition*, trans. Kisiel, T. & Murray G., Evanston: Northwestern University Press, 1971.

West, D., *An Introduction to Continental Philosophy*, Cambridge: Polity Press, 1996.

Whitehead, N. A., *Process and Reality*, New York: Free Press, 1979.

Willems, B., *Hopkins and Heidegger*, London & New York: Continuum International Publishing Group, 2009.

Winner, L., *Autonomous Technology: Technics-Out-Of-Control as a Theme in Political Thought*, Cambridge, Massachusetts: The MIT Press, 1978.

Wiplinger, F., *Wahrheit und Geschichtlichkeit: Eine Untersuchung über die Frage nach dem Wesen der Wahrheit im Denken Martin Heideggers*, Freiburg/München: Karl Alber, 1961.

Wolin, R., (ed.), *The Heidegger Controversy, A Critical Reader*, Cambridge, Massachusetts: MIT Press Edition 1991, 1993.

– *The Heidegger Controversy*, New York: Columbia University Press, 1991.

Wood, E. M., *The Origin of Capitalism: a longer view*, London & New York: Verso Publication, 2002.

WRATHALL, A. M., "Unconcealment," in *A Companion to Heidegger*, eds. Dreyfus, H. & Wrathall, M., Oxford, England: Blackwell Publishing Ltd., 2007 edition.

– *Heidegger and Concealment. Truth, Language and History*, New York: Cambridge University Press, 2011.

WYSCHOGROD, E., *Emmanuel Levinas: The Problem of Ethical Metaphysics*, New York: Fordham University Press, 2000.

YOUNG, J., "What is Dwelling? The homelessness of Modernity and the worlding of the world," In *Heidegger, Authenticity, and Modernity*, eds. Wrathall, M & Malpas, J., Cambridge, MA: MIT Press, 2000.

– *Heidegger, Philosophy, Nazism*, Cambridge: Cambridge University Press, 1997.

– *Heidegger's Later Philosophy*, Cambridge: Cambridge University Press, 2002.

– *Heidegger's Philosophy of Art*, Cambridge: Cambridge University Press, 2001.

ZAHAVI, D., *Husserl's Phenomenology*, Stanford: Stanford University Press, 2003.

ZIMMERMAN, M. E., "The Search for Heideggerian Ethics," in *Ethics and Responsibility in the Phenomenological Tradition*, Ninth Annual Symposium of the Simon Silverman Phenomenology Center, Duquesne, 1992.

– *Contesting Earth's Future, Radical Ecology and Postmodernism*, Berkeley and Los Angeles: University of California Press, 1994.

– *Eclipse of the Self, The Development of Heidegger's Concept of Authenticity*, 2nd ed. Anthens, Ohio: Ohio University Press, 1986.

– *Heidegger's Confrontation with Modernity: Technology, Politics, Art*, Bloomington and Indianapolis: Indiana University Press, 1990.

ZYGMUNT, B., *Liquid Life*, Cambridge: Polity Press, 2005.

– *Liquid Modernity*, Cambridge: Polity Press in association with Blackwell Pub. Ltd., 2000.

– *The Individualized Society*, Cambridge: Polity Press, 2002.

– *Identity*, Cambridge: Polity Press, 2004.

Selected Journals

ALDERMAN, H. G., Heidegger on being human. *Philosophy Today* 15, 1971, 16–29.

BARNES, B., Elusive memories of technoscience.*Perspectives on Science 13*(2), 2005, 142–165.

BECK, U., World at Risk: The New Task of Critical Theory, in *Development and Society*, *37*(1), 2008, 1–21.

– From industrial society to the risk society: Questions of survival, social structure and ecological enlightenment. *Theory, Culture and Society* 9, 1992, 97–123.

– World risk society as cosmopolitan society? Ecological questions in a framework of manufactured uncertainties. *Theory, Culture & Society 13*(4), 1996, 1–32.

– Living in the world risk society.*Economy and Society 35*(3), 2006, 329–345.

Beck, U., Bonss, W. & Lau, C., The theory of reflexive modernization: Problematic, hypotheses and research program. *Theory, Culutre and Society 20*(2), 2003, 1–33.

Benso, S., The time of the feminine: For a politics of material corporeality. *Continental Philosophy Review, 36*(2), 2003, 195–202.

Bernstein, R. J., Rethinking responsibility. *The Legacy of Hans Jonas*. Hastings Centre Report *25*(7 Spec No), 02/1995, 13–20.

Capobianco, R., The fate of being in Heidegger's Four Seminars 1966–1973. *Existentia: An International Journal of Philosophy* XV (3–4), 2005, 161–183.

Cooke, M., Five arguments for deliberative democracy. *Political Studies*, 48 (5), 2000, 947–969.

De Vries, Gerard de, What is political in sub-politics?: How Aristotle might help STS. *Social Studies of Sciences* 37 (5), 2007, 781–809.

Depre, O., "Ce dont la possibilité contient l'exigence de sa réalité". De l'être au devoir-être chez Hans Jonas. *Étude Phénoménologiques* XV11 (33–34), 2001, 111–129.

Dingwall, R., Risk society': The cult of theory and the millennium. *Social Policy and Administration 33*(4), 1999, 474–491.

Dotray, Z., "Heidegger's Unzuhandenheit as a Fourth Mode of Being," *Macalester Journal of Philosophy*, Vol. 19, Iss. 1, Article 12, 2010, 205–208.

Ellen, R., The errors of Thamus: An Analysis of Technology Critique. *Bulletin of Science, Technology, & Society, 23*(3), 2003, 147–156.

Elliot, A., Beck's sociology of risk: A critical assessment.*Sociology 35*(2), 2002, 293–315.

Ellul, J., Symbolic function, technology and society.*Journal of Social and Biological Structures 3*(1), 1978, 207–218.

Fédier, F., Traduire les Beiträge zur Philosophie (Vom Ereignis). *Heidegger Studies* vol. 9, 1993, 15–33.

Gandy, M., Rethinking the ecological Leviathan: Environmental regulation in an age of risk. *Global Environmental Change 9*(1), 1999, 59–69.

Giddens, A., Risk and responsibility. *The Modern Law Review, 62*(1), 1999, 1–10.

Godzinksi, R. Jr., (En)Framing Heidegger's Philosophy of Technology. *Essays in Philosophy, 6*(1), article 9, 2005.

Greisch, J., Identité et différence dans la pensé de Martin Heidegger. Le chemin de l'Ereignis. *Revue des sciences philosophiques et théologiques* 57, 1973, 71–111.

Haas, M. P., Banning chlorofluorocarbon: Epistemic community efforts to protect stratospheric ozone. *International Organization*, *46*(1), 1992, 187–224

– Introduction: Epistemic Communities and International Policy Co-ordination. *International Organization*, *46*(1), Knowledge, Power, and International Policy Coordination, 1992, 1–35.

Hans, L., Collective self-legislation as an *Actus Impurus*: a response to Heidegger's critique of European nihilism. *Continental Philosophy Review 41*(3), 2008, 55–72.

Iain, T., What's wrong with being a technological essentialist? A response to Feenberg. *Inquiry 43*(4), 2000, 429–444.

Iliffe, R., "The Masculine Birth of Time: Temporal Framework of Early Modern Natural Philosophy." *British Journal of the History of Science*, 33, 2000, 527–453.

Klein, T. & William, P. E. (trans.) Kant's Thesis about Being. *Southwestern Journal of Philosophy* 4(3), 1973, 7–33.

Lammi, W., Hans Georg Gadamer's Platonic *destruktion* of the later Heidegger. *Philosophy Today* 41(¾), 1997, 394–404.

Lash, S., Reflexivity as non-linearity. *Theory, Culture, and Society, 20*(2), 2003, 49–57.

Latour, B., Morality and technology: The end of the means. *Theory, Culture & Society, 19* (5/6), 2002, 247–260.

Mackenzie, D., Fukushima radioactive fallout nears Chernobyl levels, *New Scientist,* 24 March, 2011.

Magnette, P., European governance and civic participation: Beyond elitist citizenship. *Political Studies 51*(1), 2003, 144–160.

Marshall, B. K., Globalization, environmental degradation and Ulrich Beck's risk society. *Environmental Values 8*(2), 1999, 253–275.

Mythen, G., "Reappraising the Risk Society thesis: Telescopic Sight or Myopic Vision?" in *Current Sociology*, 2007, 55(6), 793–813.

Pada, S. R., The Paradox of Ipseity and Difference: Derrida's Deconstruction and Logocentrism. *Kritike, 1*(1), 2007, 32–51.

Peperzak, A. T., Emmanuel Levinas: Jewish experience and philosophy. *Philosophy Today* 27, 1983, 297–396.

Rivas, V. The 'Turn' to Time and the Misscarriage of Being. *Kritike, 1*(2), 2007, 65–81.

Rosentine, L., Heidegger and Plato and the Good. *Philosophy Today*, *22*(4/4), 1978, 332–354.

Rustin, M., Incomplete modernity: Ulrich Beck's risk society. *Radical Philosophy,* 67, 1994, 3–12.

Sallis, J., La différence ontologique et l'unité de la pensé de Heidegger. *Revue Philosophique de Louvain,* 65, 1967, 192–206.

Sheehan, T., A Paradigm shift in Heidegger research. *Continental Philosophy Review 34*(2), 2002, 183–202.

Shlomo, G., Living in a world risk society: A reply to Mikkel V. Rasmussen. *Millennium: Journal of International Studies 31*(1), 2002, 149–160.

Stirling, A., 'Opening up' and 'closing down': Power, participation, and pluralism in the social appraisal of technology. *Science, Technology and Human Values 33*(2), 2008, 262–294.

Taminiaux, J., Sur une éthique pour l'âge technologique. *Le Messenger Européen,* 5, 1991, 187–202.

Taylor-Gooby, P., Hastie, C. & Bromley, C., Querulous citizens: Welfare knowledge and the limits of welfare reform. *Social Policy and Administration 37*(1), 2003, 1–20.

Tsoukas, H., David and Goliath in the risk society: Making sense of the conflict between Shell and Greenpeace in the North Sea. *Organization 6*(3), 1999, 499–528.

Important Websites

http://ec.europa.eu/research/science-society/document_library/pdf_06

http://en.wikipedia.org/wiki/Hans_Jonas

http://thetechnologicalcitizen.com/?p=1975

http://en.wikipedia.org/wiki/Early-Modern

http://www.newscientist.com/article/dn20285-fukushima-radioactive-fallout-nears-chernobyl-levels-htr

Dictionary

Langenscheidt- Redaktion Wörterbücher, Berlin: Cornelsen Verlag GmbH & Co., 2002.

www.ingramcontent.com/pod-product-compliance
Ingram Content Group UK Ltd.
Pitfield, Milton Keynes, MK11 3LW, UK
UKHW041824100926
8556IPUK00002B/11

9 783631 672884